TRIBOLOGY

Friction and Wear of Engineering Materials

TRIBOLOGY

Friction and Wear of Engineering Materials

Ian M. Hutchings

Department of Materials Science and Metallurgy,
University of Cambridge

OXFORD AMSTERDAM BOSTON LONDON NEW YORK PARIS
SAN DIEGO SAN FRANCISCO SINGAPORE SYDNEY TOKYO

Butterworth-Heinemann
An imprint of Elsevier Science
Linacre House, Jordan Hill, Oxford OX2 8DP
200 Wheeler Road, Burlington, MA 01803

First published 1992
Reprinted 1999, 2001
Transferred to digital printing 2003

British Library Cataloguing in Publication Data
A catalogue record for this book is available from the British Library

ISBN 0 340 56184 X

Preface

Tribology, the study of friction, wear and lubrication, is an interdisciplinary subject which draws on the expertise of the physicist, the chemist and the mechanical engineer, as well as the materials scientist or metallurgist. It can, therefore, be approached from several different viewpoints. Most previous textbooks have focused on topics in tribology which can be modelled with mathematical precision: contact mechanics, fluid film lubrication and bearing design, for example. These are undoubtedly important. In the present book, however, the emphasis is more on an equally important subject which is less amenable to precise quantitative analysis: the behaviour of materials in the context of tribology, and in particular, friction and wear.

The book is intended for final-year undergraduate students of the physical sciences and technology, especially students of mechanical engineering, materials science and metallurgy. It should also be valuable for those taking postgraduate or post-experience courses, and since it provides numerous references to the research literature, I hope it will prove a useful source of initial information on tribology for scientists and engineers working in industry.

I am well aware that in offering a textbook on the tribological behaviour of materials I have undertaken a daunting task, particularly in tackling the subject of wear. Professor Duncan Dowson has remarked that whereas the scientific study of friction dates back some 300 years, and that of lubrication more than a century, wear has received similar attention for only 50 years. One can go even further, and suggest that our understanding of wear mechanisms has developed most rapidly only with the widespread use of electron microscopy and instrumental methods of microanalysis over the past 20 years. In a subject so young and so complex, it is inevitable that there will still be competition between theories, confusion over nomenclature and definitions, and inconsistencies between experimental observations. Nevertheless, the foundations of the subject now seem to be well established, and I have tried to present them, as well as more recent developments, in such a way that the reader will be able to appreciate the future advances which will certainly occur.

I have listed recommendations for further reading at the end of each chapter, but the sources of the numerous illustrations, which I have tried to cite accurately in the captions, should also provide valuable points of entry into the research literature. I am very grateful to numerous authors and publishers for their permission to reproduce copyright material.

I M Hutchings
Cambridge, 1992

Contents

1

Introduction

The movement of one solid surface over another is fundamentally important to the functioning of many kinds of mechanism, both artificial and natural. The subject of this book, *tribology*, is defined as 'the science and technology of interacting surfaces in relative motion', and embraces the study of friction, wear and lubrication.

In many instances, low friction is desirable. The satisfactory operation of joints, for example, whether hinges on doors or human hip joints, demands a low friction force. Work done in overcoming friction in bearings and other mechanical components of machines is dissipated as heat, and its reduction will lead to an overall increase in efficiency. But low friction is not necessarily beneficial in all cases. In brakes and clutches, friction is essential; high friction is similarly desirable between a vehicle tyre and the road surface, just as it is between shoe and floor for walking.

Whenever surfaces move over each other, *wear* will occur: damage to one or both surfaces, generally involving progressive loss of material. In most cases, wear is detrimental, leading to increased clearances between the moving components, unwanted freedom of movement and loss of precision, often vibration, increased mechanical loading and yet more rapid wear, and sometimes fatigue failure. The loss by wear of relatively small amounts of material can be enough to cause complete failure of large and complex machines. As in the case of friction, though, high wear rates are sometimes desirable; grinding and polishing, for example, employ wear processes to remove material rapidly and in a controlled manner, and a small amount of wear is often anticipated and even welcomed during the 'running-in' process in some kinds of machinery.

One method of reducing friction, and often wear, is to lubricate the surfaces in some way, and the study of lubrication is very closely related to that of friction and wear. Indeed, even when an artificial lubricant is not added to a system, components of the atmosphere (especially oxygen and water vapour) often play a similar role, and must be considered in any study of the interaction of the surfaces.

The word 'tribology' (from the Greek: τριβος = rubbing or attrition) was first coined in 1966 by a UK government committee, although friction, lubrication and wear had been studied for many years before then and have a long and fascinating history. That committee also made an estimate of the savings which could be made by UK industry if known tribological principles were widely applied. Similar exercises have subsequently been carried out in several other countries; their conclusions are all in broad agreement, that at least 1% of the gross national product might be saved with minimal further investment in research, and that the potential for even larger savings might exist with further research. The savings arise from several sources. The original (1966) survey identified the savings listed in Table 1.1. The reduction in energy consumption through reduced friction was, in money terms, rather a small component of the total; savings in maintenance and replacement costs, in losses consequential upon breakdowns, and in investments through the increased life of machinery were, at that time, much more important. A subsequent survey in the UK, taking account of the relative increase in the cost of energy since 1966, has focused on the savings in energy which can be made by improved tribological design, and identified the kinds of savings shown in Table 1.2. It is clear that as the relative price of energy continues to rise, and the need to conserve both energy and raw materials becomes more widely appreciated, correct tribological design will become increasingly important.

In the following chapters we shall explore various aspects of tribology. We shall start by examining the topography of surfaces and the way in which they interact when placed in contact (Chapter 2). The origins of friction, and the frictional response of metals and non-metals are discussed next (Chapter 3). Lubricants and lubrication form the subject of Chapter 4, and wear is the topic of Chapters 5 and 6. In these two chapters the distinction is made between wear which occurs when two relatively smooth surfaces slide over

Table 1.1 Methods by which financial savings could be made through improved tribological practice in UK industry. The percentages represent proportions of the total annual saving, which was estimated at £515 million (at 1965 prices) (from UK Department of Education and Science, Lubrication (Tribology): Education and Research, HMSO, 1966)

Reduction in energy consumption from lower friction	5%
Reduction in manpower	2%
Savings in lubricant costs	2%
Savings in maintenance and replacement costs	45%
Savings in losses resulting from breakdowns	22%
Savings in investment through greater availability and higher efficency	4%
Savings in investment through increased life of plant	20%

Table 1.2 Methods by which savings of energy could be made through improved tribological practice in UK industry (from Jost H P and Schofield J, *Proc. Inst. Mech. Engrs.* **195**, 151–195, 1981)

Direct savings of energy:		
	Primary:	saving of energy dissipated by friction
	Secondary:	saving of energy needed to *fabricate* replacement parts
	Tertiary:	saving of energy content of *materials* for replacement parts
Indirect savings of energy:		
	Savings consequential on direct savings, e.g. in investment in plant needed to compensate for frictional losses	

each other, either lubricated or unlubricated (i.e. sliding wear, covered in Chapter 5), and wear involving hard particles (i.e. abrasive or erosive wear, covered in Chapter 6). The terms 'sliding wear', 'abrasion' and 'erosion' are not intended to describe mechanisms of wear, but form broader classifications; as we shall see later, many different mechanisms can be involved in wear and it would be simplistic to attempt to list them here. Ways in which wear can be taken into account in the design process are discussed in Chapter 7, and in the final chapters we examine the important topics of surface engineering (Chapter 8) and materials for bearings (Chapter 9).

Further reading

Dowson D, *History of Tribology*, Longman 1979

Glossary of Terms and Definitions in the Field of Friction, Wear and Tribology (including eight-language index), in Peterson M B and Winer W O (Eds.), *Wear Control Handbook*, ASME, 1980, pp 1143–1303

Jost H P, Tribology – origin and future, *Wear* **136**, 1–18, 1989

Sibley L B and Kennedy F E (Eds), *Achievements in Tribology*, ASME, 1990

Standard Terminology Relating to Wear and Erosion, Standard G40, ASTM, 1991

UK Department of Education and Science, Lubrication (Tribology) Education and Research, HMSO, 1966

2

Surface topography and surfaces in contact

2.1 INTRODUCTION

When studied on a sufficiently fine scale, all solid surfaces are found to be uneven. In the limit, the surface irregularities will be on the scale of individual atoms or molecules; it is possible, for example, to prepare carefully cleaved specimens of the mineral mica which are truly smooth on a molecular scale over areas of several square centimetres. However, the surfaces of even the most highly polished engineering components show irregularities appreciably larger than atomic dimensions, and many different methods have been employed to study their topography. Some involve examination of the surface by electron or light microscopy, or by other optical methods, while others employ the contact of a fine stylus, electrical or thermal measurements, or rely on the leakage of a fluid between the surface and an opposing plane. Perhaps the highest resolution can be achieved by the techniques of scanning tunnelling microscopy or atomic force microscopy, which can resolve individual atoms; but for most engineering surfaces less sensitive methods are adequate to study their topography.

In this chapter we shall see how surface roughness is measured, and then examine what happens when two surfaces are placed in contact.

2.2 MEASUREMENT OF SURFACE TOPOGRAPHY

One of the most common methods of assessing surface topography is the *stylus profilometer*, the principles of which are illustrated in Fig. 2.1. A fine stylus is dragged smoothly and steadily across the surface under examination. As the stylus travels over the surface it rises and falls. Its vertical displacement is converted by a transducer into an electrical signal which is amplified and, in the simplest form of the instrument, moves the pen of a chart recorder. The graph drawn by the pen represents the vertical displacement of the stylus as a function of the distance travelled along the surface.

The graphical representation of the surface profile generated by a stylus

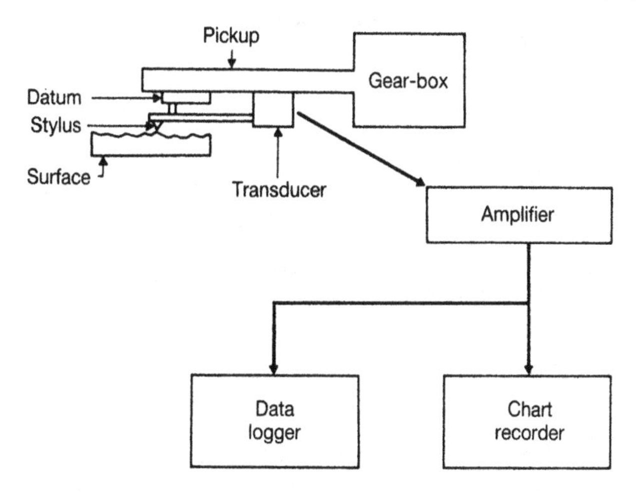

Fig. 2.1 The principles of operation of a simple stylus profilometer. The stylus moves steadily over the surface under examination, and its vertical displacement is recorded on a moving chart, or digitized for computer processing (from Thomas T R (Ed), *Rough Surfaces*, Longman, 1982)

profilometer differs from the shape of a genuine cross-section through the surface for several reasons. The major difference is due to the different magnifications employed by the instrument in the horizontal and vertical directions. The vertical extent of surface irregularities is nearly always much less than their horizontal scale. It is therefore convenient to compress the graphical record of the surface profile by using a magnification in the vertical direction which is greater than that in the plane of the surface. The ratio of the magnifications will depend on the roughness of the surface, but typically lies between 10 and 5000; with refinements, stylus instruments can be used at absolute vertical magnifications up to 10^6. In Fig. 2.2 the shape of a real surface is compared with profilometer recordings of the same surface at magnification ratios (vertical : horizontal) of 5:1 and 50:1.

Although the amplitudes and wavelengths of the irregularities are faithfully recorded on both graphs, the surface slopes appear much steeper on the profilometer records than they really are. The distorting effect of this horizontal compression, present in nearly all profilometer traces, must be recognized in interpreting them. Thus, although surface slopes may appear very steep on a profilometer trace, in reality they are very rarely steeper than 10°.

An unavoidable limitation of this method of profilometry results from the shape of the stylus. For reasons of strength, the diamond styli used are pyramidal or conical, with minimum included angles of about 60° and tip radii of 1 to 2.5 μm (Fig. 2.3). The combination of finite tip radius and included angle prevents the stylus from penetrating fully into deep narrow features of the surface. In some applications this can lead to significant error. Special styli, with chisel edges and minimum tip radii as small as 0.1 μm, can be used to examine very fine surface details where a conventional stylus would be too blunt, but all stylus methods inevitably produce some 'smoothing' of the profile due to the finite dimensions of the stylus tip.

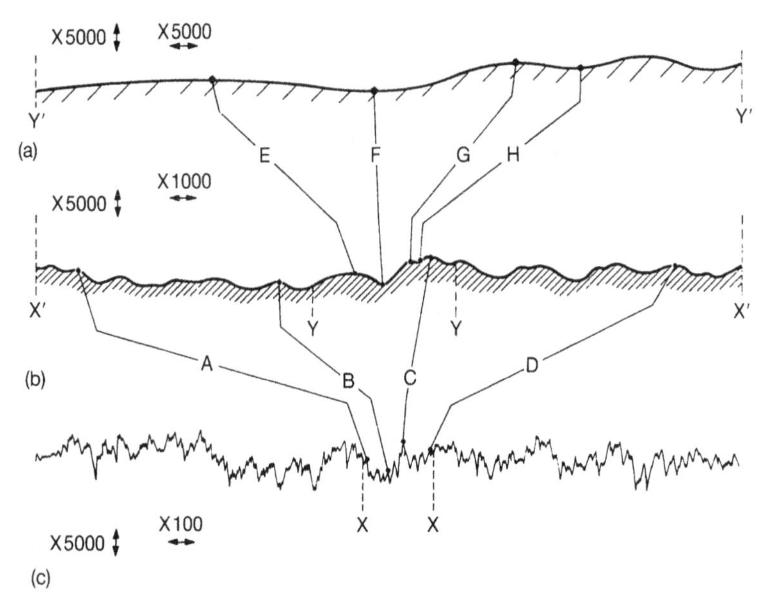

Fig. 2.2 (a) The profile of a real surface, greatly magnified; (b) the same surface, as depicted on a profilometer trace in which the vertical magnification is five times that in the horizontal plane; (c) as for (b), but with a ratio of 50:1 between vertical and horizontal magnifications (from Dagnall H, *Exploring Surface Texture*, Rank Taylor Hobson, 1980)

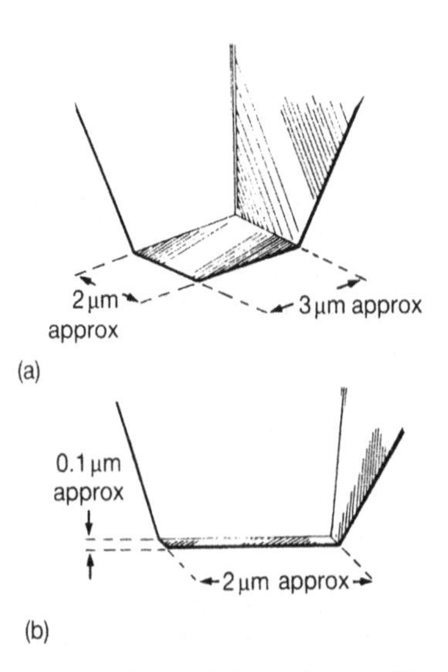

Fig. 2.3 (a) A standard diamond stylus used for surface profilometry; (b) a chisel-edge stylus, giving greater sensitivity to fine surface features than the stylus in (a) (from Dagnall H, *Exploring Surface Texture*, Rank Taylor Hobson, 1980)

A further error is introduced in examining very compliant or delicate surfaces by stylus methods, in that the load on the stylus, although small, may nevertheless distort or damage the surface. Non-contacting, optical methods of surface measurement are attractive for such applications, and recent advances in computer capabilities have enabled optical interferometers to be designed which can record surface profiles without distortion or damage. Figure 2.4 illustrates the principles of one such instrument. Interference between two beams of light, reflected from the surface under examination and from a perfectly plane reference surface, creates fringes which are digitally recorded by an array of photodiodes linked to a microprocessor. Small, accurately known displacements of the reference surface, under microprocessor control, cause changes in the fringe pattern from which the distribution of surface heights over the specimen can be computed. The instrument can be used with a linear array of photodiodes to generate a surface profile along a line on the surface under examination, just like a stylus profilometer. Alternatively, a square array of photodiodes can be used to record the variation of height over an area of the surface. Vertical resolution as small as 0.1 nm can be achieved, although the maximum height which is measurable is limited by the depth of focus of the instrument to a few micrometres. For the examination of very fine surface features, especially on compliant surfaces (e.g. polymers), optical interferometry has clear advantages over stylus methods, but for coarser surfaces the latter technique must be used. In many applications the two techniques complement each other rather than competing.

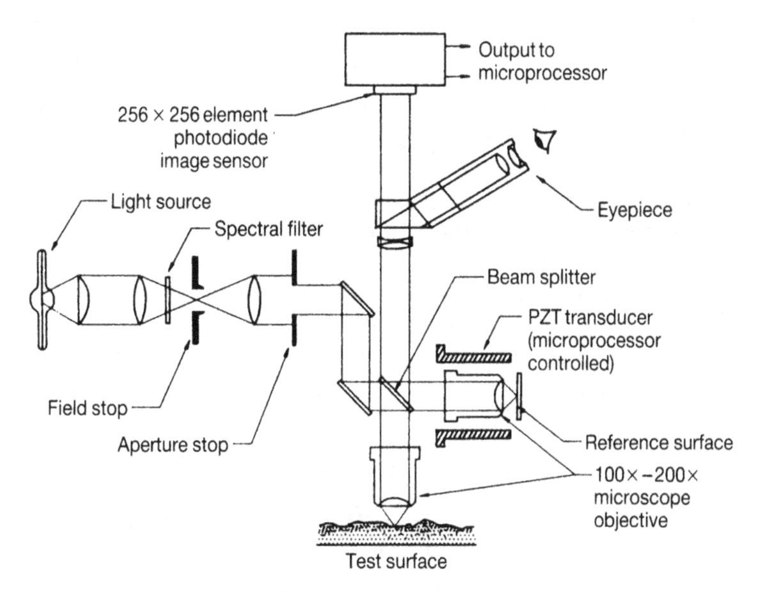

Fig. 2.4 Schematic diagram of a digital optical interferometer, used to determine surface profiles without mechanical contact (from Bhushan B, Wyant J C and Meiling J, *Wear* **122**, 301–312, 1988)

2.3 QUANTIFYING SURFACE ROUGHNESS

It is convenient to differentiate between *roughness*, meaning the small scale irregularities of a surface, and *form error*, which is a measure of the deviation of the shape of the surface from its intended ideal shape (e.g. plane, cylindrical or spherical). The distinction between roughness and form error is arbitrary, although clearly it involves the horizontal scale of the irregularity. The picture may be further complicated by the presence of *waviness*, a periodic surface undulation intermediate in scale between roughness and form error.

By various methods, form error and waviness may be subtracted from the surface profile recorded by a profilometer, so that the graph depicts only the roughness, or short wavelength irregularities. A simple mechanical method, frequently used in stylus profilometry, is to arrange for the measuring head of the instrument to be supported on a small skid which rides on the surface just behind or in front of the stylus. The profilometer then records the displacement of the stylus relative to the skid. In this way the average local level of the surface is used as a datum, and surface disturbances of wavelength rather longer than the size of the skid are not recorded. An alternative method is to filter the displacement signal during or after recording, so that components corresponding to long wavelength surface displacements (form error or waviness) are removed. Electronic filtering methods may also be used to remove the roughness signal and detect only the form error or waviness. If filters are used, the distinction between roughness and form error may be quantified by quoting the filter cut-off wavelength.

The graph of a surface profile generated by a stylus or optical profilometer contains most of the information needed to describe the topography of the surface along a single direction. The profile graph itself, however, does not provide a sufficiently simple and readily interpreted means of describing surface roughness; several quantities derived from the profile, which are often automatically computed by the profilometer instrument, are used for this purpose.

The most commonly quoted measure of surface roughness is the *average roughness* (symbols R_a, c.l.a. for 'centre line average' or AA for 'arithmetic average'). R_a is defined as the arithmetic mean deviation of the surface height from the mean line through the profile. The mean line is defined so that equal areas of the profile lie above and below it (Fig. 2.5).

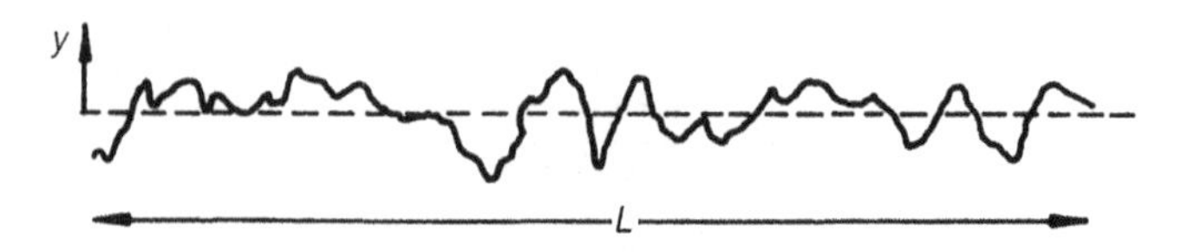

Fig. 2.5 A surface profile is a graph of surface height, y, relative to a mean line, plotted against distance. The overall length of the profile under examination is L

Formally, the average roughness R_a is defined by

$$R_a = \frac{1}{L}\int_0^L |y(x)|\,\mathrm{d}x \tag{2.1}$$

where y is the height of the surface above the mean line at a distance x from the origin, and L is the overall length of the profile under examination (Fig. 2.5).

The *r.m.s. roughness* (symbol R_q) is defined as the root mean square deviation of the profile from the mean line:

$$R_q^2 = \frac{1}{L}\int_0^L y^2(x)\,\mathrm{d}x \tag{2.2}$$

For many surfaces, the values of R_q and R_a are similar; for a Gaussian distribution of surface heights, $R_q = 1.25\ R_a$.

It is inevitable that in attempting to describe a profile by a single number, some important information about the surface topography will be lost. R_a and R_q, for example, give no information on the shapes or spacings of the surface irregularities, and convey no indication of the probability of finding surface heights within certain limits. For a fuller description of the topography of the surface, information is needed about the *probability distribution* of surface heights and the *spatial distribution* of peaks and valleys across the surface.

The need for a method of describing the distribution of surface heights is met by defining an *amplitude density function*, $p(y)$, the value of which, for any height y, is proportional to the probability of finding a point on the surface at height y above the mean line. The quantity $p(y)\Delta y$ is then the fraction of the surface profile which lies at heights between y and $y + \Delta y$ above the mean line, as shown in Fig. 2.6. A symmetrical profile, such as a sine curve, leads to an amplitude density curve which is symmetrical about the position of the mean line. Asymmetry of the surface profile leads to skewing of the amplitude density function, which therefore contains some information about the shapes of surface irregularities as well as their vertical extent. The r.m.s. roughness, R_q, is the standard deviation of the amplitude density function and, in this context, is sometimes given the symbol σ.

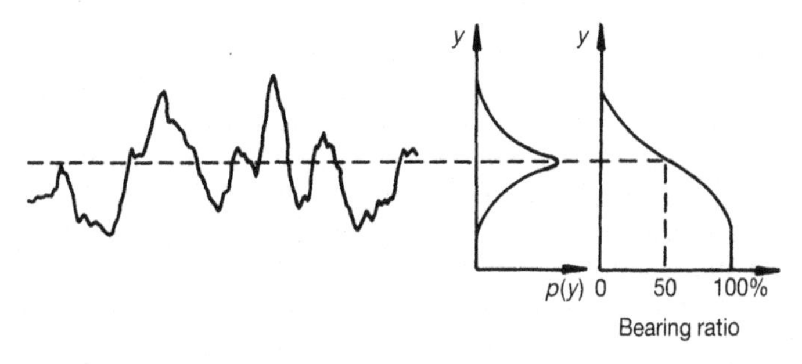

Fig. 2.6 Illustration of the derivation of the amplitude density function $p(y)$ and the bearing ratio curve from a profile trace

The shape of the amplitude density curve can be described by its *skewness*, which provides a measure of its asymmetry, and *kurtosis* (from the Greek word for a hump), a measure of the sharpness of the peak of the distribution curve. Skewness (symbol *Sk*) is defined by

$$Sk = \frac{1}{R_q^3} \int_{-\infty}^{\infty} y^3 p(y)\, \mathrm{d}y \tag{2.3}$$

and kurtosis (symbol *K*) by

$$K = \frac{1}{R_q^4} \int_{-\infty}^{\infty} y^4 p(y)\, \mathrm{d}y \tag{2.4}$$

Both quantities can be readily calculated from the surface profile by modern computerized profilometers. A Gaussian (normal) probability distribution has a skewness value of zero and a kurtosis of 3.0. A kurtosis value less than 3.0 indicates a broad, flat distribution curve, while a greater kurtosis corresponds to a more sharply-peaked distribution.

Another function sometimes used in tribological studies and closely related to the amplitude density function is the *bearing ratio* curve (or *Abbot–Firestone* curve). This curve can be understood by imagining a straight line, representing the profile of a plane smooth surface, being brought slowly down towards the profile of the surface under investigation. When the plane first touches the surface at a point, the bearing ratio (defined as the ratio of the contact length to the total length of the profile) is zero. As the line is moved further downwards, the length over which it intersects the surface profile increases, and therefore the bearing ratio increases. Finally, as the line reaches the bottom of the deepest valley in the surface profile, the bearing ratio rises to 100%. The bearing ratio curve is a plot of bearing ratio against surface height, as shown in Fig. 2.6. Differentiation of the bearing ratio curve yields the amplitude density function.

The way in which the hills and valleys are distributed across a surface is not described by either the amplitude distribution or the bearing ratio curve. Two methods may be used to extract this information from a profile: *autocorrelation* and *spectral analysis*. The *autocorrelation function* $C(\beta)$ of a profile graph is defined by:

$$C(\beta) = \frac{1}{L} \int_{0}^{L} y(x)\, y(x+\beta)\, \mathrm{d}x \tag{2.5}$$

The value of the autocorrelation function for some displacement β along the surface is therefore derived by shifting the profile a distance β along the surface, multiplying the shifted profile function by the corresponding unshifted value, and calculating the area beneath the resultant product curve. When the displacement β is zero, the value of the autocorrelation function is a maximum and is simply the mean square roughness (R_q^2).

The autocorrelation function provides a measure of the correlation between the heights of the surface at positions separated by a distance β along the surface. The shape of the curve summarizes statistical information on the characteristic spacings, if any, of the surface features. Any regular undulation of the surface will show up as an oscillation of the same wavelength in the value of the autocorrelation function. For many real surfaces the autocorrelation function decays steadily to zero as β increases, and may be approximated by an exponential function.

The *power spectral density* function $P(\omega)$, which conveys direct information about the spatial frequencies (i.e. wavelengths) present in the surface profile, is the Fourier transform of the autocorrelation function:

$$P(\omega) = \frac{2}{\pi}\int_0^{\infty} C(\beta)\cos(\omega\beta)\,\mathrm{d}\beta \tag{2.6}$$

Power spectral density is a particularly suitable function with which to study machined surfaces, since it clearly depicts and separates any strong surface periodicities which may result from the machining process. Both the autocorrelation function and the power spectral density function can be computed automatically by some modern profilometers.

2.4 THE TOPOGRAPHY OF ENGINEERING SURFACES

So far we have treated the two-dimensional graph produced by a profilometer as if it were completely representative of the topography of the surface. Yet many surface finishes are directional, and a profile graph represents a section through the surface in only one direction. Directional surfaces, which might be generated for example by turning, milling or grinding, are usually examined by stylus profilometry with the stylus moving across the 'lay' of the surface. The profile, and the measures of roughness derived from it, thus correspond to the maximum roughness of the surface, but give no information about the distribution of surface features in a direction parallel to the machining marks.

It is possible, however, by making a large number of profilometer traverses across a specimen and displacing the specimen slightly between each traverse, to generate a three-dimensional picture of the surface. Some stylus profilometers will do this automatically. As noted earlier, a digital optical interferometer equipped with a square array of photosensors can also be used to produce such a map, and examples generated by both methods are shown in Fig. 2.7.

The distinct lay of the ground and turned surfaces, and the contrasting lack of directionality of the shot-blasted surface are clearly evident. Height data gathered in this way may also be used to generate contour maps of real surfaces, to study the statistics of feature distribution across two-dimensional areas, and to examine the area of contact between surfaces by numerically modelling their approach.

The topography of a freshly-machined surface depends on the machining process used to generate it as well as on the nature of the material. Likewise, the topography of a worn surface depends on the conditions under which the wear has occurred. Much useful information is therefore contained in the amplitude density function and the autocorrelation function. Examples of

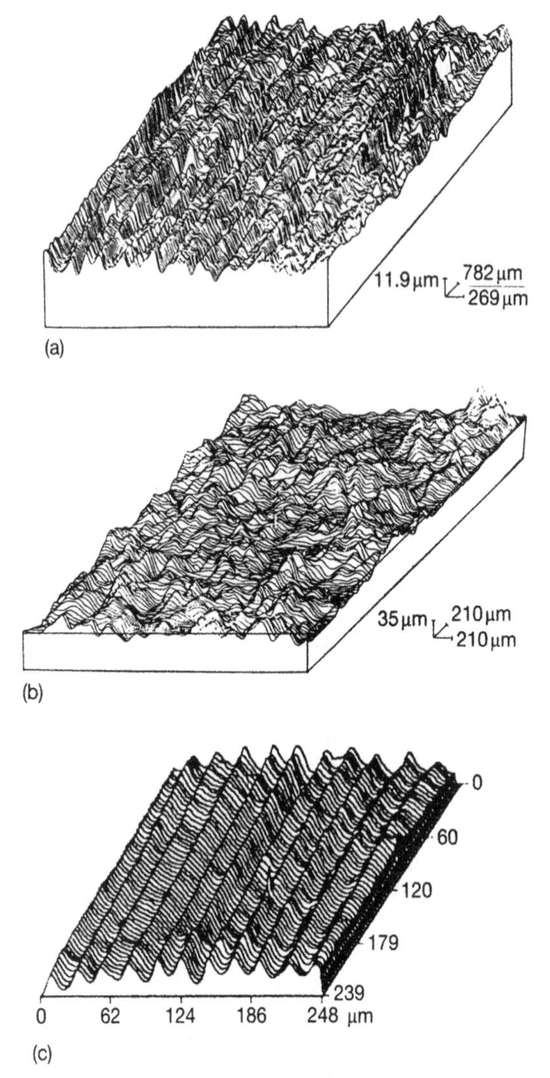

Fig. 2.7 Examples of three-dimensional profile maps: (a) a ground steel surface; (b) a shot-blasted steel surface; (c) a diamond-turned surface. (a) and (b) were generated by repeated stylus profilometry, while (c) was produced by optical interferometry (from Sayles R S and Thomas T R, *J. Phys. E: Sci. Instrum.* **9**, 855–861, 1976; and Bhushan B, Wyant J C and Meiling J, *Wear* **122**, 301–312, 1988)

amplitude density functions, in the form of experimentally determined histograms for three machined surfaces, are shown in Fig. 2.8.

The continuous curves are theoretically fitted distribution functions with the same mean, standard deviation, skewness and kurtosis as the experimental data. The mild steel surface finished by centreless grinding

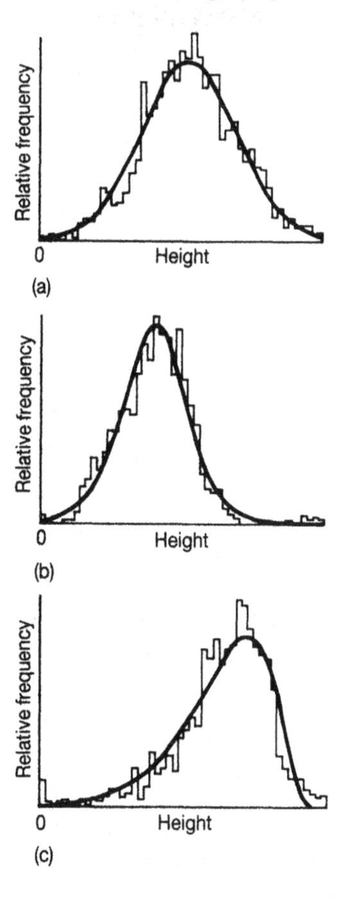

Fig. 2.8 Surface height distributions (amplitude density functions) derived from experimental measurements on (a) mild steel finished by centreless grinding; (b) lathe-turned mild steel ; (c) lathe-turned grey cast iron (from Watson W, King T G, Spedding T A and Stout K J, *Wear* **57**, 195–205, 1979)

exhibits a symmetrical distribution which is well fitted by a Gaussian function (skewness = 0, kurtosis = 3). Many surfaces, especially those resulting from wear processes rather than from deliberate machining, have height distributions which are close to Gaussian. Some combinations of machining process and material, however, can lead to appreciably non-Gaussian distributions. The curve for a turned mild steel surface (Fig. 2.8b), although nearly symmetrical, has a kurtosis value rather greater than 3. Turning of grey cast iron results in a markedly asymmetric distribution (Fig. 2.8c).

Although a great deal of information may be conveyed by the use of skewness and kurtosis values, the roughnesses of engineering surfaces are frequently specified only by average roughness (R_a) values. Table 2.1 lists typical ranges of R_a values for engineering surfaces finished by various processes.

Table 2.1 Typical average roughness values for engineering surfaces finished by different processes (data from Stout K J, *Materials in Engineering*, **2**, 287–295, 1981)

	R_a (μm)
Planing, shaping	1–25
Milling	1–6
Drawing, extrusion	1–3
Turning, boring	0.4–6
Grinding	0.1–2
Honing	0.1–1
Polishing	0.1–0.4
Lapping	0.05–0.4

2.5 CONTACT BETWEEN SURFACES

When two nominally plane and parallel surfaces are brought gently together, contact will initially occur at only a few points. As the normal load is increased, the surfaces move closer together and a larger number of the higher areas or *asperities* on the two surfaces come into contact.

Since these asperities provide the only points at which the surfaces touch, they are responsible for supporting the normal load on the surface and for generating any frictional forces which act between them. If an electrical current passes from one surface to the other, then it must travel through the points of contact. An understanding of the way in which the asperities of two surfaces interact under varying loads is therefore essential to any study of friction, wear or electrical contact resistance. In the rest of this chapter, the behaviour of surfaces under normal loading will be discussed; tangential forces, which are responsible for friction, will be discussed in Chapter 3.

2.5.1 Deformation of a single asperity

Before examining the behaviour of two rough surfaces in contact, in which large numbers of asperities of different shapes and sizes are pressed against each other, we shall first consider the simpler idealized case of a single asperity loaded against a rigid plane surface. Results derived for this geometry will then be used in discussing the more complex case.

Asperities are revealed by the study of surface profiles to be blunt; as we have seen, surface slopes are very seldom steeper than about 10°, and are usually much shallower. In examining the behaviour of single asperities, it is convenient to model them as perfectly smooth protuberances of spherical, conical or pyramidal shape.

Elastic deformation

The elastic contact between a sphere and a plane has been particularly closely studied. When a sphere of an elastic material is pressed against a plane (Fig. 2.9) under a normal load w, contact will occur between the two over a circular area of radius a, given by the following equation due to Hertz (1881):

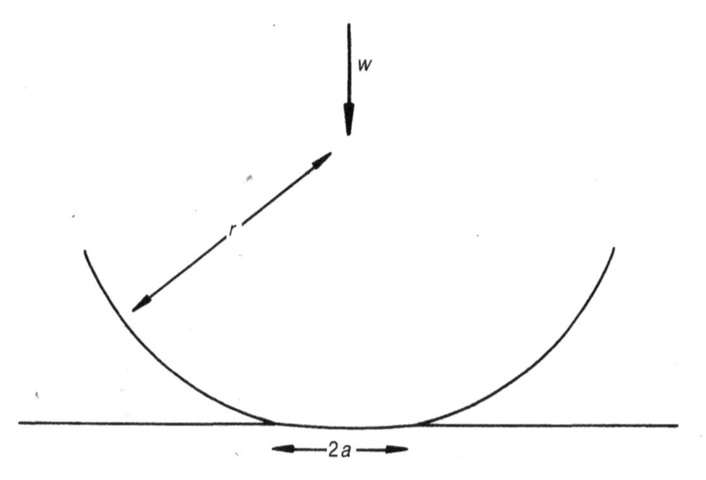

Fig. 2.9 Elastic deformation of a sphere of radius r, pressed against a plane surface under a load w. The radius of the contact circle is a

$$a = \left(\frac{3wr}{4E}\right)^{1/3} \tag{2.7}$$

Here r is the radius of the sphere and E is an elastic modulus which depends on Young's moduli, E_1 and E_2, and on the Poisson's ratios, ν_1 and ν_2, for the materials of the sphere and of the plane in the following way:

$$\frac{1}{E} = \frac{(1-\nu_1^2)}{E_1} + \frac{(1-\nu_2^2)}{E_2} \tag{2.8}$$

The area of contact between the sphere and the plane, πa^2, is then given by:

$$\pi a^2 \approx 0.83\pi\left(\frac{wr}{E}\right)^{2/3} \tag{2.9}$$

For this case, in which the deformation is purely elastic, the area of contact is therefore proportional to $w^{2/3}$. The mean pressure (normal stress) P_{mean} over the contact area is $w/\pi a^2$, and thus varies as $w^{1/3}$. This stress is not uniform over the circular area of contact, but has a maximum at the centre and falls to zero at the edge. The distribution of normal stress is shown in Fig. 2.10; the

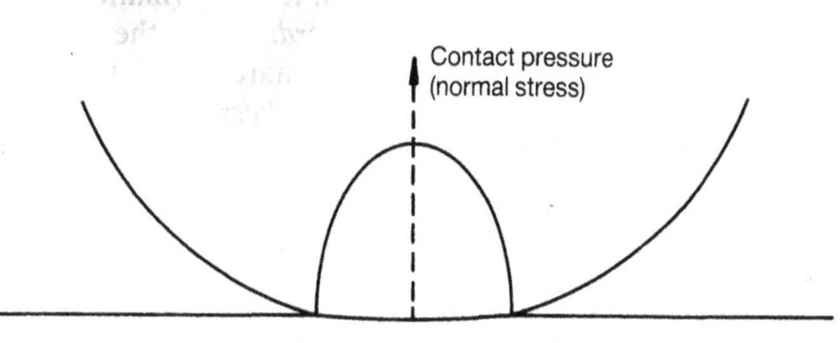

Fig. 2.10 The distribution of normal stress (contact pressure) under a sphere loaded elastically against a plane

maximum stress, at the centre of the contact circle, is 3/2 times the mean pressure.

Plastic deformation

As the normal load between the sphere and the plane is increased, one or other of the components may start to deform plastically. The situation may be simplified by considering two cases: first, that in which the sphere is assumed to be rigid, and plastic flow is confined to the plane, and second, that in which the plane does not deform and plastic flow occurs only in the sphere.

The indentation of a plastic half-space by a rigid sphere has been extensively investigated in connection with indentation hardness testing. Hertz's analysis of the elastic stress field due to a spherical indenter on a flat surface shows that the maximum shear stress beneath the indenter occurs at a depth of about $0.47a$ (where a is the radius of the contact circle). Plastic flow first occurs at this point when the yield criterion is satisfied; if the Tresca criterion is applicable, this will be at a maximum shear stress of $Y/2$ where Y is the uniaxial yield stress of the material. For a material in which Poisson's ratio has a value of about 0.3, the shear stress at a depth $0.47a$ beneath the sphere is 0.47 times the mean contact pressure. Plastic deformation therefore initiates at a mean contact pressure of $1.1Y$.

As the normal load is further increased, the zone of plastic deformation extends from beneath the indenter until it eventually reaches the surface. At this point, which may be shown theoretically (by finite element analysis) to occur in a metal at a load some 50 to 100 times the load at which plastic flow initiates, the contact area is still extremely small, with a radius typically less than 1% of the radius of the sphere. The mean pressure over the contact area has risen by this stage to about $3Y$ and remains at nearly the same value for subsequent increases in load. The independence of the mean contact pressure from the load once full plasticity has been reached, and the constant proportionality between that pressure and the yield stress of the indented material, provide the basis for *indentation hardness testing*.

Similar results are also found for indentation by strong indenters of other shapes, in particular for cones and pyramids of different angles, and for flat-ended punches. The constant of proportionality between indentation pressure and yield stress in these cases depends on the geometry of the indenter, but it is never far from the value of about 3 found for the case of a sphere. The mean indentation pressure when the deformation is fully plastic is used as a measure of the *indentation hardness* of the material. If the indenter has the Vickers geometry of a square-based pyramid with an included angle between its sides of 136°, the *Vickers hardness* is often quoted in units of kg force per mm^2 ($1\ kgf\ mm^{-2} = 9.8$ MPa), but with the symbol for the units omitted, followed by the symbol HV, VPH or VPN.

For the case of a soft sphere or a blunt cone, pressed against a rigid plane, in which yield occurs in the sphere or cone rather than in the plane surface, similar results are obtained. Provided the extent of the deformation is not too large, plastic constraint again raises the mean contact pressure to about three times the uniaxial yield stress of the sphere or cone.

We might therefore expect that when an asperity, of whatever shape, is pressed on to an opposing surface, it does not much matter which component yields; the mean pressure over the contact area will always be of the order of three times the uniaxial yield stress of the softer material. More important, the area of contact should be directly proportional to the load.

2.5.2 Simple theory of multiple asperity contact

We can use the results of the previous section in an elementary treatment of the contact between two rough surfaces. Although the statistical approach discussed in the next section describes the real case more closely, a simpler model can, nevertheless, provide a valuable physical picture of the dependence of contact area on normal load.

From consideration of the plane of symmetry between two identical rough surfaces brought into contact, it is clear that the problem may be treated as one of contact between a single rough surface and a rigid frictionless plane. If we make the simplifying assumption that the rough surface consists of an array of spherical asperities of constant radius and height, and that each asperity deforms independently of all the others, then we can immediately apply the results of the previous section.

Each asperity will bear the same fraction of the total normal load, and each will contribute the same area to the total area of contact. By summing the contributions from all the asperities over the whole area of contact, we may show that the total real area of contact A will be related to the total load W in exactly the same way as the individual area of contact for each asperity, πa^2 is related to the load borne by each asperity, w. For the case of purely elastic contact,

$$A \propto W^{2/3} \tag{2.10}$$

and for perfectly plastic behaviour of the asperities,

$$A \propto W \tag{2.11}$$

Real surfaces, however, are not composed of uniform asperities of a single radius and height; both the radii and heights of the surface irregularities are statistically distributed. As the load on a real surface is increased, not only will the area of contact for each individual asperity increase, but more asperities will come into contact and start to carry some load. Under these circumstances, if the average area of contact for each contacting asperity remains constant, and the increase in load is borne by a correspondingly increased number of asperities in contact, then even for purely elastic contact the total area will be directly proportional to the load. To examine whether this is a valid picture of the behaviour of real surfaces, we must look in more detail at the behaviour of a statistically distributed set of asperities.

2.5.3 Statistical theories of multiple asperity contact

One of the first statistical theories for the contact of rough surfaces was presented by Greenwood and Williamson in 1966, and is still widely cited.

Although later theories have progressively removed some of the simplifications made by Greenwood and Williamson, they support the broad conclusions of this model which we shall discuss here in some detail. Subsequent developments of the theory of contact of rough surfaces are discussed in some of the references listed at the end of this chapter, to which the reader is referred for further details.

In the Greenwood and Williamson model it is assumed that all the contacting asperities have spherical surfaces of the same radius r, and that they will deform elastically under load according to Hertz's equations (Section 2.5.1). Figure 2.11 illustrates the contact between the rough surface and a rigid plane surface assumed in the model.

The height of an individual asperity above the reference plane is z. If the separation between the reference plane and the flat surface, d, is less than z then the asperity will be elastically compressed and will support a load w which can be predicted from Hertz's theory:

$$w = \frac{4}{3} E r^{1/2} (z - d)^{3/2} \tag{2.12}$$

The heights of the asperities will be statistically distributed. The probability that a particular asperity has a height between z and $z + \mathrm{d}z$ will be $\varphi(z)\mathrm{d}z$ where $\varphi(z)$ is a probability density function describing the distribution of asperity heights.

It is important to distinguish between the distribution of surface heights, which we have met in Section 2.3 as the amplitude density function $p(y)$, and the distribution of asperity or peak heights, $\varphi(z)$, the definition of which depends upon a criterion for identifying individual asperities. The method of identification employed by Greenwood and Williamson in their experimental work involved the comparison of each value of surface height, taken from a digitized profilometer trace, with the two neighbouring values, and therefore required the definition of a sampling length.

The probability that an asperity makes contact with the opposing plane surface is the probability that its height is greater than the plane separation, d:

$$\text{prob}\,(z>d) = \int_d^\infty \varphi(z)\,\mathrm{d}z \tag{2.13}$$

If there are a total of N asperities on the surface, then the expected number of contacts, n, will be given by

$$n = N \int_d^\infty \varphi(z)\,\mathrm{d}z \tag{2.14}$$

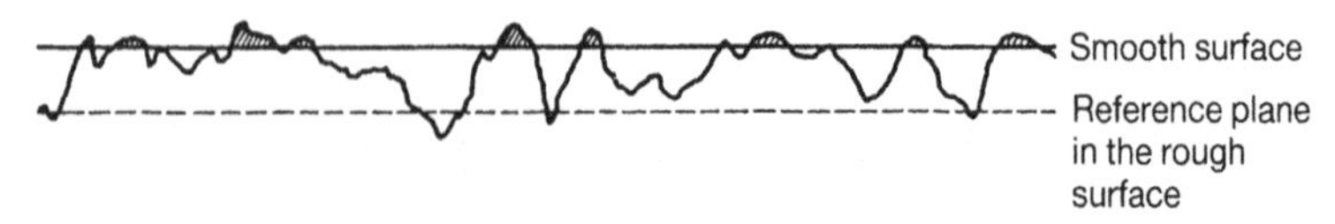

Fig. 2.11 Model for contact between a rough surface and a smooth rigid plane (from Greenwood J A and Williamson J B P, *Proc. Roy. Soc. Lond.* **A295**, 300–319, 1966)

and the total load carried by all the asperities, W, will be

$$W = \frac{4}{3} NEr^{1/2} \int_d^\infty (z-d)^{3/2} \varphi(z)\, dz \tag{2.15}$$

Greenwood and Williamson studied the behaviour of their model for two different distributions of asperity heights. If the asperity heights follow an exponential distribution, then equation 2.15 may be integrated analytically. The load W is found to be linearly proportional to the total real area of contact. As the load is increased, the size of each individual contact spot increases, but more asperities come into contact so that the *mean* size of each asperity contact remains constant.

Although the exponential distribution provides a fair description of perhaps the highest tenth of all asperities on many real surfaces, the Gaussian (normal) distribution is found experimentally to provide a better model for the height distribution. For a normal distribution of asperity heights, equation 2.15 must be integrated numerically; if physically reasonable quantities are employed, then the results are not greatly different from those obtained with the exponential distribution. The real area of contact is no longer exactly proportional to the load, although as Fig. 2.12 illustrates for a typical ground steel surface, the relationship is close to linear. The real area of contact predicted by the theory is also effectively independent of the nominal contact area.

The theory of Greenwood and Williamson was derived for purely elastic contact, but it does allow the onset of plastic flow at asperities to be predicted. It is found that the proportion of asperity contacts at which plastic flow has occurred depends on the value of a *plasticity index*, ψ, given by

$$\psi = \frac{E}{H}\left(\frac{\sigma^*}{r}\right)^{1/2} \tag{2.16}$$

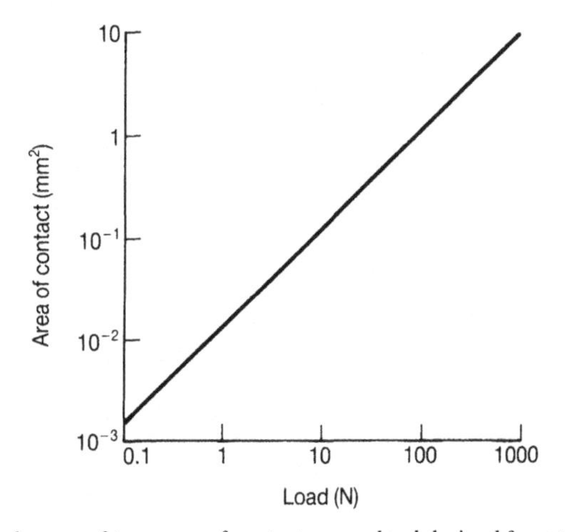

Fig. 2.12 Theoretical curve of true area of contact versus load derived from the Greenwood and Williamson model for steel flats of 10 cm^2 nominal area (from Greenwood J A and Williamson J B P, *Proc. Roy. Soc. Lond.* **A295**, 300–319, 1966)

where E is defined by equation 2.8, H is the indentation hardness of the rough surface (a measure of the plastic flow stress of the asperities) and σ^* is the standard deviation of the distribution of asperity heights. The quantity $(\sigma^*/r)^{1/2}$ is approximately equal to the average slope of the asperities.

In principle, the proportion of asperity contacts which are plastic is determined by both the value of the plasticity index and the nominal pressure, but in practice the plasticity index dominates the behaviour. For values of ψ less than about 0.6, plastic flow at the asperities would be caused only by extremely high nominal pressures, while for ψ greater than about 1 most asperities will deform plastically under even the lightest of loads. For metal surfaces, produced by normal engineering methods, ψ lies typically in the range 0.1 to 100; as can be seen from Fig. 2.13, only for very finely polished surfaces will asperity contact between metals remain elastic. For ceramics and polymers, however, elastic contact is much more likely than plastic since for these materials the value of E/H is typically one tenth that of metals, leading to a proportionate reduction in the value of ψ.

Theories of the contact between rough surfaces, and also experimental observations, thus suggest that in many practical cases of contact between metals, the majority of asperity contacts will be plastic. The load supported by each asperity is directly proportional to its contact area, and provided that the asperities deform independently, the total real area of contact for the whole surface will be proportional to the normal load and independent of the detailed statistical distribution of asperity heights. For ceramics and polymers, which have lower values of the ratio E/H, contact is more likely to be elastic. It may also be predominantly elastic for metals if their surfaces are very smooth. But even for elastic contact, as we have seen above, for realistic distributions of surface heights the total real area of contact will remain roughly proportional to the normal load, since as the load is increased the number of asperity contacts grows but the mean area of each remains effectively constant.

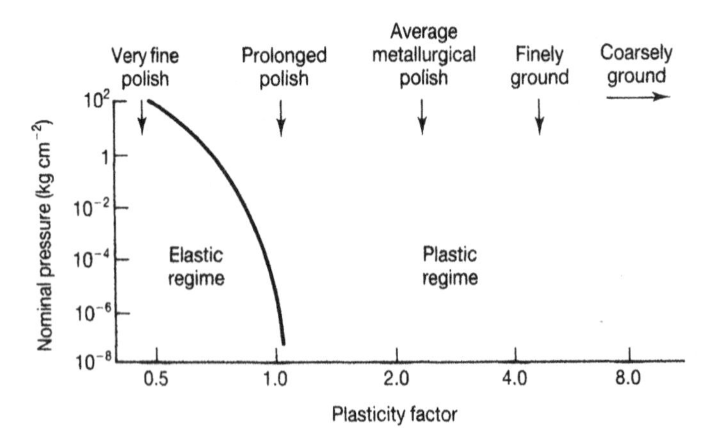

Fig. 2.13 The dependence of asperity deformation mode on plasticity index ψ for aluminium surfaces with different roughnesses (from Tabor D, Friction, lubrication and wear, in Matijevic E (Ed), *Surface and Colloid Science* vol. 5, John Wiley, 1972, pp. 245–312)

Further reading and references

Greenwood J A and Williamson J B P, Contact of nominally flat rough surfaces, *Proc Roy Soc Lond* **A295**, 300–319, 1966
Johnson K L, *Contact Mechanics*, Cambridge University Press, 1985
McCool J I, Comparison of models for the contact of rough surfaces, *Wear* **107**, 37–60, 1986
Thomas T R, *Rough Surfaces*, Longman, 1982

3

Friction

3.1 INTRODUCTION

We have seen in Chapter 2 that when two solid surfaces are placed together, contact will generally occur only over isolated parts of the nominal contact area. It is through these localized regions of contact that forces are exerted between the two bodies, and it is these forces which are responsible for friction. In this chapter we shall examine the origins of the frictional force, and try to understand the magnitude of the frictional interaction between metals, polymers, ceramics and other materials.

3.2 DEFINITION OF FRICTION

The force known as *friction* may be defined as the resistance encountered by one body in moving over another. This broad definition embraces two important classes of relative motion: sliding and rolling. The distinction between rolling and sliding friction is useful, but the two are not mutually exclusive, and even apparently 'pure' rolling nearly always involves some sliding (see Section 9.2.4).

In both ideal rolling and sliding, as illustrated in Fig. 3.1, a tangential force F is needed to move the upper body over the stationary counterface. The ratio between this frictional force and the normal load W is known as the

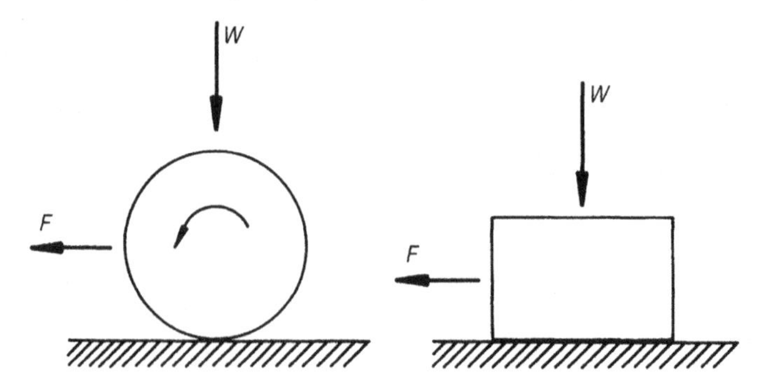

Fig. 3.1 A frictional force, F, is needed to cause motion by (a) rolling or (b) sliding

coefficient of friction, and is usually denoted by the symbol μ:

$$\mu = F/W \tag{3.1}$$

The magnitude of the frictional force is conveniently described by the value of the coefficient of friction, which can vary over a wide range: from about 0.001 in a lightly loaded rolling bearing to greater than 10 for clean metals sliding against themselves in vacuum. For most common materials sliding in air, however, the value of μ lies in the narrower range from about 0.1 to 1.

3.3 THE LAWS OF FRICTION

Under some conditions of sliding, μ for a given pair of materials and fixed conditions (or absence) of lubrication may be almost constant. This observation led to the formulation of two empirical *Laws of Sliding Friction*, often called after Amontons who rediscovered them in 1699; Leonardo da Vinci, however, had been the first to describe them some 200 years earlier. The Laws of Friction may be stated as follows:

(1) the friction force is proportional to the normal load;
and

(2) the friction force is independent of the apparent area of contact.

To these is sometimes added a third law, often attributed to Coulomb (1785):

(3) the friction force is independent of the sliding velocity.

These three Laws of Friction are of varying reliability, but except in some important cases they do provide useful summaries of empirical observations.

The First Law, which may be expressed as

$$F = \mu W \tag{3.2}$$

amounts to the statement that the coefficient of friction, μ, is independent of the normal load. For many materials under conditions of lubricated or unlubricated sliding, this is true. Amontons, in the experiments from which his Laws were deduced, used several metals and wood, all lubricated with pork fat; his materials therefore experienced *boundary lubrication* (see Section 4.6). The First Law, however, is also often obeyed for unlubricated sliding. Figure 3.2 shows typical results for the unlubricated sliding, in air, of steel on polished aluminium. The coefficient of friction remains effectively constant although the load was varied by a factor of nearly 10^6.

Although most metals, and many other materials, obey the First Law well, polymers often do not. The reasons for this difference in behaviour are discussed in Section 3.8.

The Second Law of Friction has not been so widely explored as the First, but it is nevertheless well attested for most materials, with the exception again of polymers. Figure 3.3 shows the coefficient of friction for wooden sliders on an unlubricated steel surface. The normal load was held constant, while the

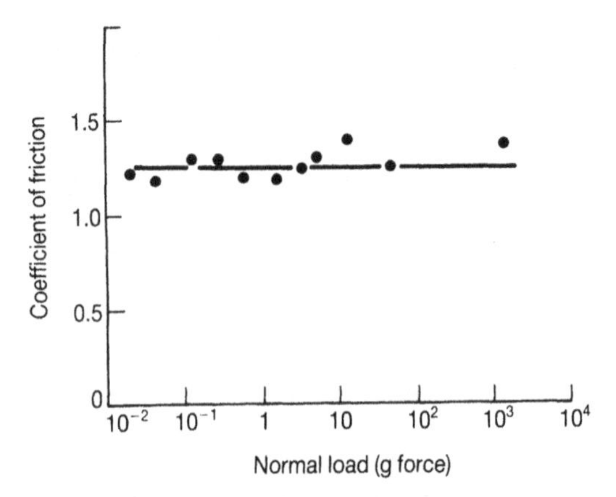

Fig. 3.2 The variation of the coefficient of friction, μ, with normal load, W, for the unlubricated sliding of steel on aluminium in air (from Bowden F P and Tabor D, *The Friction and Lubrication of Solids*, Clarendon Press, Oxford, 1950)

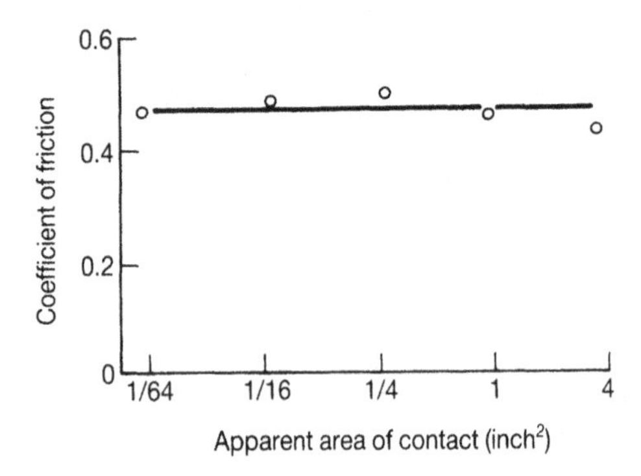

Fig. 3.3 The variation of the coefficient of friction, μ, with apparent area of contact for wooden sliders on an unlubricated steel surface (from Rabinowicz E, *Friction and Wear of Materials*, John Wiley, 1965)

apparent area of contact was varied by a factor of about 250; the value of μ is seen to be effectively constant.

The Third Law of Friction is rather less well founded than the first two. It is a matter of common observation that the frictional force needed to initiate sliding is usually greater than that necessary to maintain it, and hence that the *coefficient of static friction* (μ_s) is greater than the *coefficient of dynamic friction* (μ_d). But once sliding is established, μ_d is found for many systems to be nearly independent of sliding velocity over quite a wide range, although at high sliding speeds, of the order of tens or hundreds of metres per second for metals, μ_d falls with increasing velocity (see Section 3.5.5).

3.4 THEORIES OF FRICTION

3.4.1 Introduction

Many early investigators, among them Amontons and Coulomb, envisaged that the major contribution to the frictional force arose from mechanical interaction between rigid or elastically deforming asperities.

Figure 3.4 illustrates a simple version of this model, often called the *Coulomb model*, in which the action of the wedge-shaped asperities causes the two surfaces to move apart as they slide from position A to position B. It may then be readily shown, by equating the work done by the frictional force to that done against the normal load, that μ is equal to $\tan \theta$. It is in considering the next phase of the motion, from B to C, however, that a fundamental defect of this model becomes apparent; now the normal load does work *on* the system, and all the potential energy stored in the first phase of the motion (from A to B) is recovered. No net energy dissipation occurs in the complete cycle, and one must therefore conclude that no frictional force should be observable on a macroscopic scale if the interaction between real surfaces followed the Coulomb model exactly.

Some mechanism of energy dissipation is clearly essential in any satisfactory model for friction; in metals and ceramics, that mechanism is usually plastic deformation, sometimes in interfacial films rather than in the bulk material, while in polymers it is often viscoelastic behaviour. We shall next consider models for sliding friction. Although these have been developed primarily for metals, they are also applicable with some modification to other materials. The behaviour of ceramics, lamellar solids and polymers is discussed, in particular detail, in later sections.

3.4.2 Sliding friction

Most current theories of sliding friction stem from the important work of Bowden and Tabor carried out, mainly in Cambridge, between the 1930s and the 1970s. The Bowden and Tabor model for sliding friction, in its simplest

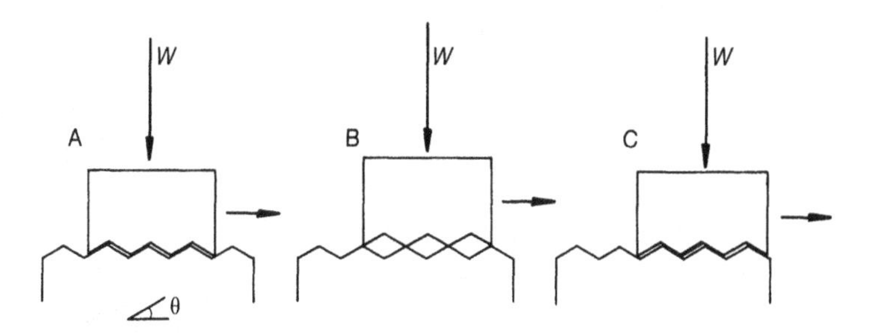

Fig. 3.4 A schematic diagram illustrating the principles behind the Coulomb model for sliding friction. The surface roughness is assumed to have a sawtooth geometry. As sliding occurs from position A to B work is done against the normal load W. The normal load then does an equal amount of work as the surfaces move from B to C

form, assumes that the frictional force arises from two sources: an *adhesion force* developed at the areas of real contact between the surfaces (the asperity junctions), and a *deformation force* needed to plough the asperities of the harder surface through the softer. Although in later developments of the theory it has become clear that these two contributions cannot be treated as strictly independent, it is convenient and illuminating to consider them separately; the resultant frictional force F is then taken to be the sum of the two contributing terms, F_{adh} due to adhesion and F_{def} due to deformation.

The adhesion term arises from the attractive forces which are assumed to operate at the asperity contacts. At first sight, this assumption may appear implausible; it is, after all, a matter of common experience that when two metal surfaces are pressed against each other they do not generally adhere. However, if the surfaces are *clean*, free from oxide and other surface films and from adsorbed gases, then significant adhesion *is* observed between metals; such conditions can be achieved under ultra-high vacuum (UHV) (at gas pressures typically $<10^{-8}$ Pa). Strong adhesion, with adhesive forces sometimes greater than the load used to press the surfaces together, is seen under UHV in ductile metals such as copper and gold. In less ductile materials, for example the hexagonal metals with a small number of slip systems, or ceramics, the adhesion is found to be weaker.

In very soft and ductile metals such as lead and indium adhesion can readily be demonstrated under ordinary laboratory conditions (Fig. 3.5).

If the rounded end of a brass or steel rod is degreased and abraded to remove some of the surface contamination, and then pressed on to the freshly-scraped surface of a block of indium, strong adhesion will occur. Furthermore, when the rod is detached fragments of indium adhere to the rod, showing that the adhesive forces at the junctions are stronger than the cohesive strength of the indium itself. Similar effects are seen in UHV experiments: if a clean iron surface is lightly pressed against a copper surface and then removed, examination of the iron surface reveals traces of copper.

Recent computer simulations of the interaction between dissimilar clean metals in contact, employing models for the relevant interatomic forces, have also shown that significant adhesion and transfer of material will occur. Figure 3.6 shows the result of a molecular dynamics simulation of the contact between a nickel indenter and an initially plane gold surface. As the indenter

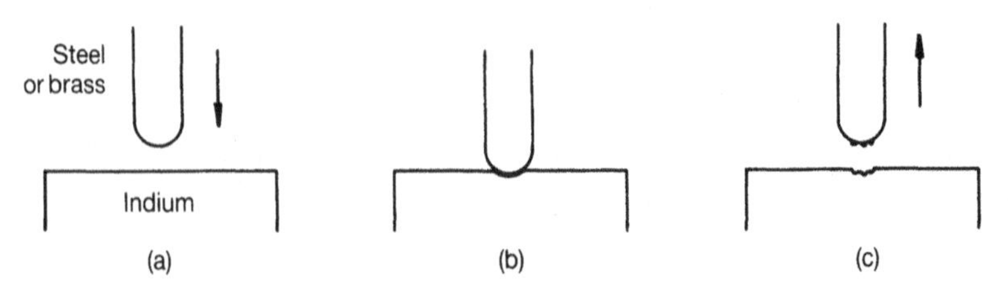

Fig. 3.5 An experiment to illustrate adhesion between metals. A clean steel or brass rod is pressed with a slight twisting motion on to the freshly-scraped surface of an indium block. Appreciable force is needed to detach the rod from the block, and fragments of indium adhere to the steel surface

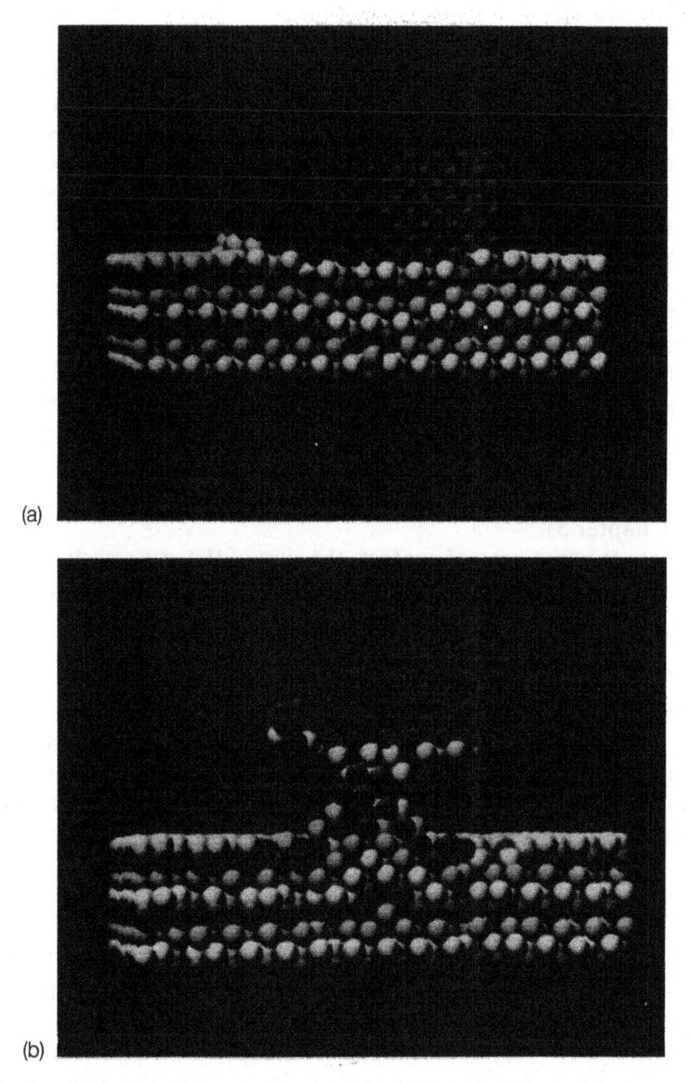

Fig. 3.6 Atomic configurations predicted by theoretical modelling of the contact between a nickel indenter (upper body) and an initially plane gold surface. No consideration is given in the model to adsorbed or reacted oxygen atoms, so that the system modelled corresponds to the contact of atomically clean metals in a perfect vacuum: (a) shows the system under normal load at the point of maximum penetration; (b) shows adhesion of gold atoms to the indenter, and a 'neck' of gold atoms drawn out as the indenter is raised from the surface (from Landman U, Luedtke W D, Burnham N A and Colton R J, *Science* **248**, 454–461, 1990)

approaches the surface, the attractive forces are sufficient to cause distortion of the gold, so that intimate contact occurs when the nickel tip is still some distance above the original level of the surface. Figure 3.6(a) shows the indenter at the point of maximum indentation into the gold. When the indenter is raised from the surface (Fig. 3.6b), the junction breaks within the gold and significant transfer of gold atoms to the nickel tip occurs.

All this evidence points to the establishment of strong interfacial bonds across asperity junctions. The crucial assumption of the Bowden and Tabor model, that adhesive forces will be present, is therefore justifiable. The reasons why significant adhesion is not observed when two metal surfaces are placed together under ordinary circumstances are twofold: in the first place the surfaces will be covered with oxide and adsorbed films which will weaken the adhesion, and secondly the elastic strains around the asperities when under load generate enough stress to break the asperity junctions during the unloading process, unless the metal is particularly ductile. Thus, only in soft ductile metals, and where oxide films are at least partially removed, can appreciable adhesion be demonstrated under ordinary conditions.

The evidence of adhesion experiments and of theoretical modelling further suggests that when two dissimilar metals slide against each other the asperity junctions formed will, in fact, be stronger than the weaker of the two metals, leading to the plucking out and transfer of fragments of the softer metal on to the harder. This is indeed observed, and gives rise to severe wear of the softer metal (see Chapter 5).

If we denote the true area of contact, the sum of the cross-sectional areas of all the asperity junctions, by A, and assume that all junctions have the same shear strength s, then the friction force due to adhesion will be given by:

$$F_{adh} = As \tag{3.3}$$

We saw in Section 2.5.3 that to a first approximation, whether the contact between asperities is predominantly elastic or predominantly plastic, the real area of contact is almost linearly proportional to the applied normal load. For contact between metal surfaces finished by conventional engineering processes, the initial asperity contacts will be effectively plastic (see Section 2.5.3), and we can write

$$W \approx AH \tag{3.4}$$

since the normal stress which the asperities are capable of supporting, if they deform plastically, will be close to the indentation hardness H of the softer material (see Section 2.5.1).

The contribution to the coefficient of friction from the adhesive forces is therefore

$$\mu_{adh} = F_{adh}/W \approx s/H \tag{3.5}$$

Since the asperity junctions fail by rupture within the weaker of the two materials, we can take s, to a first approximation, to be the shear strength of that material; H is the indentation hardness of the same material. For metals, the indentation hardness is about three times the uniaxial yield stress:

$$H \approx 3Y \tag{3.6}$$

The yield stress Y will be about 1.7 or 2 times the yield stress in pure shear, s, the precise factor depending on the yield criterion. We therefore expect that

$$H \approx 5s \tag{3.7}$$

and thus that

$$\mu_{adh} \approx s/H \approx 0.2 \tag{3.8}$$

The frictional force due to the ploughing of harder asperities through the surface of a softer material, the *deformation* term, may be estimated by considering a simple asperity of idealized shape.

If a rigid conical asperity of semi-angle α (Fig. 3.7) slides over a plane surface, the tangential force needed to displace it will be some flow pressure, which we may take as the indentation hardness H of the surface material, multiplied by the cross-sectional area of the groove:

$$F_{def} = Hax = Hx^2 \tan \alpha \tag{3.9}$$

The normal load supported by the asperity is given by

$$W = H\pi a^2/2 = \tfrac{1}{2}H\pi x^2 \tan^2 \alpha \tag{3.10}$$

The coefficient of friction due to the ploughing term will therefore be

$$\mu_{def} = F_{def}/W = (2/\pi) \cot \alpha \tag{3.11}$$

A plane strain model, where the asperity is taken to be a wedge of semi-angle α, leads in a similar way to

$$\mu_{def} = \cot \alpha \tag{3.12}$$

These relationships are supported by the evidence of experiments in which macroscopic model asperities are dragged across softer metal surfaces.

The slopes of real surfaces are nearly always less than 10° (i.e. $\alpha > 80°$), and therefore from equations 3.11 and 3.12, we would expect μ_{def} to be less than about 0.1.

We therefore conclude from our simple model that, even for a hard metal sliding on a softer one, the total coefficient of friction, representing the contributions from both ploughing and adhesion terms, should not exceed 0.3 or so. For a metal sliding against a counterface of the same material, where the contribution from ploughing should be negligible, μ should be slightly lower, of the order of 0.2.

If we look at the experimental measurements of μ for the unlubricated sliding of metals listed below in Table 3.1, we find that the measured values are, in fact, typically several times these estimates. The discrepancy is so large that other effects must be playing a role. Two such effects dominate: *work-hardening* and *junction growth*.

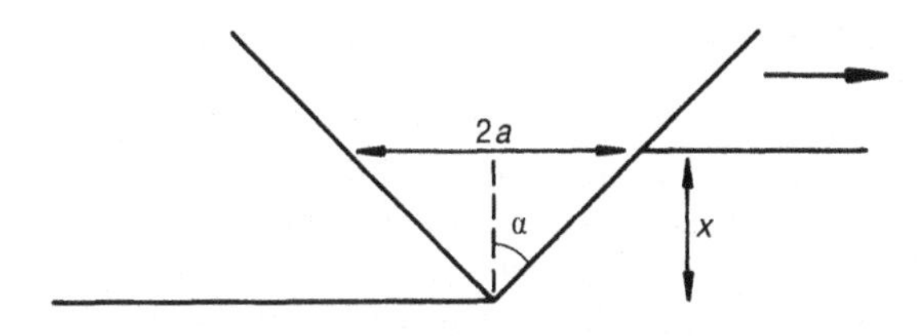

Fig. 3.7 Model for the deformation component of friction, in which a conical asperity of semi-angle α indents and slides through the surface of a plastically deforming material

Table 3.1 Typical values of the coefficient of static friction, μ_s, for combinations of metals in air and without lubrication. It must be appreciated that the value of μ_s depends on the precise conditions of the test, and these values should be taken as only broadly representative. Where two values are quoted, they refer to conditions under which the oxide films are intact, and in which they are penetrated (data from Bowden F P and Tabor D, *The Friction and Lubrication of Solids, Part II*, Clarendon Press, Oxford, 1964, and Tabor D, in Neale M J (Ed.), *Tribology Handbook*, Butterworths, 1973)

(a) Self-mated metals in air	μ_s
Gold	2
Silver	0.8–1
Tin	1
Aluminium	0.8–1.2
Copper	0.7–1.4
Indium	2
Magnesium	0.5
Lead	1.5
Cadmium	0.5
Chromium	0.4

(b) Pure metals and alloys sliding on steel (0.13% C) in air	μ_s
Silver	0.5
Aluminium	0.5
Cadmium	0.4
Copper	0.8
Chromium	0.5
Indium	2
Lead	1.2
Copper–20% lead	0.2
Whitemetal (Sn based)	0.8
Whitemetal (Pb based)	0.5
α-brass (Cu–30% Zn)	0.5
Leaded α/β brass (Cu–40% Zn)	0.2
Grey cast iron	0.4
Mild steel (0.13% C)	0.8

In the simple model developed above, the material is assumed to have a constant flow stress. However, nearly all metals strain-harden to some extent, and although the normal load is supported by plastic flow some distance from the immediate vicinity of the asperity junctions, the junctions themselves will work-harden significantly, which will tend to raise the relative value of s in comparison with that of H. Work-hardening will therefore tend to increase μ_{adh}, although the phenomenon is difficult to quantify. Probably more important is the effect of junction growth, which will be considered in the next section.

3.4.3 Junction growth

In the model just described, we have assumed that the true area of contact is determined solely by the normal load, and that it is unaffected by tangential forces. This is, in fact, a considerable oversimplification. Whether the metal flows plastically or not is determined by a yield criterion, which takes account of both the normal and the shear stresses acting; a simple illustration will indicate that incorporation of a yield criterion into the model markedly affects its predictions.

Figure 3.8 shows a slab of material loaded against a rigid plane surface, representing in very idealized form an asperity contact. The element of material just inside the slab at (a) is subjected to uniaxial compression by a normal stress p_o and we can assume it to be on the point of yielding, since we know that nearly all asperity contacts between metals are plastic. When a tangential stress is then applied to the asperity junction, as at (b), the element of material experiences an additional shear stress τ. For the material to remain at the point of yielding, the normal stress on the element must be reduced to a value p_1. If the normal load remains constant, then the area of contact must grow: the phenomenon is therefore known as *junction growth*.

The relationship between p_o, p_1 and τ is determined by the yield criterion. For Tresca's criterion, in which plastic flow occurs at a critical value of the maximum shear stress,

$$p_1^2 + 4\tau^2 = p_o^2 \tag{3.13}$$

For the von Mises yield criterion,

$$p_1^2 + 3\tau^2 = p_o^2 \tag{3.14}$$

Whether we use equation 3.13 or 3.14 does not much matter; both lead to the same qualitative conclusions. Let us examine these for equation 3.13. The normal and shear stresses are given by:

$$p_1 = W/A \tag{3.15}$$

and

$$\tau = F/A \tag{3.16}$$

where A is the true area of contact. Note that F here denotes the tangential force, and does not necessarily imply that sliding is actually occurring. We can now substitute in equation 3.13 and obtain

$$W^2 + 4F^2 = A^2 p_o^2 \tag{3.17}$$

In a typical sliding experiment under dead-weight loading, W is constant, while p_o is a property of the material (its yield stress in compression). The real

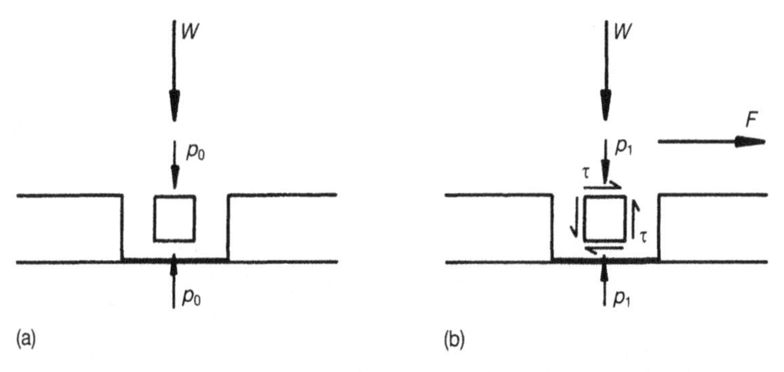

Fig. 3.8 Stresses acting on an element within an idealized asperity pressed against a counterface (a) with no tangential force, and (b) with a tangential force applied

area of contact will therefore increase with increasing tangential force, and the ratio F/W, the instantaneous value of μ, will also increase steadily.

There is nothing in this model to limit the growth process; in theory it could continue until the whole area of the specimen was actually in contact, and the coefficient of friction would reach a very high value. Under certain conditions junction growth in metals can indeed proceed this far (see Section 3.5.2), but in most practical cases it is limited by the ductility of the material, and by the presence of weak interfacial films. We can model the effect of a weak interface if we assume that it will fail at some shear stress τ_i, less than the shear strength of the bulk material. The maximum possible tangential force is then given by

$$F_{max} = \tau_i A_{max} \tag{3.18}$$

and the coefficient of friction is

$$\mu = F_{max}/W \tag{3.19}$$

If the shear yield stress of the bulk asperity material is τ_o, then from the Tresca yield criterion

$$p_o = 2\tau_o \tag{3.20}$$

An expression for μ can now be derived:

$$\mu = \frac{F_{max}}{W} = \frac{1}{2((\tau_o/\tau_i)^2 - 1)^{1/2}} \tag{3.21}$$

When the interface has the same shear strength as the bulk material ($\tau_o/\tau_i = 1$), equation 3.21 shows that μ becomes infinite, since junction growth is unlimited. For weaker interfaces μ is finite, and drops rapidly as

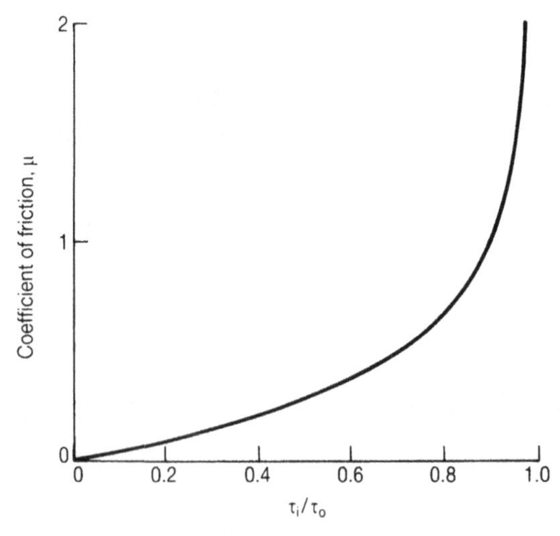

Fig. 3.9 The variation of coefficient of friction, μ, with the ratio beween the shear strength of the interface and that of the bulk material (τ_i/τ_o). The function plotted is that given by equation 3.21

τ_i/τ_o falls. The dependence of μ on interfacial shear strength is shown in Fig. 3.9.

An interface only 10% weaker than the bulk is sufficient to reduce μ to about unity, while for an interface with half the bulk strength, μ falls to about 0.3. If the interface is very weak, then μ can have an extremely low value: for an interfacial strength one tenth of the bulk value, $\mu = 0.05$.

Weak interfacial films lead to very limited junction growth, and we would then expect essentially the same value of μ predicted from equation 3.21 as from a model which takes no account of junction growth at all. Such a theory can be readily developed: if we assume the idealized asperity in Fig. 3.8 to be separated from the substrate by a weak film of shear strength τ_i, then the frictional force is determined by the shear strength of the film:

$$F = A\tau_i \tag{3.22}$$

and the normal load is supported by the plastic flow stress of the bulk of the asperity:

$$W = Ap_o \tag{3.23}$$

Hence we predict that

$$\mu = F/W = \tau_i/p_o \tag{3.24}$$

Equation 3.21 reduces to the same simple form if $\tau_i \ll \tau_o$ and equation 3.20 is used to relate p_o and τ_o.

Equation 3.24 is important, since it suggests a way of reducing friction; if a film of material of low shear strength can be interposed between two surfaces, then the coefficient of friction can be lowered. This principle underlies the action of lubricants, discussed in Chapter 4, and is also exploited in the design of some journal bearing materials (Section 9.4).

3.4.4 Better models for asperity deformation

The picture of a slab-like asperity illustrated in Fig. 3.8 is very idealized, and several attempts have been made to provide a better model for asperity deformation. The asperities involved in the contact of real surfaces can be approximated more accurately by spherical, conical or wedge-shaped protuberances: for all these cases, as a result of plastic constraint, the normal stress needed to cause plastic deformation will be higher than that for the frictionless compression of a slab, and the arguments put forward in the previous section need amendment.

Bowden and Tabor made a simple modification to enable them to treat asperities of general shape, by introducing a more general form of equation 3.13:

$$p_1^2 + a\tau^2 = p_o^2 \tag{3.25}$$

where a is a numerical factor, determined empirically, with a value of the order of 12. The conclusions from this approach, which is not rigorous, are essentially the same as those from the simple model described in the previous section; in particular, equation 3.21 is changed only by the introduction of a numerical factor.

A different approach, the results of which are also in qualitative agreement with those of the simpler theory, has been to analyse by plasticity theory the contact between a rigid-plastic plane and a rigid wedge-shaped asperity. Even this is a considerable simplification of the contact of many real materials, but the model nevertheless provides some valuable insights into the processes of friction and wear.

Figure 3.10 shows slip-line fields for the behaviour of a rigid wedge loaded against a deformable (rigid-plastic) plane. Under a purely normal load the wedge will indent the plane to a certain depth, as in a hardness test. When a tangential force is then superimposed, the wedge will sink further into the plane, pushing a prow of material ahead of itself and leading to an increase in the area of contact; this is junction growth. Eventually, as the sideways force is further increased, continuous tangential motion of the wedge will occur. Under these conditions, which correspond to steady-state sliding, the prow of material is pushed ahead of the wedge. Let us consider two extreme cases: one where perfect adhesion occurs between the wedge and the substrate, and one where there is no adhesion.

Full adhesion leads to a large prow, as shown in Fig. 3.10(a). Here μ, the ratio between the tangential and normal forces, is close to unity. With zero

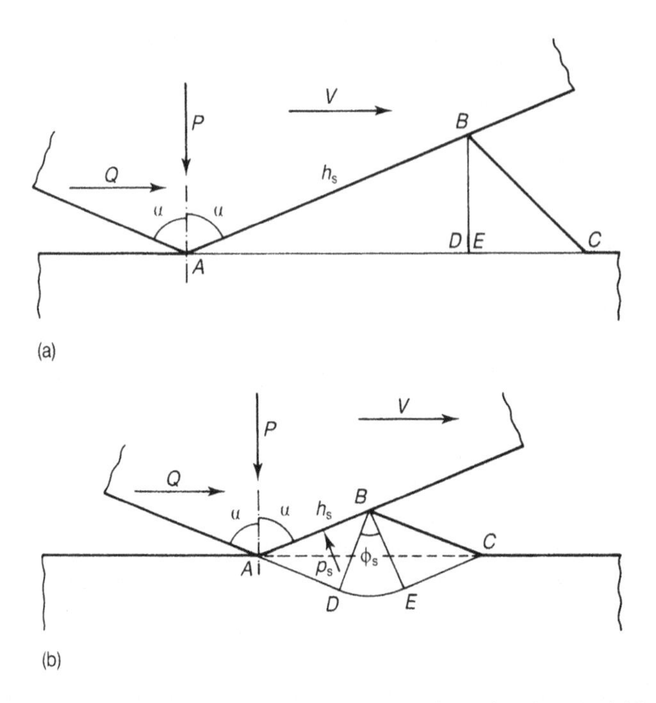

Fig. 3.10 Predictions from slip-line field analysis of the deformation due to a rigid wedge sliding under steady-state conditions over an ideal-plastic surface with (a) perfect interfacial adhesion, and (b) no adhesion (from Johnson K L, *Contact Mechanics*, Cambridge University Press, 1985)

adhesion (Fig. 3.10(b)), a much smaller prow is formed, the plastic strains in the substrate are less, and the coefficient of friction is given by

$$\mu = \cot \alpha \tag{3.26}$$

as in the simple plane-strain ploughing model discussed in the previous section (equation 3.12).

The approach used in this theory is attractive, in that it integrates the two influences of adhesion and ploughing, separated by Bowden and Tabor in their simpler picture, into one model. The theory can be further developed by considering the interface to have a shear strength less than that of the bulk, and the effects of interfacial films can then be treated.

Figure 3.11 illustrates results from this model. For typical asperity angles ($\alpha > 80°$) the qualitative picture is very similar to that of Fig. 3.9: the high value of μ predicted for $\tau_i = \tau_o$ drops rapidly as the interface is weakened. In this case, however, there is a small residual ploughing component of friction (where $\mu = \cot \alpha$) even when the interface has no strength at all.

3.5 FRICTION OF METALS

3.5.1 Introduction

Table 3.1 lists values of the coefficient of static friction μ_s, taken from the literature, for several combinations of metals in air and without lubrication.

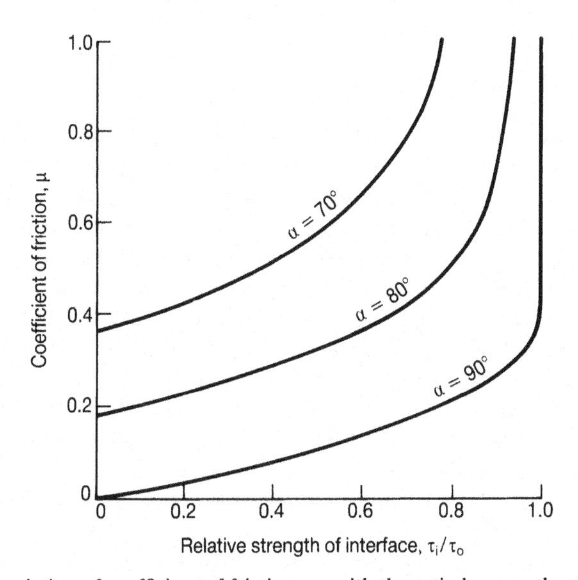

Fig. 3.11 The variation of coefficient of friction, μ, with the ratio beween the shear strength of the interface and that of the bulk material (τ_i/τ_o), derived from slip-line field analyses of wedge sliding (from Johnson K L, in Dowson D, Taylor C M, Godet M and Berthe D (Eds), *Friction and Traction*, Westbury House, 1981, pp. 3–12)

As was noted in Section 3.3, the coefficient of dynamic friction μ_d, applicable once sliding is established, would be expected to be somewhat lower.

It must be borne in mind in all quantitative discussions of friction that the precise value of μ depends critically on the experimental conditions under which it is measured; the usefulness of the figures in Table 3.1 lies more in their relative magnitudes than in their absolute values. In some cases, as shown by the examples below, to quote a single value for μ is very misleading, since the coefficient depends so strongly on the conditions. We shall discuss some of the underlying reasons for the values of μ listed in Table 3.1 in the following sections.

3.5.2 Clean metals in high vacuum

If metal surfaces are cleaned in high vacuum and then placed in contact, strong adhesion is usually observed. The surfaces do not need to be atomically clean for them to adhere: merely heating them in vacuum to drive off adsorbed gases is often sufficient. The coefficient of friction under these conditions has a very high value, typically 2 to 10 or even more, and gross seizure frequently follows when sliding is attempted. Strong metallic bonds form across the interface; when the surfaces are pulled apart, metal is transferred from one body to the other. With little or no interfacial contamination present, the extent of junction growth is limited only by the ductility of the asperity material. The coefficient of friction is therefore very high. Sliding friction under ultra-high vacuum conditions occurs in space engineering, and special measures must be taken in designing sliding components to operate in this demanding environment. Solid lubricants and thin soft metallic films (see Section 4.7) can provide valuable protection.

3.5.3 Self-mated metals in air

In most practical applications, metals slide against one another in air. Coefficients of friction are then much lower than in vacuum, and lie typically, for unlubricated sliding, in the range from 0.5 to 1.5. Table 3.1(a) lists some representative values of μ, although it must be remembered that frictional behaviour depends, sometimes markedly, on the composition and microstructure of the materials and on the conditions of measurement.

Gold has a particularly high value of μ among the metals listed. It forms no oxide film although adsorbed gases will be present in air, and the asperity junctions therefore tend to be strong. Gold is also ductile, so that considerable junction growth can occur. Both these factors contribute to a high frictional force, although not so high as that found in high vacuum.

All the other metals oxidize in air to some extent, forming oxide films typically between 1 and 10 nm thick within a few minutes exposure of an atomically clean surface. These films play a critical role in determining the sliding behaviour, since friction between oxide surfaces, or between oxide and bare metal, is almost always less than between surfaces of bare metal.

Figure 3.12 illustrates the effect of oxygen on the sliding friction of pure iron. In high vacuum, strong adhesion and seizure occur. The admission of

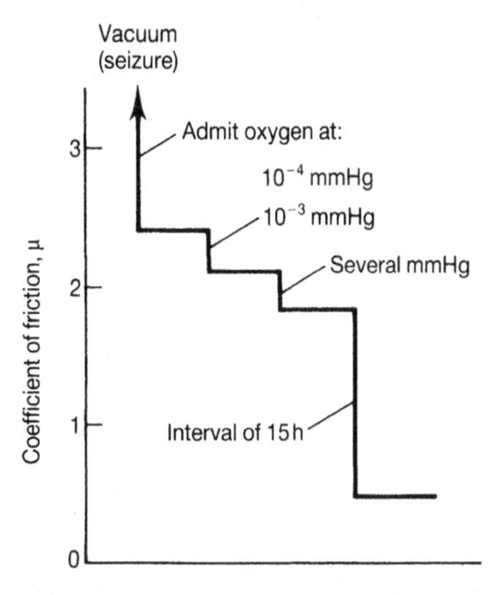

Fig. 3.12 The effect of oxygen on the friction of pure iron. In vacuum, the value of μ is very high; as the oxygen concentration is increased, so the coefficient of friction falls (from Buckley D H, *Surface Effects in Adhesion, Friction, Wear and Lubrication*, Elsevier, 1981)

only a small pressure of oxygen allows sliding, although with a high coefficient of friction. As more oxygen is admitted to the system, the value of μ drops, until it eventually reaches the value normally measured in air at ambient pressure.

The effect of an oxide film in mitigating friction can be destroyed to some extent, however, if the film is penetrated during sliding. Figure 3.13 shows a common form of behaviour, here seen in copper.

At low normal loads, the oxide films effectively separate the two metal surfaces and there is little or no true metallic contact. The electrical resistance of the interface is high, and the track formed by the slider appears smooth and

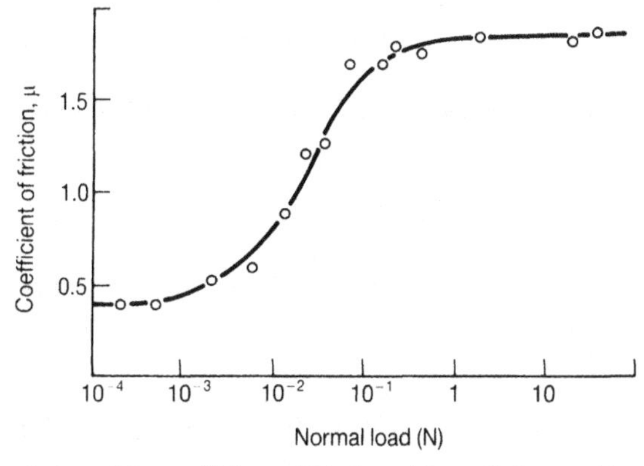

Fig. 3.13 The variation of the coefficient of friction with applied normal load for copper sliding against copper in air, unlubricated (from Whitehead J R, *Proc. Roy. Soc. Lond.* **A201**, 109–124, 1950)

polished. The coefficient of friction is low, perhaps because the oxide acts as a low shear strength film, but more probably because its low ductility limits junction growth. As the normal load is increased, a transition occurs to a higher value of μ: in the case of copper in these experiments from less than 0.5 to greater than 1.5. The track of the slider now exhibits considerable surface damage with evidence of metallic transfer, the electrical resistance of the contact falls to a low value, and we deduce that metallic contact is occurring between the copper asperities.

Transitions of this kind are common in other metals also, although the change in μ may not be so great as in copper. In aluminium, for example, μ typically increases from about 0.8 to 1.2 as the oxide film breaks down. Cadmium shows almost no change in μ, although electrical resistance measurements confirm that an insulating film is present at low loads which is broken down at higher loads. In all these cases, when the surfaces are separated by protective oxide films the friction is essentially being measured between oxide surfaces. At higher loads the surface films may deform and fracture allowing true metallic contact to occur, and the friction often (but not always) rises. As might be expected, the wear rates associated with these two regimes are markedly different, and are further discussed in Chapter 5.

For some very soft and ductile metals, notably tin and indium, metallic contact between the surfaces occurs even at the lightest loads, and the coefficient of friction is consequently high and does not change markedly with load. Here the oxide films, although formed, are penetrated easily since the soft substrate offers little mechanical support. At the other extreme of behaviour, chromium forms a thin but very strong oxide film, and exhibits over a wide range of load no metallic contact (as determined by electrical resistance measurements) and a constant low coefficient of friction.

In summary, the friction of pure metals sliding against themselves in air, unlubricated, is often determined by the presence of surface oxides. If the oxide film is not broken during sliding, surface damage is slight and the oxide itself determines the coefficient of friction. The value of μ is often (but not always) lower than at higher loads where the oxide film is penetrated and metallic contact between asperities occurs. Under these conditions considerable surface damage and rapid wear occur. In several metals no transition occurs because the protective oxide coating is retained over a wide range of load, while in others, where the oxide is penetrated even at very light pressure, the contact is always metallic.

3.5.4 Dissimilar metals, and alloys

Oxide films are also important in the friction of dissimilar metals and alloys in air. In general, the coefficient of friction for an alloy tends to be rather less than that for its pure components.

The sliding friction of steels has been widely studied; μ varies with both composition and microstructure, and also often depends on load. Figure 3.14 shows typical behaviour.

The solid curve, for a 0.4% carbon steel, shows a transition from a relatively high value of μ at low loads to a lower value at high loads. It has

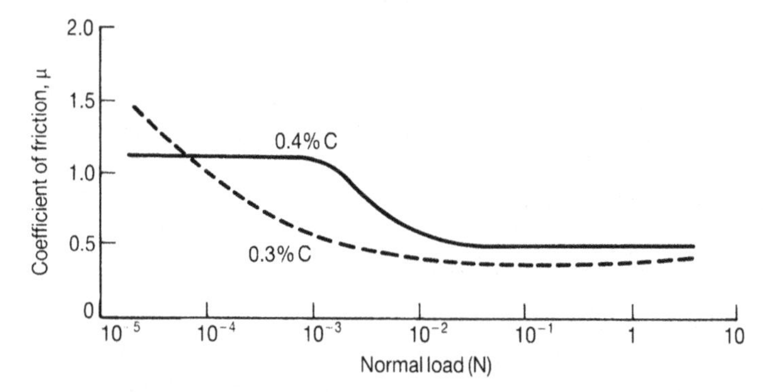

Fig. 3.14 The variation of coefficient of friction with normal load for steels sliding against themselves in air, unlubricated. Results are shown for two different plain carbon steels, with carbon contents of 0.4% (full curve) and 0.3% (broken curve) (from Wilson R W, 1952, cited by Bowden F P and Tabor D, *The Friction and Lubrication of Solids, Part II*, Clarendon Press, Oxford, 1964)

been suggested that the transition results from the stratified structure of the oxides present on the steel. The uppermost layer is Fe_2O_3, while beneath this lie layers of Fe_3O_4 and finally FeO next to the metal itself. According to this model, the transition then results from penetration of the Fe_2O_3 layer at the higher loads. Some other steels do not show such a pronounced transition: the broken line in Fig. 3.14 (for a steel of lower carbon content) illustrates behaviour commonly observed.

The effects of minor constituents on the frictional behaviour of alloys can be marked, as a result of surface segregation. Aluminium in steels, for example, segregates to the surface and in vacuum tends to increase the coefficient of friction. Oxidation of the surface, however, produces a layer of aluminium oxide which promotes a reduction in friction. The effects of surface segregation are complex, but can provide a way of changing the frictional properties of an alloy through relatively small modifications to its composition.

Table 3.1(b) lists representative values of μ for various alloys sliding against a low-carbon steel. Some alloys, notably leaded α/β brass (copper–zinc), grey cast iron and copper–lead, stand out as showing particularly low coefficients of friction. The reason in all three cases is the same: all contain phases which provide thin interfacial films of low shear strength. Leaded brass and copper–lead both contain dispersions of metallic lead, which has very little solubility in the matrix phases. On sliding, a thin weak film of lead is formed on the surface and results in a low value of μ by the mechanism expressed in equation 3.24. In grey cast iron the low shear strength film is provided by the graphite constituent (see Section 3.7). These alloys with intrinsically low coefficients of friction in dry sliding against steel, which do not depend on the formation of a protective oxide layer, have wide engineering use: copper–lead and other copper-based alloys containing lead are used in journal bearings (see Sections 9.3 and 9.4), while the good tribological properties of grey cast iron add to the attractions of its low cost and damping capacity for applications such as heavy machine tool slideways. It is worth noting that the traditional bearing alloys, the tin- and lead-based whitemetals, do not have intrinsically low coefficients of friction against steel (Table 3.1(b)). Other factors, as discussed in Section 9.3.2, determine their suitability for this purpose.

3.5.5 Effect of Temperature

When the temperature of a sliding metal is increased, several effects will occur: its mechanical properties will change, its rate of oxidation will increase, and phase transformations may take place. All these will influence its frictional behaviour.

Figure 3.15 shows measurements of μ in ultra-high vacuum for metals of the three common crystal structures sliding against themselves, and illustrates how the influence of temperature on their plastic deformation behaviour changes the measured coefficient of friction. For both the cubic close-packed (c.c.p.) and body-centred cubic (b.c.c.) metals, transitions in friction occur. In the c.c.p. metals these are associated with a marked change in the work-hardening rate (which is higher at low temperatures), while in the b.c.c. metals the transition coincides with the ductile–brittle transition.

The hexagonal close-packed (h.c.p.) metals show no change in friction with temperature over this range, since their mechanical properties do not alter significantly, but there is a clear correlation between the ductilities of the individual metals and their values of μ. Titanium and zirconium, being fairly ductile, resemble the c.c.p. metals in their friction behaviour, whereas beryllium and cobalt, with their ductility limited by the small number of slip systems operating at these low temperatures, exhibit lower values of μ. In these experiments, where the influence of interfacial films was deliberately excluded by performing them in vacuum, it is clear that μ was largely determined by the ductility of the metal at the points of contact. In general, as the ductility of a metal increases, so does the value of μ.

When metals are heated in air, their rate of oxidation will increase and so the thickness, and possibly also the nature, of the oxide films will change. This provides a second mechanism through which friction can vary with temperature.

Figure 3.16 illustrates this effect, for the sliding of an austenitic stainless steel against pure nickel. As the temperature is increased, the rising ductility of the metals leads to a steady rise in μ, until suddenly at 750 °C a thick film of nickel oxide forms on the nickel and the friction drops sharply to a lower value. On cooling, the oxide continues to separate the asperities, and the low friction is maintained down to much lower temperatures. Similar behaviour is commonly observed in steels. A surface layer of Fe_3O_4 provides a lower coefficient of friction than Fe_2O_3; the predominant oxide which forms is determined by the temperature and by the compositions of the alloy and of the gas phase. Other surface films formed by gas–solid reactions can change the frictional behaviour in suitable systems: chlorine, for example, reacts strongly with many metals and the resultant chloride films can reduce friction.

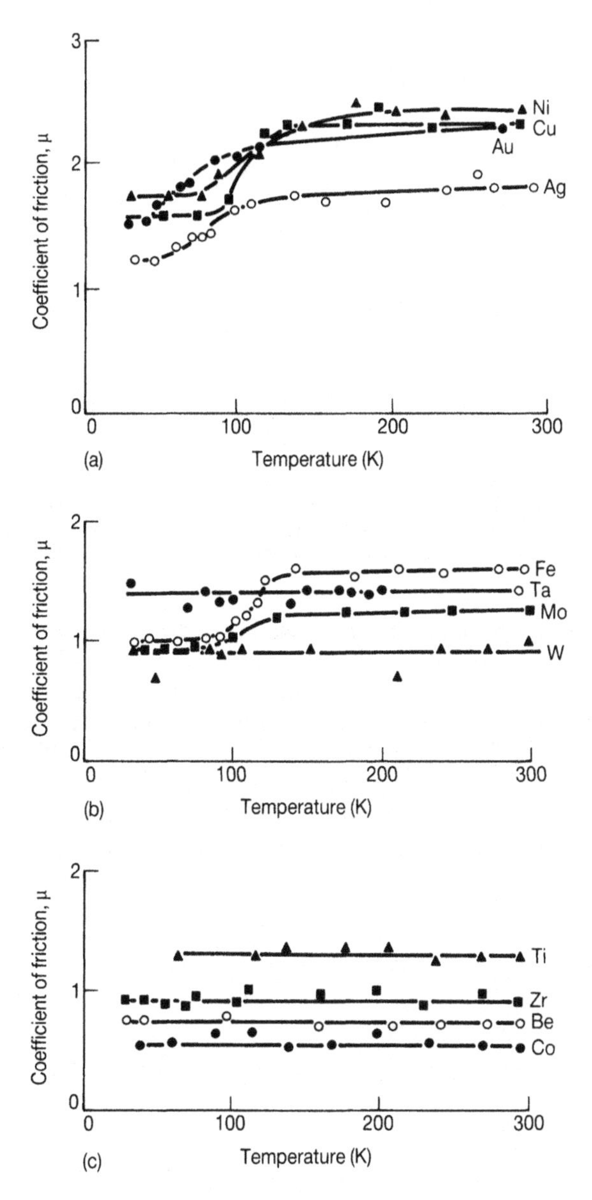

Fig. 3.15 The variation of coefficient of friction with temperature for various pure metals sliding against themselves in ultra-high vacuum: (a) c.c.p. metals; (b) b.c.c. metals; (c) h.c.p. metals (from Bowden F P and Childs T H C, *Proc. Roy. Soc. Lond.* **A312**, 451–466, 1969)

Phase transformations, through their influence on the mechanical properties of the materials, can result in large changes in friction. The most drastic effect is that due to melting; as a metal approaches its melting point its strength drops rapidly, and thermal diffusion and creep phenomena become more important. The resulting increased adhesion and ductility at the points

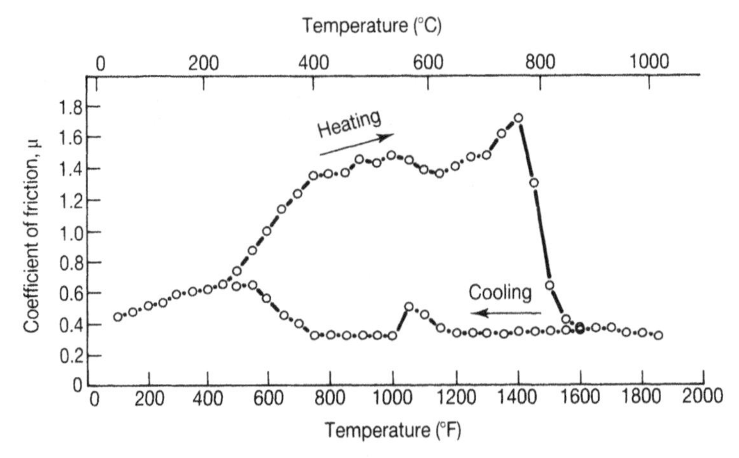

Fig. 3.16 The dependence of coefficient of friction on temperature for an austenitic stainless steel sliding against pure nickel in air. The two curves relate to data gathered while the specimens were being heated (upper curve) and subsequently cooled (lower curve) (from Peterson M B, Florek J J and Lee R E, *Trans. Am. Soc. Lubrication Eng.* **3**, 101–115, 1960)

of contact lead to a pronounced increase in friction. When one of the sliding surfaces actually becomes molten, however, and therefore loses its shear strength, the friction force drops to a low value determined by viscous forces in the liquid layer. This occurs in the sliding of metals at very high speeds (typically $>100\ \mathrm{m\ s^{-1}}$: see Fig. 3.17) and also in the sliding of a ski over ice or snow. In both these cases, the dissipation of frictional work generates enough local heat to raise the temperature at the interface to the melting point, and sliding then takes place under conditions of effectively hydrodynamic lubrication (see Section 4.4). At low sliding speeds, or at sufficiently low ambient

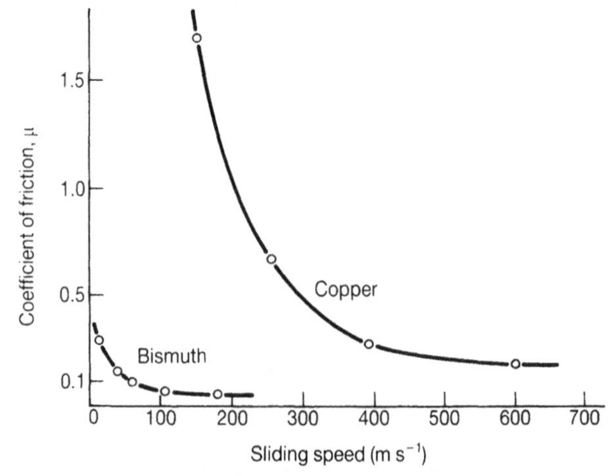

Fig. 3.17 The variation of coefficient of friction with sliding speed for pure bismuth and copper sliding against themselves (from Bowden F P and Tabor D, *The Friction and Lubrication of Solids, Part II*, Clarendon Press, Oxford, 1964)

temperatures, a molten film does not form and the friction is controlled by the interaction of the solid surfaces.

Solid-state phase transformations also influence friction. Figure 3.18 shows the variation of the friction of cobalt with temperature. Cobalt exhibits a rather sluggish transformation (equilibrium temperature 417 °C) from its low-temperature hexagonal close-packed structure, with limited slip ductility, to a cubic close-packed structure which is fully ductile. The coefficient of friction associated with the c.c.p. structure is high, whereas that of the h.c.p. is low. The change in frictional behaviour is first observed at an ambient temperature below the equilibrium temperature for the transformation

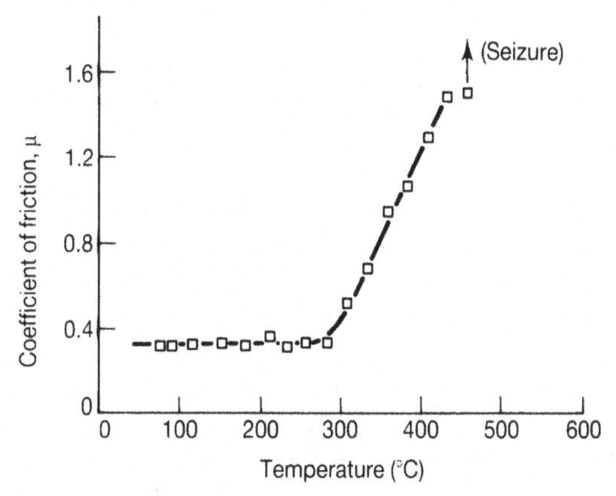

Fig. 3.18 The variation of coefficient of friction with temperature for cobalt sliding against itself in vacuum (from Buckley D H, *Surface Effects in Adhesion, Friction, Wear and Lubrication,* Elsevier, 1981)

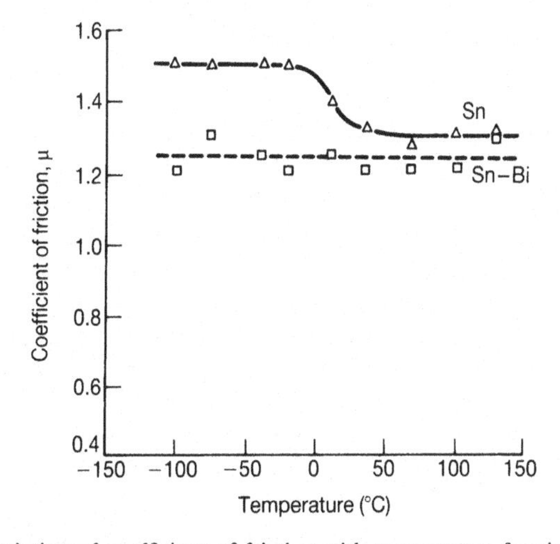

Fig. 3.19 The variation of coefficient of friction with temperature for tin sliding against itself (solid curve). The addition of a small amount of bismuth suppresses the allotropic transformation to grey tin (broken curve) (from Buckley D H, *Surface Effects in Adhesion, Friction, Wear and Lubrication,* Elsevier, 1981)

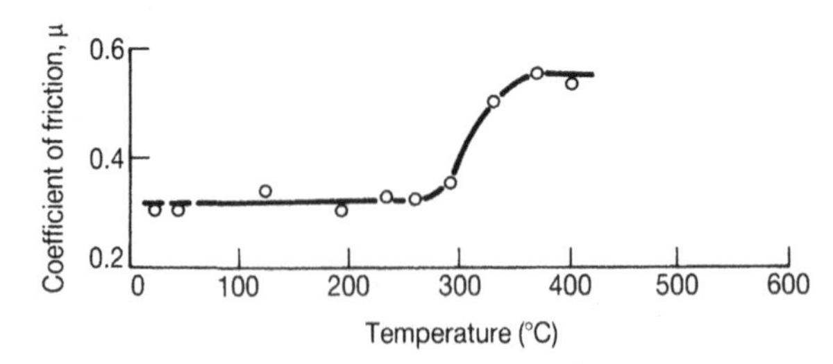

Fig. 3.20 The variation of coefficient of friction with temperature (in vacuum) for the alloy Cu_3Au which exhibits an order–disorder transformation at 390 °C (from Buckley D H, *Surface Effects in Adhesion, Friction, Wear and Lubrication*, Elsevier, 1981)

because frictional heating raises the local interfacial temperature.

Tin is another metal exhibiting a polymorphic transformation which leads to a change in friction, as seen in Fig. 3.19. Below the transformation temperature of 13 °C grey tin (cubic, with the diamond structure) is the stable allotrope, while above, white tin (body-centred tetragonal) is stable. The change in μ shown in Fig. 3.19 is fully reversible with temperature. The addition of a small amount of bismuth to tin suppresses the transformation from the white to the grey allotrope, and the coefficient of friction for the tin–bismuth alloy therefore does not change with temperature.

Another class of phase transformation affecting friction in metals is the order–disorder transformation, which occurs in many alloy systems. The copper–gold alloy of composition Cu_3Au has been investigated (Fig. 3.20). At low temperatures Cu_3Au has a long-range ordered cubic structure; an order–disorder transformation occurs at 390 °C. The mechanical properties change markedly with the transformation: both the elastic modulus and the hardness increase on ordering and, as can be seen from Fig. 3.20, there is also a pronounced decrease in friction.

3.6 FRICTION OF CERAMIC MATERIALS

The ceramic materials of most interest to the tribologist are the so-called *engineering ceramics*, which combine low density with excellent mechanical properties (e.g. high strength, hardness and stiffness) up to high temperatures. Typical examples of engineering ceramics are silicon nitride (Si_3N_4), silicon carbide (SiC), alumina (Al_2O_3) and zirconia (cubic and/or tetragonal ZrO_2). Although these materials may be nominally pure, they usually contain additives, which may be present either in small amounts to assist the fabrication process (e.g. as sintering aids), or in more substantial proportions as alloying additions. There is also considerable interest in the use of ceramic materials as thin coatings on substrates of other materials: titanium nitride (TiN) and diamond, in particular, have attractive tribological properties when used in this way.

The major differences in mechanical behaviour between the engineering ceramics and metals arise from the different nature of the interatomic forces:

ionic or covalent in ceramics, rather than metallic bonding. Ionic bonding, in ceramics such as MgO or Al_2O_3, leads to crystal structures with only a small number of independent slip systems available for dislocations, fewer than the five necessary to accommodate a general plastic strain. Covalent bonding, as in SiC, TiC or diamond, leads to very narrow dislocations which move only under high stress, even though five independent slip systems may exist.

Ceramic materials of either bond type thus show only limited plastic flow at room temperature, and correspondingly, much less ductility than metals. Also, as we saw in Section 2.5.3, asperity contact between ceramics is more likely to be elastic than in metals. The large plastic strains associated with junction growth in metals do not therefore occur in ceramics except at high temperature, and although adhesive forces (of covalent, ionic or van der Waals origin) are present between ceramic materials in contact, the coefficient of friction never reaches the very high values seen in clean metals sliding in the absence of oxygen (see Section 3.5.2). The value of μ for ceramic–ceramic contacts lies typically in the range from 0.25 to 0.8, although some lower values will be discussed below. These are similar to the values seen for metallic couples sliding in air in the presence of intact oxide films, and indeed, there is much similarity between the contact of oxidized metal surfaces and that of bulk oxide ceramics.

Wide variability is typically seen in reported values of μ for engineering ceramics, and it would therefore be misleading to state even representative coefficients of friction for particular materials. Environmental factors are responsible for a good part of this variation. Despite their reputation for chemical inertness, the surfaces of most ceramics are susceptible to tribochemical reactions which lead to the formation of surface films and thus modify their frictional behaviour. These reactions occur much more rapidly at a sliding contact than on a free surface at the same bulk temperature; local high 'flash' temperatures at the asperity contacts (discussed further in Section 5.5), the exposure of atomically clean surfaces by the wear process, and direct mechanical stimulation of the reaction (e.g. in the highly strained region at an asperity junction or crack tip) are all mechanisms which can accelerate surface reactions at sliding contacts.

Non-oxide ceramics in air commonly form oxide films on their sliding surfaces. For example, silicon nitride, silicon carbide, titanium nitride and titanium carbide have all been found after sliding to have reacted significantly with oxygen, derived from oxygen in the air or from water vapour. Oxide ceramics will react with water, whether it is present as liquid or as vapour: alumina and zirconia, for example, both form hydrated surface layers on sliding in humid air. In the case of non-oxide ceramics, oxidation can be followed by hydration, so that the nature of the surface films formed on silicon nitride in humid air is controlled by both the reactions

$$Si_3N_4 + 6H_2O = 3SiO_2 + 4NH_3$$

and

$$SiO_2 + 2H_2O = Si(OH)_4$$

These tribochemical effects are responsible for the influence of atmospheric

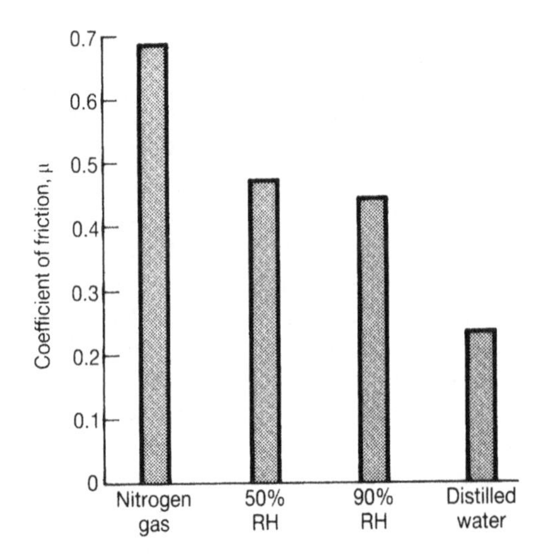

Fig. 3.21 Illustration of the effect of environment on the friction of hot-pressed silicon nitride. The values of μ are derived from pin-on-disc tests at a sliding speed of 150 mm s^{-1}, carried out in the environments indicated (from Ishigaki H, Kawaguchi I, Iwasa M and Toibana Y, in Ludema K C (Ed.), *Wear of Materials 1985*, ASME, 1985, pp. 13–32)

composition on friction which is commonly observed in ceramics. Figure 3.21, for example, shows how the value of μ falls with the increasing availability of water for silicon nitride sliding against itself in dry nitrogen gas, air of two different humidity levels, and liquid water. The reactions outlined above lead to the formation and hydration of a silica film at the interface, lowering its shear strength and thereby reducing the coefficient of friction.

In some ceramics the effects of surface films can be very marked: diamond and titanium nitride provide good examples. Both show high friction when sliding against themselves in vacuum: with diamond, μ reaches ~1 after cleaning or repeated sliding in vacuum. Yet in air, much lower friction is measured: $\mu = 0.05$ to 0.15 for diamond, and typically 0.1 to 0.2 for titanium nitride. Significant surface oxidation has been reported in titanium nitride and appears to be responsible for its low friction in air, but in diamond the surface modification is more likely to be due to adsorbtion of a gaseous species, rather than the formation of a reaction product. The particularly low friction of diamond on diamond in air is due to the very low adhesive force between the surfaces in the presence of the adsorbed contamination, together with a small contribution from other dissipative processes, probably plastic deformation at the asperity contacts on a very fine scale.

As well as surface chemical reactions, a second factor which can be important in the friction of ceramics is the extent of fracture on the sliding surfaces. As we shall see in considering the wear of these materials (in Section 5.10), under some conditions widespread brittle fracture can occur in the contact zone: often intergranular in polycrystalline ceramics, but under more severe conditions, transgranular. The occurrence of fracture leads to increased friction, since it provides an additional mechanism for the dissipation of energy at the sliding contact.

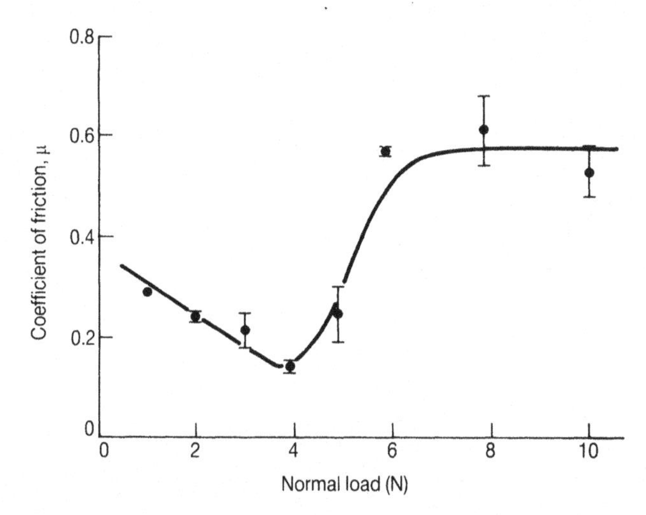

Fig. 3.22 Variation of coefficient of friction with normal load for a 60° diamond cone sliding over the (0001) face of a silicon carbide single crystal. The increase for loads above ~4 N is associated with fracture along the sliding path (from Adewoye O O and Page T F, *Wear* **73**, 247–260, 1981)

Fracture is readily produced in concentrated contacts, such as a hard sharp pin or stylus sliding against a flat. Figure 3.22, for example, shows results from experiments in which a diamond cone was slid over a single crystal of silicon carbide. At low loads, plastic grooving of the silicon carbide occurred, with no fracture; the value of μ was relatively low. As the load was increased brittle fracture occurred around the sliding track, similar to that shown in Fig. 6.15, leading to a higher coefficient of friction.

The effects of atmospheric composition, temperature, load, sliding speed and time of sliding on the friction of ceramics can usually be interpreted in terms of changes in the tribochemical surface films and the extent of fracture in the contact region. Both load and sliding speed will also affect the rate of frictional energy dissipation and hence the temperature at the interface; at sufficiently high temperatures increased plasticity will occur in most ceramics and will also affect the frictional behaviour. The influence of these factors can be large. Figure 3.23, for example, shows the variation of coefficient of friction with temperature for alumina and zirconia, sliding in self-mated couples in air. The initial marked rise in friction with temperature has been attributed to the removal of adsorbed water from the interface.

The effects of interfacial temperature on tribochemical processes are also often assumed to be responsible for the changes in friction with sliding speed commonly observed in many ceramic systems: Figure 3.24 shows representative results for silicon nitride and silicon carbide in self-mated sliding, and similar behaviour has also been reported in alumina and zirconia. Fracture may also play an important role in some cases, as seen in Fig. 5.31 for silicon nitride, where, above a critical sliding velocity, the onset of fracture coincides with a sharp increase in both friction and sliding wear rate.

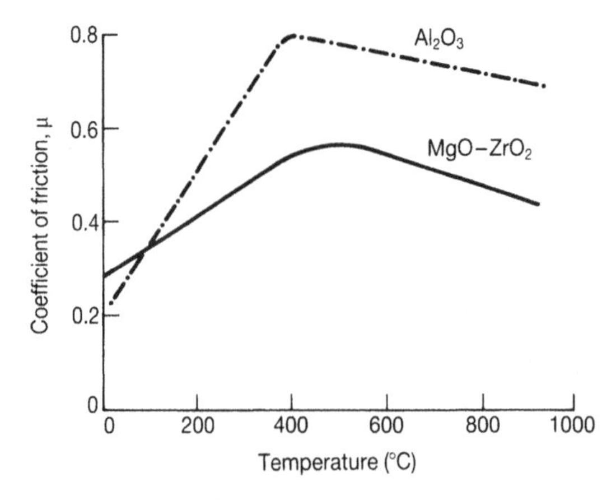

Fig. 3.23 Variation of coefficient of friction with temperature for magnesia-partially stabilized zirconia and alumina sliding against themselves (from Hannink R H J, *Wear* **100**, 355–366, 1984)

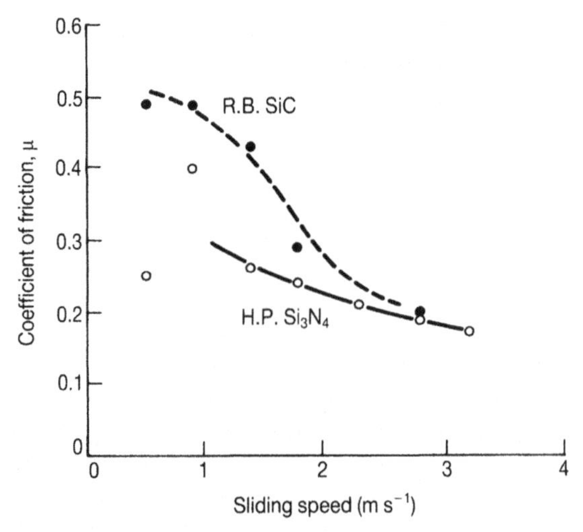

Fig. 3.24 Variation of coefficient of friction with sliding speed for reaction-bonded silicon carbide and hot-pressed silicon nitride samples sliding in self-mated couples in air (data from Cramer D C, *J. Mat. Sci.* **20**, 2029–2037, 1985)

3.7 FRICTION OF LAMELLAR SOLIDS

Several materials with lamellar structures exhibit low values of friction under certain conditions, and are therefore of interest as solid lubricants; this application is discussed in detail in Section 4.7. Foremost among these materials are graphite (an allotrope of carbon) and molybdenum disulphide (MoS_2), the structures of which are shown in Fig. 3.25.

In each case the bonding between the atoms within the layers of the structure is covalent and strong, while that between the layers is considerably

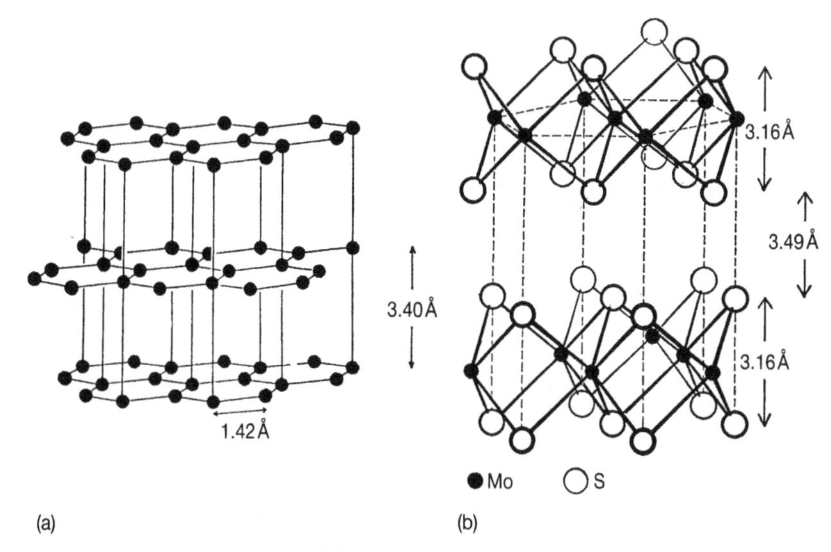

Fig. 3.25 The structures of (a) graphite and (b) molybdenum disulphide (MoS_2)

weaker. In graphite the interplanar bonding is primarily from van der Waals forces, with a weak covalent contribution resulting from interaction between the π-electron orbitals of the carbon atoms. The interplanar bond energy is about one tenth to one hundredth of that between atoms within the layers. In molybdenum disulphide the bonding between the layers of sulphur atoms is due only to van der Waals forces, and is rather weaker than in graphite. Both materials are strongly anisotropic in their mechanical and physical properties: in particular, they are much less resistant to shear deformation in the basal plane (i.e. parallel to the atomic planes) than in other directions.

The low friction of both graphite and molybdenum disulphide is associated with their lamellar structures and weak interplanar bonding, but by no means all compounds with similar structures show low friction, and the low friction values cannot therefore be ascribed to these factors alone.

The sliding friction of graphite against itself or other materials in air is low; typically, $\mu \approx 0.1$. If the surface of graphite is examined by electron diffraction after sliding, it is found that the basal planes have become oriented nearly parallel to the plane of the interface, with a misalignment of the order of 5°. The friction of graphite depends strongly on the nature of the ambient atmosphere. In vacuum or in dry nitrogen, μ is typically ten times greater than in air, and graphite under these conditions wears very rapidly. Controlled addition of gases and vapours reveals that the low friction and wear of graphite depends on the presence of oxygen, water vapour, or other condensable vapours.

Figure 3.26 illustrates the effect on the sliding wear rate of graphite of varying the partial pressure of several gases. Oxygen shows an effect similar to that of water vapour, but at pressures about 100 times greater. Figure 3.27 (upper curve) shows how the coefficient of friction of graphite increases

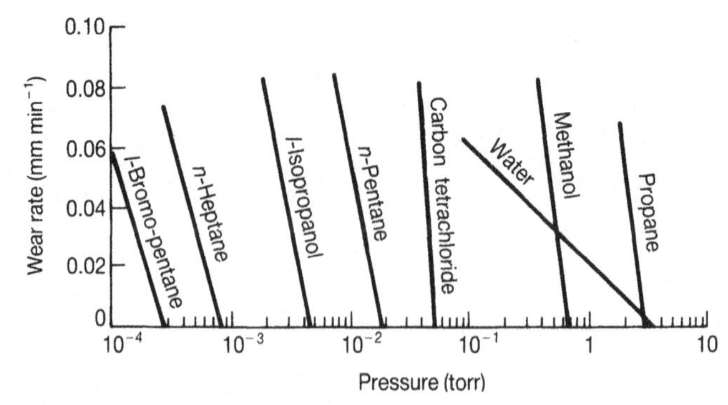

Fig. 3.26 The wear rate of graphite in sliding experiments under different partial pressures of various gases (from Savage R H and Schaefer D L, *J. Appl. Phys.* **27**, 136–138, 1956)

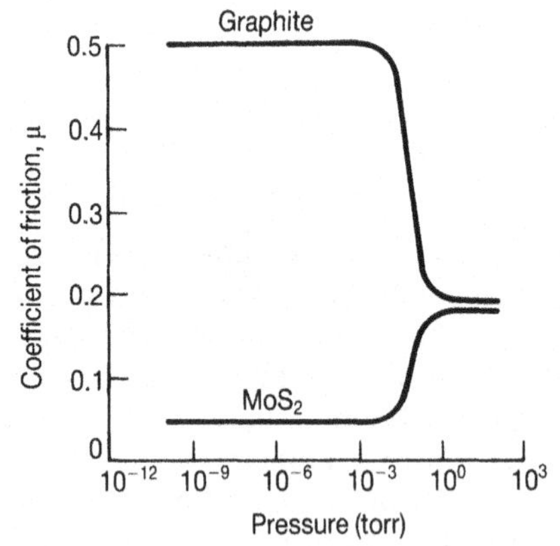

Fig. 3.27 The variation of the coefficient of friction for graphite and molybdenum disulphide with oxygen concentration (air pressure) (from Buckley D H, *Surface Effects in Adhesion, Friction, Wear and Lubrication*, Elsevier, 1981)

dramatically with decreasing air pressure. Although some features of the behaviour of graphite are still not well understood, it is clear that the adsorption of certain molecules is necessary for a low value of μ. The friction between graphite lamellæ in their planes appears always to be low; they are low-energy surfaces and show little adhesion. But the edges of the lamellæ are high-energy sites, and bond strongly to other edge sites or to basal planes. In sliding friction some edge sites will always be exposed, and so the friction of graphite in vacuum is high. Condensable vapours lower the friction by adsorbing selectively to the high-energy edge sites, saturating the bonds, reducing the adhesion, and thus lowering the friction. Only a small concentration of adsorbed molecules is needed to produce this effect.

Molybdenum disulphide, unlike graphite, exhibits an intrinsically low coefficient of friction. Figure 3.27 (lower curve) shows how the friction depends on air pressure; MoS_2 shows a low value of μ in air which, in contrast to the behaviour of graphite, falls even lower in vacuum. For MoS_2 the addition of condensable vapours *raises* the coefficient of friction. Like graphite, MoS_2 forms an oriented film on a sliding surface, with the basal planes tending to be aligned parallel to the surface; bonding of the crystallites to the surface is probably aided by the internal polarization of the lamellæ which results from their 'sandwich' structure. A coefficient of friction of ~0.1 is typically found for sliding between basal planes; for edge-oriented crystallites sliding against basal planes, μ is two or three times higher. The low coefficient of friction of MoS_2 observed in practice can therefore be ascribed to the strong orientation of the films of MoS_2 produced by sliding, and the intrinsically low adhesion and shear strength between the basal planes of the MoS_2 structure.

Many other compounds with lamellar structures exist. Some, like talc, tungsten disulphide, graphite fluoride, cadmium chloride and lead iodide, do show low friction and are potentially useful as solid lubricants; these are discussed further, together with practical applications of graphite and molybdenum disulphide as lubricants, in Section 4.7. Others, such as mica and calcium hydroxide, do not show low values of μ.

3.8 FRICTION OF POLYMERS

3.8.1 Introduction

Contact between polymers, or between a polymer and a metal, is often predominantly elastic. In this important respect the friction of polymers differs fundamentally from that of metals. As we saw in Section 2.5.3, the ratio E/H, where E is Young's modulus and H is the hardness of the material, determines the extent of plasticity in the contact region through equation 2.16; the surface topography is also important. For metals, the value of E/H is typically 100 or greater, while for many of the softer (i.e. low modulus) polymers E/H is only about 10. The plasticity index ψ (see Section 2.5.3) for a soft polymer thus has only about one tenth of the value for a metal, and the contact is therefore almost completely elastic, except against very rough surfaces. Polymers for which this is true include polyethylene (high and low density - HDPE and LDPE), polytetrafluoroethylene (PTFE), nylons, polymethylmethacrylate (PMMA) and some epoxy resins. Rubbers, too, make contact elastically.

A second factor which plays an important role in the friction of polymers is the strong time-dependence of their mechanical properties: most polymers are viscoelastic, and also show a marked increase of flow stress with strain rate.

Coefficients of friction between polymers sliding against themselves, or against metals or ceramics, commonly lie in the range from 0.1 to 0.5, although values outside this range are also seen. In these materials, for which Amontons' Laws are not even broadly applicable, μ varies so much with normal load, sliding speed and temperature that a list of coefficients of friction for specific materials would be of little value.

The friction of polymers, like that of metals, can be attributed to two

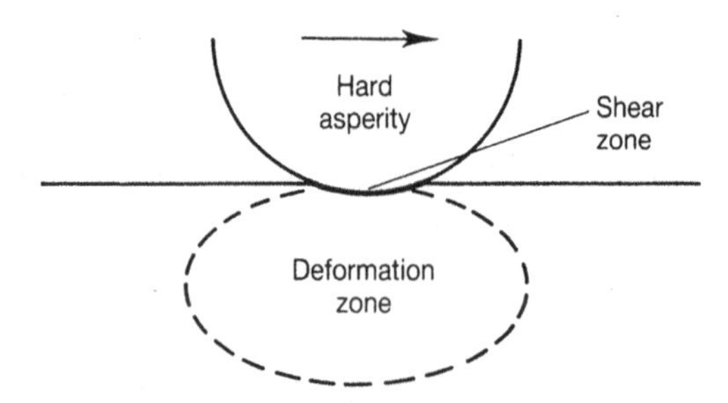

Fig. 3.28 The origins of the friction associated with the sliding of a hard smooth asperity over a polymer surface (from Briscoe B J, in Dowson D, Taylor C M, Godet M and Berthe D (Eds), *Friction and Traction*, Westbury House, 1981, pp. 81–92)

sources: a *deformation* term, involving the dissipation of energy in quite a large volume around the local area of contact, and an *adhesion* term originating from the interface between the slider and the counterface. The regions where these two sources of friction originate are illustrated in Fig. 3.28, for the sliding of a hard asperity over a polymer surface. As for metals, the distinction between the deformation and adhesion components of friction is somewhat artificial, and under many circumstances no clear demarcation can be made. However, in some experiments, and in some practical applications, one term dominates and can then be examined in isolation.

In the next two sections we shall discuss the contributions to polymer friction from these sources. Since in most practical applications polymers slide against more rigid counterfaces, we shall concentrate on this situation; much of our discussion is also applicable to the case of polymers sliding against other polymers.

3.8.2 Friction due to deformation

We can isolate the *deformation* component of friction by eliminating adhesion; this can effectively be achieved by rolling a hard sphere or cylinder over a polymer surface treated with lubricant to reduce the interfacial adhesion, or by dragging a slider across it under well-lubricated conditions. Figure 3.29 illustrates the case of rolling.

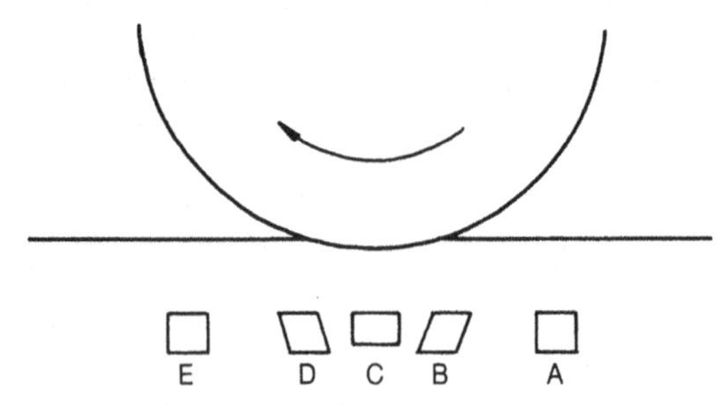

Fig. 3.29 Deformation of elements in a polymer under a rolling sphere (from Briscoe B J, in Dowson D, Taylor C M, Godet M and Berthe D (Eds), *Friction and Traction*, Westbury House, 1981, pp. 81–92)

The resistance to rolling arises from energy dissipation in the bulk of the polymer beneath the sphere, and is a direct result of its viscoelastic response. As the sphere in Fig. 3.29 rolls towards the right, elements of the polymer in its path become progressively deformed, while those behind it recover their undeformed shape. An individual element will experience the deformation cycle illustrated by the sequence ABCDE. In a viscoelastic material, energy will be dissipated during this cycle as heat; the frictional force will be equal to the energy dissipated per unit distance moved by the force.

It can be shown that if a fraction β of the total energy input is dissipated (i.e. the fraction (1 − β) is recovered mechanically) then, for a sphere of radius R rolling under a normal load W, the frictional force F_{def} will be given by

$$F_{def} = 0.17\,\beta\, W^{4/3}\, R^{-2/3}\, (1-\nu^2)^{1/3}\, E^{-1/3} \tag{3.27}$$

Here ν is Poisson's ratio for the polymer, and E is the real part of Young's modulus. For a well-lubricated conical slider of semi-angle α, rather than a rolling sphere, the friction force is given by

$$F_{def} = \left(\frac{\beta W}{\pi}\right) \cot \alpha \tag{3.28}$$

For deformation by a sphere or a cone, the fraction β is found to be about two to three times the fraction of the energy lost in a single cycle of pure shear deformation (the 'loss factor' α), reflecting the more complex cycle of deformation experienced by the polymer under rolling or sliding contact. The values of α and E vary with loading frequency and temperature in a polymer, and can be measured independently from F_{def} in separate experiments.

Good agreement is generally found between the predictions of equation

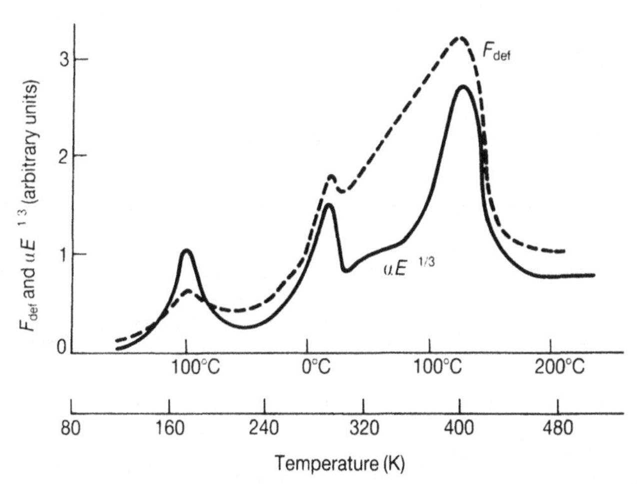

Fig. 3.30 Dependence of the deformation component of the frictional force (F_{def}) and the quantity $\alpha E^{-1/3}$ on temperature, for a steel sphere rolling over PTFE (48% crystalline). The units employed for the ordinate of the graphs are arbitrary (from Ludema K C and Tabor D, *Wear* **9**, 329–348, 1966)

3.27 and experimental measurements of the rolling friction force for polymers. Figure 3.30 shows results for the rolling of a steel sphere on PTFE over a range of temperature. The correlation between F_{def} and the quantity $\alpha E^{-1/3}$, as measured in independent experiments, is good; the peaks in the curves are associated with specific energy-absorbing motions of the polymer molecules. Similar dependence on the rate of loading (i.e. on the speed of rolling or sliding) is also found.

3.8.3 Friction due to adhesion

If a smooth polymer surface slides against a relatively smooth rigid counterface so that the contribution of deformation to friction is negligible, the observed frictional force then originates primarily in *adhesion* between the two surfaces, and can be analysed in the same way as the adhesion component of friction for metals (Section 3.4). The analysis differs, however, in that the contact of a polymer against a typical engineering surface will be predominantly elastic rather than plastic.

At low loads and for moderately rough surfaces, where the true area of contact is a small fraction of the apparent area, the contact area remains effectively proportional to the normal load (Section 2.5.3) and the coefficient of friction is therefore constant and independent of the normal load. Figure 3.31(a) illustrates this case, for the friction of crossed PMMA cylinders at low loads, when the surfaces have been lathe-turned to produce moderate surface roughness.

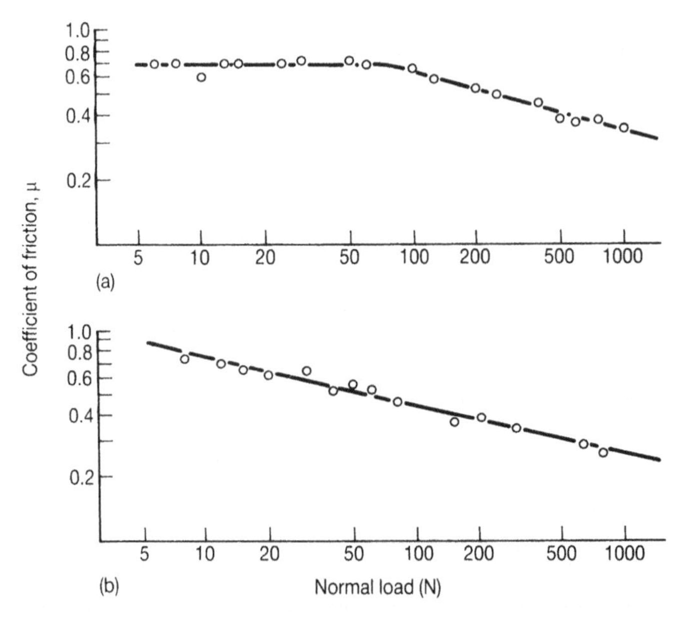

Fig. 3.31 Variation of coefficient of friction with normal load for sliding of crossed cylinders of polymethylmethacrylate (PMMA) with different surface roughness: (a) lathe turned, and (b) smooth polished (from Archard J F, *Proc. Roy. Soc. Lond.* **A243**, 190, 1957)

When the load is increased however (Fig. 3.31(a)), or if the surfaces are smooth and polished (Fig. 3.31(b)), the coefficient of friction is found to fall with increasing load. Under these conditions, although the frictional force remains proportional to the real area of contact, that area is no longer proportional to the normal load.

At high loads, or for very smooth surfaces, the elastic deformation at the points of contact between the cylinders is so great that individual asperities on the contacting surfaces are 'ironed out', and the situation approximates to the contact of a single giant asperity. We saw in Section 2.5.1 that for a single spherical asperity deforming elastically the area of contact A will be proportional to the load W raised to the power 2/3. The same exponent applies for the contact of elastic cylinders.

We would therefore expect under these conditions

$$\mu \propto A/W \propto W^{2/3}/W \propto W^{-1/3} \tag{3.29}$$

In practice, fair agreement is found with this relationship, although the exponent of normal load is usually slightly less than $-1/3$. The slope of the graph in Fig. 3.31(b) is about $-1/4$; the difference is probably due to the presence of some plastic flow around the contact, although it remains predominantly elastic. Similar dependence of the coefficient of friction on load is found for the heavily loaded contact of a rounded polymer body against a hard smooth substrate.

The adhesion responsible for this component of polymer friction results primarily from the weak bonding forces (e.g. hydrogen bonding and van der Waals forces) which are also responsible for the cohesion between the polymer chains themselves in the bulk of the material. The relative magnitude of the forces may be described in terms of the surface energies of the polymer and substrate. Surface energies of non-polar polymers are typically 20 to 30 mJ m^{-2}, while those of polar polymers have up to twice these values. Polar polymers therefore tend to show stronger adhesion. The junctions formed by adhesion show many similarities to those responsible for the friction of metals. Failure of the junction tends to occur within the bulk of the polymer, rather than at the interface; however, only very limited junction growth occurs, and the high values of μ associated with extensive junction growth in metals are never observed in polymers.

Many polymers sliding against hard counterfaces (e.g. metals) transfer detectable films of polymer on to the counterface. The formation and behaviour of the transfer films are important factors in the friction and wear of these polymers. Once a transfer film has formed, subsequent interaction occurs between the polymer and a layer of similar material, irrespective of the composition of the substrate. On further sliding the polymer may continue to wear by adding material to the transfer film, since the interfacial bond to the counterface is often stronger than that within the bulk of the polymer itself. The sliding wear of polymers is discussed further in Section 5.11.

The properties of thin films of polymers may be investigated by placing the film between a rigid slider and a hard substrate. If sliding occurs under a sufficiently high load, the real area of contact will be equal to the apparent area and a direct comparison can then be made between the measured coefficient of friction and the value predicted from independent measurements of the shear yield strength of bulk samples of the polymer. Agreement is generally good. The coefficient of friction depends on the load, since the yield stress of a polymer, unlike that of a metal, depends strongly on the hydrostatic component of stress. The shear yield stress τ varies with the applied hydrostatic pressure P in the following way:

$$\tau = \tau_o + \alpha P \tag{3.30}$$

where τ_o and α are constants for a particular polymer. For the case of a thin polymer film sandwiched between rigid surfaces, $P = W/A$ and $F = \tau A$. It therefore follows that

$$\mu = \frac{F}{W} = \frac{\tau}{P} = \frac{\tau_o}{P} + \alpha \tag{3.31}$$

Good agreement is generally found between measured values of μ for polymers which form transfer films and those calculated from this equation, when the values of τ_o and α are measured in conventional mechanical tests.

From equation 3.31, it can be seen that the coefficient of friction is expected to fall with increasing hydrostatic pressure (and hence with increasing normal load), and this is indeed observed in practice for polymers which form transfer films. For very high pressures, the value of μ tends to α. Equation 3.31 is also found to predict the friction of *bulk* samples of many polymers if P is taken to be the bulk hardness of the polymer.

Two polymers, linear (i.e. unbranched: high density) polyethylene (HDPE) and polytetrafluoroethylene (PTFE) exhibit appreciably lower coefficients of friction than most others and have important applications as solid lubricants and bearing materials. Their molecular structures are characterized

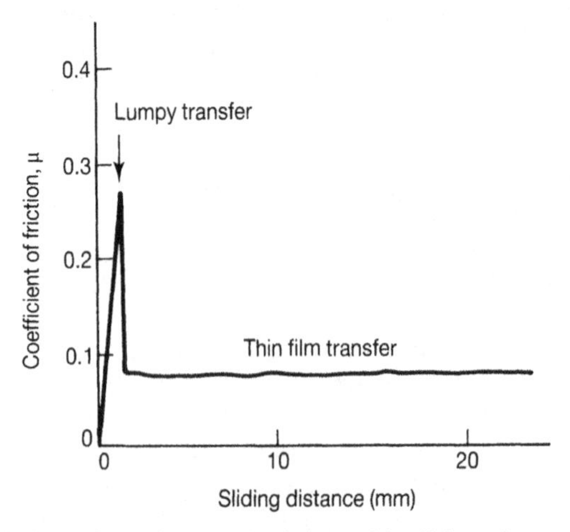

Fig. 3.32 The variation of coefficient of friction with sliding distance for high density polyethylene (HDPE) sliding against glass (from Pooley C M and Tabor D, *Proc. Roy. Soc. Lond.* **A329**, 251–274, 1972)

by linear, unbranched chains without bulky or polar side groups, giving weak intermolecular bonding and a high degree of crystallinity. Although these polymers form transfer films on a hard counterface, subsequent sliding in the same direction tends to occur at the interface between the bulk polymer and the film, leading to low wear rates. The coefficient of friction for initial sliding on a clean hard substrate is not particularly low (typically $\mu = 0.2$ to 0.3) and the transfer film is of the order of micrometres thick. As sliding progresses, the coefficient of friction drops to a much lower value (perhaps as low as 0.05 for PTFE); the transfer film becomes much thinner and contains molecular chains strongly oriented parallel to the sliding direction. This behaviour is illustrated for HDPE sliding on glass in Fig. 3.32.

For subsequent sliding with the same orientation of slider on the same track, the friction remains low. But if the slider is turned, destroying the molecular alignment between the slider and the transfer film, the friction and wear rate rise substantially.

The behaviour of HDPE and PTFE has been attributed to the smoothness of the molecular chains in these materials, although it seems that a low surface energy, associated with weak intermolecular forces, is also needed to allow the very thin oriented films to be drawn out from the bulk.

Further reading

Arnell R D, Davies P B, Halling J and Whomes T L, *Tribology: Principles and Design Applications*, Macmillan, 1991

Bowden F P and Tabor D, *The Friction and Lubrication of Solids, Part I*, 1950 and *Part II*, 1964, Clarendon Press, Oxford

Briscoe B J and Tabor D, Friction and wear of polymers, in Clark D T and Feast W J (Eds), *Polymer Surfaces*, John Wiley, pp 1–23, 1978

Buckley D H, *Surface Effects in Adhesion, Friction, Wear and Lubrication*, Tribology Series No. 5, Elsevier, 1981

Buckley D H and Miyoshi K, Tribological properties of structural ceramics, in Wachtman J B (Ed.), *Structural Ceramics, Treatise on Materials Science and Technology* **29**, pp 293–365, 1989

Chaudhri M M, J. Mater. Sci Lett. **3**, pp 565–568, 1984

Clauss F J, *Solid Lubricants and Self-lubricating Solids*, Academic Press, 1972

Lancaster J K, Friction and wear, in Jenkins A D (Ed.), *Polymer Science*, Vol. 2, North-Holland, pp. 959–1046, 1972

Ling F F and Pan C H T (Eds), *Approaches to Modeling of Friction and Wear*, Springer-Verlag, 1988

Rabinowicz E, *Friction and Wear of Materials*, John Wiley, 1965

Rigney D A (Ed.), *Fundamentals of Friction and Wear of Materials*, ASM, 1981

Singer I L and Pollock H M (Eds), *Fundamentals of Friction*, Kluwer, 1992

Tabor D, Friction, lubrication and wear, in Matijevic E (Ed.), *Surface and Colloid Science*, Vol. 5, John Wiley, pp 245–312, 1972

4

Lubricants and lubrication

4.1 INTRODUCTION

We have seen in the previous chapter that the coefficient of sliding friction for dry metals, and for many ceramics and some polymers, is rarely lower than 0.5, and in many cases is rather higher. Such high values of μ in an engineering application would often lead to intolerably high friction forces and frictional energy losses. For most practical uses, therefore, *lubricants* are used to reduce the frictional force between surfaces.

A lubricant functions by introducing between the sliding surfaces a layer of material with a lower shear strength than the surfaces themselves. In some lubricated systems, the lubricant may not completely prevent asperity contact, although it will reduce it and may also reduce the strengths of the junctions formed. In other cases, the lubricant completely separates the surfaces and no asperity junctions are formed at all. Thus, to a greater or lesser extent, the use of a lubricant will always reduce the rate of sliding wear, and this is another substantial benefit of lubrication.

A wide variety of materials, gases, liquids or solids, may be used as lubricants. It is convenient in discussing the subject to distinguish between various types of lubrication. In *hydrodynamic* lubrication the surfaces are separated by a fluid film, which is usually thick in comparison with the asperity heights on the bearing surfaces. The hydrostatic pressure in the film causes only small elastic distortion of the surfaces which, to a first approximation, can be treated as rigid. *Elastohydrodynamic* describes the case where the local pressures are so high and the lubricant film so thin that elastic deformation of the surfaces can no longer be neglected; indeed, it is a vital factor in this regime of lubrication. In *boundary* lubrication, the surfaces are separated by adsorbed molecular films, usually laid down from an oil or grease containing a suitable boundary lubricant; appreciable asperity contact and junction formation may nevertheless occur. *Solid lubricants* function by providing a solid interfacial film of low shear strength.

We shall discuss these four types of lubrication below, but since in the first three the lubricant is in most cases an oil or grease, we shall first describe the properties and compositions of lubricating oils and greases, starting with a brief discussion of *viscosity*.

4.2 VISCOSITY

The most important property of an oil for lubricating purposes is its *viscosity*. Viscosity provides a measure of the resistance of a fluid to shearing flow, and may be defined as the shear stress on a plane within the fluid, per unit velocity gradient normal to that plane.

Figure 4.1 illustrates this definition. We consider two planes within the fluid, one of which is moving parallel to the other with a relative velocity v_o (Fig. 4.1(a)). The velocity gradient dv/dy in the fluid is assumed to be constant and uniform between the two planes (Fig. 4.1(b)) and is given by

$$dv/dy = v_o/h \tag{4.1}$$

The shear stress acting on the planes is τ. The viscosity η is defined by the relationship

$$\tau = \eta \, dv/dy \tag{4.2}$$

An exactly equivalent definition, which may be more readily understood by those with a background in solid mechanics, is provided by the equation

$$\tau = \eta \, d\gamma/dt \tag{4.3}$$

where γ is the shear strain in the fluid. A fluid, since it has no shear strength, can support no static shear stress, but its viscous nature leads to a dynamic shear stress τ which is proportional to the shear strain rate $d\gamma/dt$. The constant of proportionality is the viscosity of the fluid, η.

Viscosity, as defined above, is termed *dynamic viscosity*, and has dimensions mass $\times$ length^{-1} $\times$ time^{-1}. The SI unit of dynamic viscosity is therefore the Pascal second ($1\ \text{Pa s} = 1\ \text{kg m}^{-1}\ \text{s}^{-1}$). Dynamic viscosities are rarely quoted in these units, however, the *centipoise* being more usually used ($1\ \text{cP} = 10^{-3}\ \text{Pa s}$). Typical lubricating oils have dynamic viscosities in the range 2 to 400 cP; the viscosity of water at room temperature is about 1 cP.

Frequently in calculations the ratio of dynamic viscosity to fluid density (η/ρ) occurs. This quantity is known as the *kinematic viscosity*; the SI unit is metre^2 second^{-1}, and the commonly used unit the *centistokes* ($1\ \text{cSt} = 10^{-6}\ \text{m}^2\ \text{s}^{-1}$).

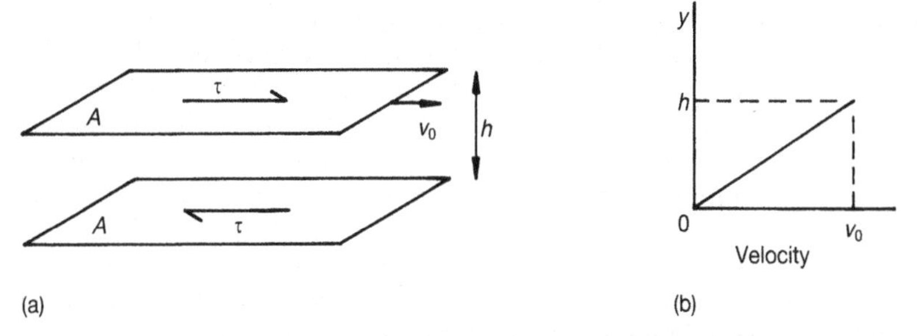

Fig. 4.1 Viscosity is defined by considering the shear stress τ acting on two planes within the fluid, of area A and separated by a distance h, moving with relative velocity v_o

4.3 COMPOSITIONS AND PROPERTIES OF OILS AND GREASES

Most lubricating oils are of mineral origin, and contain several different hydrocarbon species with mean molecular weights between about 300 and 600. Predominant are saturated long-chain hydrocarbons (*paraffins*) with straight or branched chains containing 20 to 30 carbon atoms, and saturated 5 or 6-membered hydrocarbon rings with attached sidechains up to 20 carbon atoms long (*naphthenes*). Aromatic components are also present in small proportions, consisting of one or more benzene rings with saturated sidechains. Many other constituents are usually present in smaller quantities. Mineral oils may be broadly classified as *paraffinic*, in which the naphthenes have long paraffinic sidechains and most of the carbon atoms are present in paraffin chains; or *naphthenic*, in which the naphthene sidechains are short and the proportion of carbon atoms in the rings is only slightly less than that in the sidechains; or *mixed*.

Synthetic oils are used as lubricants under conditions where mineral oils would be unsuitable, for example at very high or low temperatures, or where low flammability is vital. Among typical synthetic oils are *organic esters* which lubricate over a wide temperature range and are used for example in gas turbine engines, *polyglycols* which have excellent boundary lubricating properties (see Section 4.6) and decompose at high temperature without leaving solid residues, and *silicones* which are chemically very stable, can operate at high and low temperatures, and are electrically insulating. All synthetic oils, however, are more expensive than mineral oils. They are therefore used only in specialized and demanding applications.

The viscosity of a mineral oil depends strongly on its composition, with higher viscosity being associated with higher molecular weight. A property which is also important in determining an oil's suitability as a lubricant is the temperature dependence of its viscosity. The behaviour of two different oils is illustrated in Fig. 4.2. Both oils have the same viscosity at room temperature, but the viscosity of oil A drops off less rapidly at higher temperatures than B. Oil A is said to have a higher *viscosity index* than B. Viscosity index (VI) is quantified by comparing the behaviour of the oil with that of two reference oils of known VI. The reference oils chosen when the scale was originally established were arbitrarily assigned VI values of 0 and 100, but viscosity indexes outside this range are now commonly encountered. Most commercial lubricants have a VI of about 100; higher values are needed for oils which will experience a wide range of operating temperature, such as multigrade oils for motor vehicle engines which typically have a VI of about 150. Suitable polymer additives, such as polybutene or polyacrylics, increase the viscosity index of an oil and are often used as *viscosity index improvers* in lubricating oils. Under conditions of high shear rate, however, these additives may lose their effectiveness, and the viscosity can then drop sharply. Figure 4.3 illustrates this effect for two mineral oils which have the same value of viscosity at low shear rates. Oil A is a plain mineral oil, while B is a multigrade oil containing viscosity index improvers.

The viscosity of an oil also depends on hydrostatic pressure, following fairly

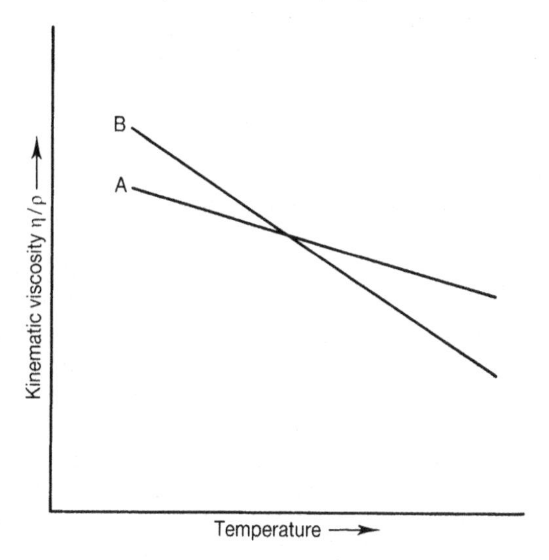

Fig. 4.2 The variation of kinematic viscosity with temperature for two mineral oils. Oil A has a higher viscosity index than oil B

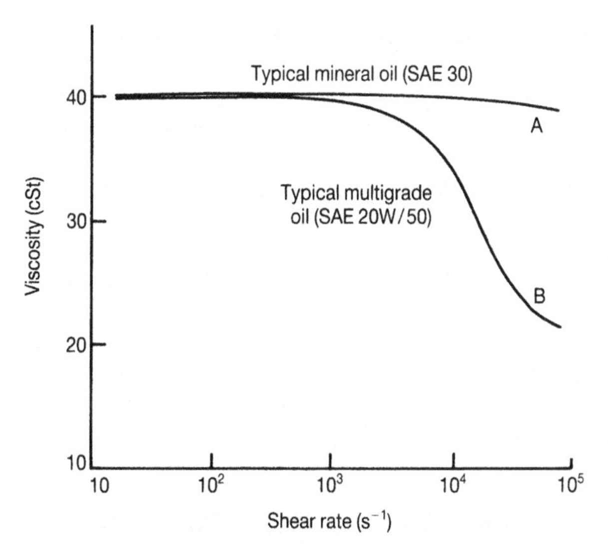

Fig. 4.3 The variation of viscosity with shear rate for two mineral oils with the same viscosity at low shear rates. Oil A is a plain mineral oil, while oil B contains viscosity index improvers which lose their effectiveness at high shear rates (from Lansdown A R, *Lubrication*, Pergamon Press, 1982)

closely an exponential dependence:

$$\eta = \eta_o \exp(\alpha P) \tag{4.4}$$

where η_o and α are constants for a particular oil and P is the hydrostatic pressure. α has been shown empirically for mineral oils to be related to the viscosity at zero pressure, η_o, by the relationship

$$\alpha \approx (0.6 + 0.965 \log_{10} \eta_o) \times 10^{-8} \tag{4.5}$$

where η_o is in centipoise and the units of α are Pa^{-1}.

A variety of additives may be used to improve the lubricating qualities of a mineral oil or to prolong its life. Viscosity index improvers have been mentioned above. Extreme pressure (EP) and boundary lubricants are sometimes added and are discussed in Section 4.6; other 'anti-wear' additives also reduce the incidence of sliding wear. Paraffinic oils at low temperatures thicken and become 'waxy'; certain complex polymers can be added as 'pour-point depressants' to improve the flow properties when cold. Oils deteriorate in service by oxidation; although naturally occurring oxidation inhibitors are present in mineral oils, antioxidants are also frequently added, together with additives to neutralize the acidic products of oxidation. Detergents and dispersants may also be added to reduce or prevent deposits resulting from oxidation or thermal degradation of the oil.

For some applications, grease is preferable to oil as a lubricant. A *grease* is a solid or semi-fluid compound consisting of an oil and a thickening agent. Most greases incorporate mineral oils, although greases based on synthetic oils are used in specialized applications (for example, silicone greases for use at very high or low temperatures). The thickening agent is often a soap (i.e. a metallic salt of a carboxylic acid) or a clay mineral (e.g. bentonite). Greases based on calcium or lithium soaps are commonly used; calcium greases are cheaper, while lithium greases will tolerate a wider range of temperature. Similar additives to those used in oils can be incorporated; EP additives may be used, together with solid lubricants such as graphite and molybdenum disulphide.

With its semi-solid consistency, grease is less easily displaced from bearing surfaces than oil. It will therefore not drain from the bearing under gravity, and also forms a very effective 'squeeze film' lubricant (see below). Grease also provides a good seal against contamination by dirt or moisture. However, unlike oil, grease provides a very poor means of transferring frictional heat away from the sliding surfaces, and also usually leads to higher friction than an oil, due to viscous forces.

4.4 HYDRODYNAMIC LUBRICATION

Under conditions of *hydrodynamic lubrication* the sliding surfaces are separated by a relatively thick film of fluid lubricant, and the normal load is supported by the pressure within this film, which is generated hydrodynamically. For hydrodynamic lubrication the opposing surfaces must be *conformal*; that is, they must be so closely matched in dimensions that they are separated by only a small gap over a relatively large area. Examples of conformal surfaces are shown in Fig. 4.4. The simplest geometry is provided by opposed planes (Fig. 4.4(a)); a common geometry for rotating shaft bearings is the plain journal bearing illustrated in Fig. 4.4(b). The gap

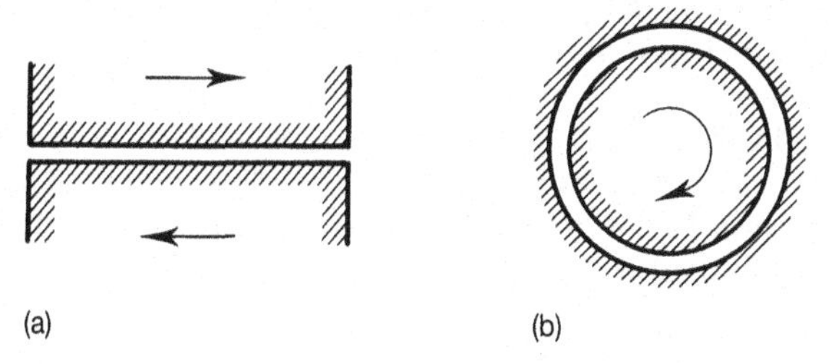

Fig. 4.4 Examples of conformal bearing geometries: (a) opposed planes, (b) plain journal bearing

between the two surfaces is filled with the lubricating fluid: oil or grease or, less commonly, water, air, or other liquid or gas.

The pressure which supports the normal load in hydrodynamic lubrication results from viscous forces within the lubricant, which in turn result from the relative motion of the two surfaces. For a hydrodynamic film to form between sliding surfaces, the gap between them must converge. The geometry must therefore be as shown in Fig. 4.5, although the degree of convergence has been considerably exaggerated here for clarity. The separations of the surfaces, and the angles of convergence, are typically very small: a plain journal bearing lubricated with oil will have a mean lubricant film thickness of the order of one thousandth of the journal diameter, while the maximum and minimum film thicknesses may differ by a factor of four or five.

The distribution of pressure within a hydrodynamic lubricant film is described by the *Reynolds equation*, which is derived from the general (Navier–Stokes) equations for fluid flow by assuming that the flow is laminar, that the fluid film is thin compared with its other dimensions, and that the dominant forces are due to viscosity. Even with these simplifications, the general form of the Reynolds equation is not readily solved analytically, but further simplifications can be made for many problems of interest. For detailed discussion and derivation of the Reynolds equation, together with its

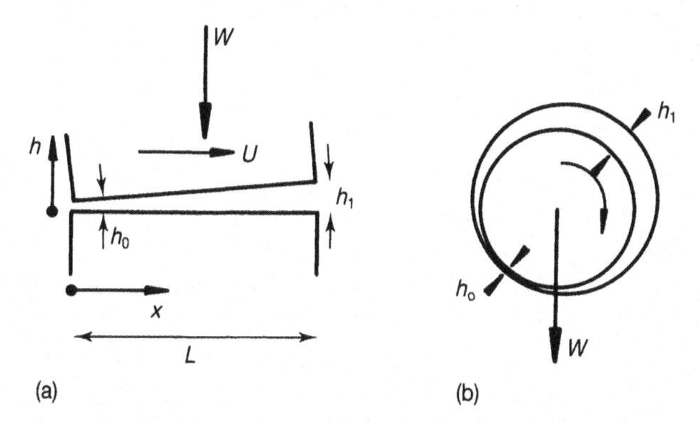

Fig. 4.5 Bearings with the geometries shown in Fig. 4.4, under conditions of hydrodynamic lubrication. The degree of convergence of the surfaces has been exaggerated for clarity

application to bearing design, the reader is referred to the specialized texts on lubrication theory listed at the end of the chapter.

The problem is simpler when the fluid is incompressible (e.g. a liquid rather than a gas) and variations in its viscosity are neglected. We shall here consider the important case of uniform tangential sliding of two planes, forming a wedge as shown in Fig. 4.5(a). The variation of pressure p with distance x is given by the simplified Reynolds equation

$$\mathrm{d}p/\mathrm{d}x = -6\eta U(h - h^*)/h^3 \tag{4.6}$$

where h^* is the separation of the surfaces at the point of maximum pressure ($\mathrm{d}p/\mathrm{d}x = 0$), U is the sliding velocity and the other quantities are as defined in the diagram.

This type of sliding bearing, the tilted pad, can be used in linear slideways. Alternatively, by attaching a ring of pads to a collar surrounding a shaft a rotary thrust bearing can be constructed to carry an axial load. The pressure distribution can be found by integrating equation 4.6, and employing the boundary conditions that $p = 0$ at $x = 0$ and $x = L$. Further integration gives the normal load W carried by the bearing (per unit width) as

$$W = 6\eta KU\left(\frac{L^2}{h_o^2}\right) \tag{4.7}$$

Here

$$K = \frac{\ln(1+n)}{n^2} - \frac{2}{n(2+n)} \tag{4.8}$$

where $n = h_1/h_o - 1$. The numerical value of K therefore depends on the ratio of the inlet and outlet film thicknesses only and is, in fact, rather insensitive to that ratio. The maximum load capacity of a tilted pad bearing occurs when h_1 is about 2.2 h_o; then $K = 0.027$.

A steadily loaded rotating journal bearing, as illustrated in Fig. 4.5(b), can be readily visualized as similar to a tilted pad slider which is 'wrapped round' on itself. Solution of the Reynolds equation is then more complicated than for the tilted pad bearing, but the forms of the equations for the load capacity are, not surprisingly, similar. The load per unit width carried by a journal bearing is given by

$$W = S\eta U\left(\frac{R^2}{h^2}\right) \tag{4.9}$$

where h is the mean film thickness (radial clearance) in the bearing, R is the journal radius, and U is its peripheral speed. S is a dimensionless number, the Sommerfeld number, which is determined by the width to diameter ratio of the bearing and by the eccentricity of the shaft within its housing. If the shaft is concentric in the housing, then no convergent film is formed and clearly no load can be carried by the bearing; under these conditions $S = 0$. Numerical solutions to the Reynolds equation are available which allow the value of S to be found for any shaft eccentricity and width of bearing. A typical design might have $h_1/h_o = 4$ (see Fig. 4.5(b)); then, for a bearing with its width equal to its diameter, S would be about 2. For an infinitely wide bearing under the same conditions, $S \approx 7$. As the eccentricity h_1/h_o increases, so S rises and the load capacity of the bearing will rise. The maximum value of h_1/h_o which can be allowed depends on the roughness of the journal and housing: too thin a lubricant film would allow asperity contact to occur, the friction would rise, wear would be rapid, and the bearing would no longer be efficient.

The frictional force in a journal bearing can also be predicted. The total tangential force F per unit width acting on the periphery of the journal is given to a fairly good approximation for an infinitely wide bearing by:

$$F = 2\pi\eta UR/h \tag{4.10}$$

The mean coefficient of friction is therefore given approximately by:

$$\mu = F/W = \left(\frac{2\pi}{S}\right)\left(\frac{h}{R}\right) \tag{4.11}$$

This simple expression provides a useful estimate of μ, although it is most accurate when the shaft and housing are nearly concentric; μ becomes higher than the value given by equation 4.11 as the bearing becomes more eccentric. From this equation it is clear that in a well-designed hydrodynamic journal bearing the coefficient of friction can be very low: typically down to 0.001 for a clearance ratio (h/R) of about the same value.

As the load on the bearing is reduced, or its rotational speed increased, the eccentricity and hence the Sommerfeld number S will decrease, leading to a rise in the coefficient of friction. On the other hand, if the load is increased or the speed is dropped then S and the eccentricity ratio h_1/h_o will rise. The local pressure in the oil film will increase, leading eventually through equation 4.4 to a significant rise in its viscosity. Elastic deformation of the shaft and housing may also become important, and as the film thickness diminishes so asperity contact between the two surfaces will become more probable. All

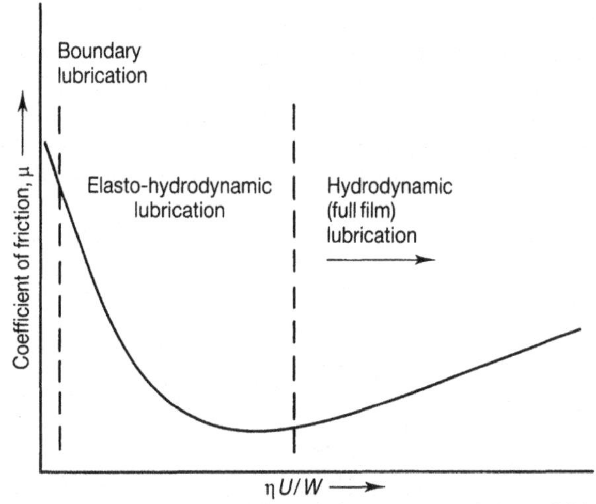

Fig. 4.6 The variation in frictional drag (expressed as the coefficient of friction μ) with the quantity $\eta U/W$ for a lubricated sliding bearing: the Stribeck curve

these effects contribute to an eventual rise in the coefficient of friction at very high loads or very low sliding speeds.

The behaviour observed in practice is illustrated in Fig. 4.6, which shows the *Stribeck curve* relating coefficient of friction to the product $\eta U/W$. At very low values of $\eta U/W$, μ rises for the reasons indicated above. The system is then in the regime known as *elastohydrodynamic lubrication* which will be discussed in the next section.

We have so far considered hydrodynamic effects associated with the wedge or entraining flow generated when two surfaces slide past each other. In some important practical applications, however, the surfaces have appreciable relative motion in the perpendicular direction. This type of motion results in *squeeze-film lubrication*, which occurs in pad or journal bearings subjected to varying normal loads. The lubrication of a big-end bearing on the crankshaft of an internal combustion engine, which is subjected to large cyclic loading by the varying force on the piston, is very largely controlled by squeeze-film phenomena rather than by the effects of wedge flow.

The simple geometry shown in Fig. 4.7 leads to the following form of Reynolds equation for squeeze-film action:

$$\frac{\mathrm{d}p}{\mathrm{d}x} = 12\eta V\left(\frac{x - x^*}{h^3}\right) \tag{4.12}$$

where V is the velocity of approach of the surfaces and x^* is the position of maximum pressure (i.e the point at which $\mathrm{d}p/\mathrm{d}x = 0$). This equation can be integrated for suitable boundary conditions to give an expression for the normal load W per unit width carried by infinitely wide plates separated by a squeeze-film:

$$W = \eta V\left(\frac{L^3}{h_o^3}\right) \tag{4.13}$$

Comparison with equation 4.7 reveals that the normal load supported by a squeeze film varies as $1/h_o^3$ whereas that due to sliding of a wedge varies as $1/h_o^2$. The normal velocity of approach of the surfaces V needs therefore be no more than about h_o/L times the sliding velocity U for squeeze film effects to become important in a tilted pad bearing. Similar conclusions apply to the case of a journal bearing. In many cases, however, one effect dominates; the general Reynolds equation, which takes account of both wedge and squeeze

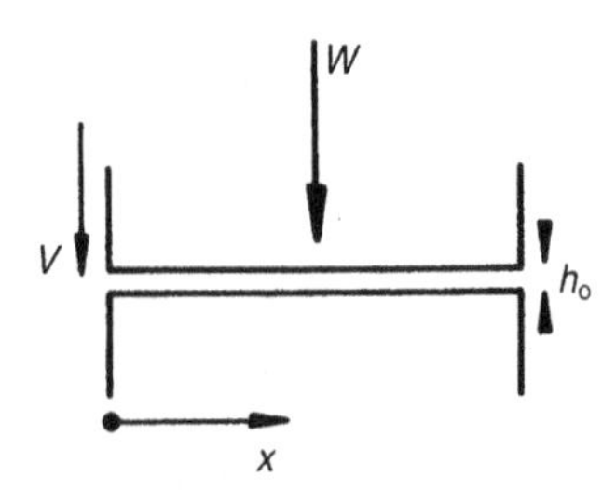

Fig. 4.7 Two opposing surfaces separated by a lubricant film and moving towards each other give rise to squeeze-film lubrication

film forces, is complicated to solve, and in many practical cases the problem can be simplified by neglecting one effect or the other.

It was mentioned above that gases can be used as lubricants in fluid film bearings. *Gas bearings* have some advantages over oil-lubricated bearings at high speeds; they offer very low friction and can operate over an extremely wide temperature range. However, since the lubricant film thickness tends to be less than that for an oil-lubricated bearing, the manufacturing tolerances and surface finish must be appreciably more accurate. The form of Reynolds equation used to analyse the behaviour of a gas bearing must take account of the compressibility of the gas; the reader is referred to specialized texts on lubrication and bearing design for further discussion.

Mean bearing pressures (normal load/area) in oil-lubricated hydrodynamic bearings running under steady load are usually less than 2 MPa, although in impulsively loaded systems such as internal combustion engine crankshaft bearings the pressures can reach a transient maximum of ~50 MPa, with local maxima in the oil film perhaps five times higher still. Gas bearings operate at much lower pressures, typically up to 100 kPa. The materials from which hydrodynamically lubricated bearings are made must be carefully selected; this aspect is discussed in Chapter 9.

4.5 ELASTOHYDRODYNAMIC LUBRICATION

If the contacting surfaces are *counterformal* (i.e. non-conforming), involving nominally line or point contact, then the local pressures in the contact zone will generally be much higher than those encountered in hydrodynamic lubrication. Examples of counterformal geometries are illustrated in Fig. 4.8.

The contact between gear teeth, between a cam and its follower, or between a ball and its track in a ball bearing, all involve concentrated contacts of very small area. Here the local pressures between steel components typically range up to several GPa. Under these conditions the dependence of the lubricant viscosity on pressure plays an important role, as does the elastic deformation of the bearing surfaces. Lubrication in these circumstances is known as *elastohydrodynamic* (often abbreviated as EHL).

The lubricated contact of soft elastic bodies, such as rubbers, also involves a type of EHL, although then the pressures are much lower and have no effect on the viscosity of the lubricant; only the elastic deformation of the

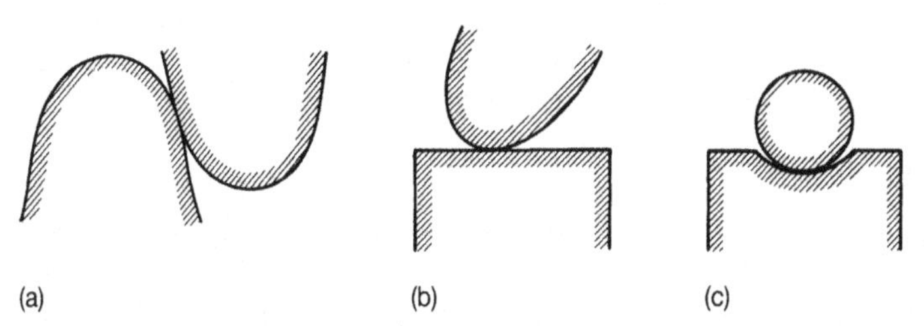

Fig. 4.8 Examples of counterformal contacts: (a) gear teeth; (b) a cam and its follower; (c) a ball in a bearing race

contacting surfaces needs be considered. This type of lubrication is sometimes known as *soft EHL* to distinguish it from *hard EHL* which occurs when both surfaces have high elastic modulus.

In hard EHL, the lubricant film is typically a few micrometres or tenths of a micrometre thick. According to the classical hydrodynamic theory discussed in the previous section, films much thinner than this should be associated with the prevailing loads, and asperity contact should occur. The classical hydrodynamic theory needs to be modified in two ways to explain why thicker films are formed in practice and why, in EHL, the surfaces can still be separated by a continuous fluid film.

The very high pressures in the film locally increase the viscosity of the lubricant, and this tends to increase the film thickness over the predictions of the hydrodynamic theory. We have seen that the pressure dependence of viscosity of an oil can be described by equation 4.4. For a typical mineral lubricating oil, the pressure coefficient of viscosity α may be $\sim 2 \times 10^{-8}\ \mathrm{Pa}^{-1}$. At the mean pressures occurring in a hydrodynamically lubricated bearing, the increase in viscosity will be only a few percent, but in EHL it can be very high indeed. At a pressure of 500 MPa, for example, the viscosity of the oil will be over 20 000 times that at atmospheric pressure, and it will behave much more like a solid than a liquid within the contact zone.

The second effect, which is also the only important effect in soft EHL, is that due to elastic distortion of the surfaces. When an elastic sphere is pressed against a rigid plane, initial contact occurs at a point (Fig. 4.9(a)). As the normal load is increased (Fig. 4.9(b)), the contact region expands and the contact area can be calculated from Hertz's theory (see Section 2.5.1). If a lubricant film is present, and the sphere slides over the flat, the pressure distribution and surface deformations predicted by Hertz's equations must be modified. Theoretical investigation of EHL involves solving Reynolds equation while taking account of the variation of lubricant viscosity with pressure, and allowing for the elastic distortion of the bounding surfaces caused by the hydrodynamically generated pressure distribution. The film predicted by this theory has a thickness profile which is shown schematically in Fig. 4.9(c). It is nearly parallel for most of its length, and then develops a sharp constriction in the exit region, in which its thickness is reduced typically by about one quarter. Associated with this constriction is a local sharp spike in the pressure distribution.

Unless drastic simplifications are introduced in this theory, numerical methods are necessary to solve the equations. A useful empirical power-law dependence can be derived from the numerical results for the minimum EHL film thickness, an important quantity. Several different power law expressions have been published, which differ only slightly. One version is as follows:

$$h_{\mathrm{min}} = 1.79 R^{0.47} \alpha^{0.49} \eta_o^{\,0.68} U^{0.68} E^{-0.12} W^{-0.07} \qquad (4.14)$$

Here h_{min} is the minimum film thickness for the contact of a sphere on a plane, E is the reduced modulus of the surfaces (given by equation 2.8), R is the radius of the sphere, η_o and α are as defined for equation 4.4, U is the sliding velocity and W is the normal load. Similar expressions can be derived for other contact geometries.

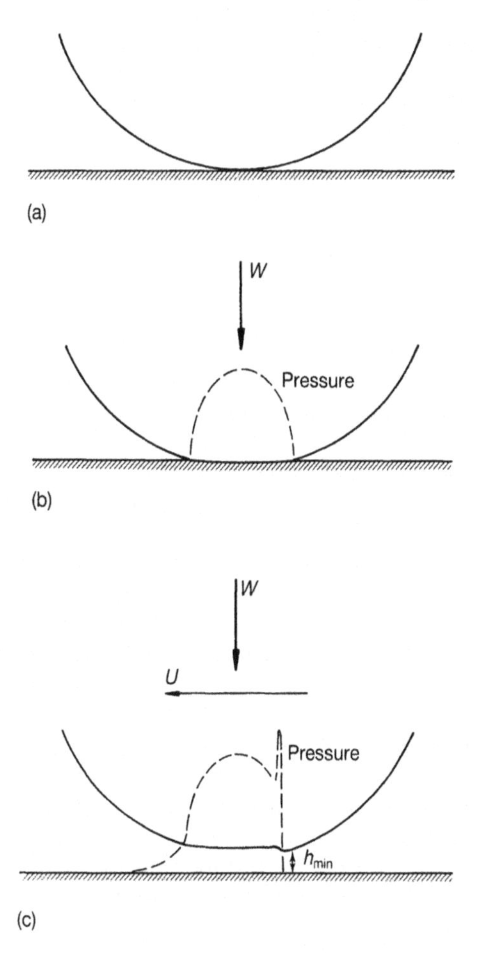

Fig. 4.9 Contact between a sphere and a plane under conditions of elastohydrodynamic lubrication: (a) point contact under zero normal load and with no sliding motion; (b) area of contact due to the normal load only; (c) elastic distortion of the sphere (shown exaggerated for clarity) under combined normal load and sliding motion

It is instructive to compare equation 4.14 with the relationship predicted for classical hydrodynamic conditions (e.g. equation 4.7). In EHL the dependence of film thickness on normal load is very slight, whereas for the hydrodynamic case the dependence is much stronger ($h \propto W^{-0.5}$). The exponents for sliding velocity and viscosity (η_o) are slightly higher in EHL than for hydrodynamic lubrication (0.68 compared with 0.5).

A power law relationship can be similarly used to indicate how film thickness varies in soft EHL. Here the pressure dependence of viscosity is not important. An empirical equation, again for a sphere on a plane, is:

$$h_{min} = 2.8 R^{0.77} \eta_o^{\ 0.65} U^{0.65} E^{-0.44} W^{-0.21} \tag{4.15}$$

The dependence of film thickness on load is stronger in this case than for hard

EHL, and the elastic modulus has a much greater effect. The dependence on viscosity and sliding speed is much the same.

The minimum lubricant film thickness, together with the surface roughness, determines when full fluid film lubrication begins to break down. It is useful to define the ratio

$$\lambda = h_{min}/\sigma^* \quad (4.16)$$

where σ^* is the r.m.s. roughness of the two surfaces, defined by

$$\sigma^{*2} = R_{q1}{}^2 + R_{q2}{}^2 \quad (4.17)$$

R_{q1} and R_{q2} are the r.m.s. roughness values for each surface (see Section 2.3).

The value of λ provides a measure of how likely, and how severe, asperity interactions will be in lubricated sliding. For $\lambda > 3$, a full fluid film separates the two surfaces, asperity contact is negligible and both the friction and wear should be low. However, many counterformal contacts in machinery operate with $\lambda < 3$. The regime $1 < \lambda < 3$ is termed *partial EHL*, or *mixed EHL*; under these conditions some contact between asperities occurs.

The breakdown of the lubricant film with decreasing film thickness in the partial EHL regime is associated with a sharp rise in friction as the quantity $\eta U/W$ falls, as illustrated in Fig. 4.6. The mechanisms of film breakdown are the subject of active research. Thermal effects undoubtedly play an important role under these conditions: a typical rate of viscous energy dissipation in a thin EHL oil film has been estimated at 100 TW m^{-3}, equivalent to dissipating the entire electrical power output of the USA in a volume of 5 litres.

At extremely high loads or low sliding speeds, when the film thickness falls further and λ drops below 1, increasingly severe surface damage (*scuffing*, discussed in Section 5.8) occurs on sliding, and the behaviour of the system depends critically on the properties of *boundary lubricants*, if present. These will be discussed in the next section.

4.6 BOUNDARY LUBRICATION

Under very high contact pressures, or at very low sliding speeds, hydrodynamic forces are insufficient to maintain even a thin EHL film between sliding surfaces, and direct contact will occur between the asperities. High friction and wear rates will then prevail unless the surfaces are protected by a suitable *boundary lubricant*. Boundary lubricants act by forming adsorbed molecular films on the surfaces; repulsive forces between the films then carry much of the load, and intimate contact between unprotected asperities, with associated adhesion and junction growth, is prevented or limited.

Figure 4.10 illustrates the mechanism of operation of a typical boundary lubricant, a long-chain carboxylic acid, on metal surfaces. The lubricant molecules are adsorbed with the polar end-groups adhering strongly to the oxide layer present on the metal. The molecular chains tend to align perpendicularly to the surface, stabilized by their mutual repulsion, and form

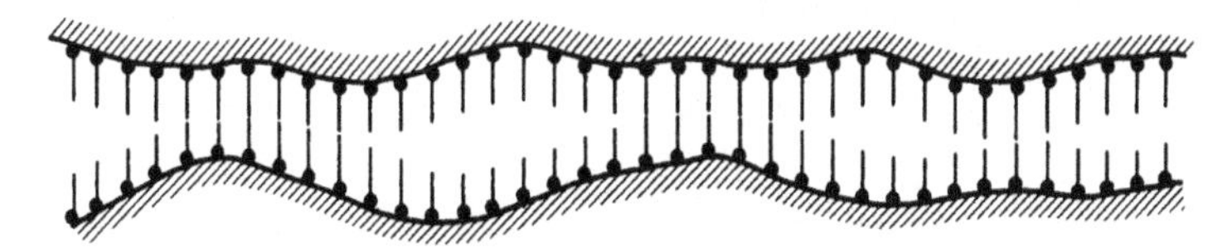

Fig. 4.10 The mechanism of operation of a boundary lubricant. Polar end-groups on the hydrocarbon chains bond to the surfaces, providing layers of lubricant molecules which reduce direct contact between the asperities

dense layers of hydrophobic chains typically 2 to 3 nm long. When the two layers come together, most of the normal load is carried by the interaction of the hydrocarbon chains, and there are only small areas of naked asperity contact. The frictional force is lower than for unlubricated sliding, and although some wear does occur, it is substantially less severe than if the surfaces were unprotected.

The lowering of friction produced by a boundary lubricant of this type is in direct proportion to its molecular weight, and hence to the length of the hydrocarbon chain. Figure 4.11 shows how the coefficient of friction for very slow sliding of steel on steel varies with the size of the boundary lubricant molecule (in these experiments, straight-chain alcohols and carboxylic acids). For these molecules, appreciable chemical affinity exists between the acid or hydroxyl group and the metal oxide, and the adsorption process involves a degree of chemical reaction (*chemisorption*). Molecular chain length has a similar influence on friction for straight-chain saturated hydrocarbons adsorbed from the gas phase in vacuum. Figure 4.12 shows that a significant effect occurs even with very short chains, only one or two carbon atoms long.

Many oils naturally contain some molecular species with boundary lubricating properties; vegetable oils, such as castor oil and rapeseed oil, contain more natural boundary lubricants than mineral oils. The boundary lubricating

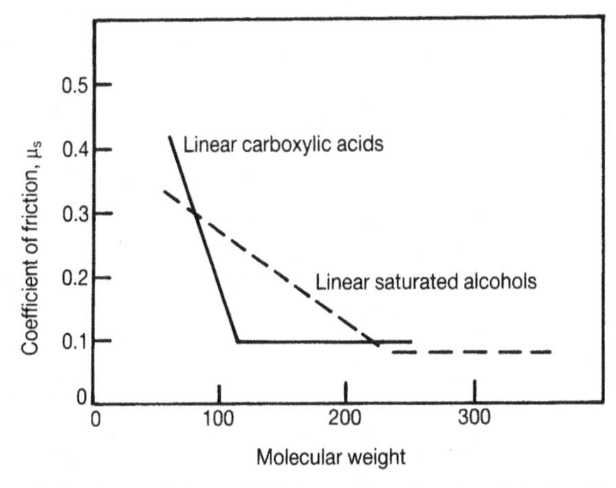

Fig. 4.11 The variation in coefficient of friction with molecular chain length for steel surfaces lubricated with carboxylic acids and alcohols (from Bowden F P and Tabor D, *The Friction and Lubrication of Solids*, Clarendon Press, Oxford, 1950)

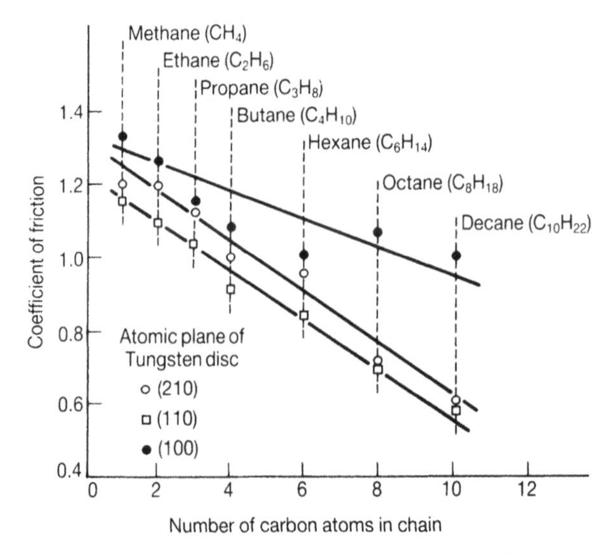

Fig. 4.12 The variation in coefficient of friction with molecular chain length for tungsten surfaces exposed to hydrocarbon vapours in vacuum (from Buckley D H, *Surface Effects in Adhesion, Friction, Wear and Lubrication*, Tribology Series no. 5, Elsevier, 1981)

action of an oil is associated with the qualitative terms *oiliness* and *lubricity*. In practical applications, additional boundary lubricant additives are usually incorporated in the oil. These might typically be long-chain carboxylic acids (fatty acids) such as stearic acid (= octadecanoic acid, $C_{17}H_{35}COOH$), carboxylic acid amides (e.g. oleic acid amide, $C_8H_{17}CHCHC_7H_{14}CONH_2$), esters such as butyl stearate ($C_{17}H_{35}COOC_4H_9$), or alcohols (e.g. cetyl alcohol = hexadecanol, $C_{16}H_{33}OH$). Complete coverage of a surface by a layer of boundary lubricant only one molecule thick is, in principle, sufficient for protection. Boundary lubricants are therefore effective in oils at very low overall concentrations; about 0.1% to 1% is typically added.

Another class of molecules with boundary lubricating properties consists of the so-called *extreme pressure* (EP) and *anti-wear* additives. These are substances which react with the sliding surfaces under the severe conditions in the contact zone to give compounds with low shear strength, thus forming a lubricating film at precisely the location where it is needed. Originally, the high pressures in the contact region were considered to play a role in the action of these additives, hence the name 'extreme pressure'; it is now clear, however, that the transient high temperatures associated with sliding in a very thin EHL film, or with the rapid plastic deformation of asperities (see Section 5.5), are more important.

Additives of this type usually contain sulphur, phosphorus or, less commonly, chlorine, and function by adsorbing on the metal surface and then reacting in the contact zone to form local sulphide, phosphide or chloride films which are easily sheared. Probably the most widely used EP additives are zinc dialkyl dithiophosphate (ZDP), tricresyl phosphate (TCP) and dibenzyl disulphide. The structures of these three compounds are shown in

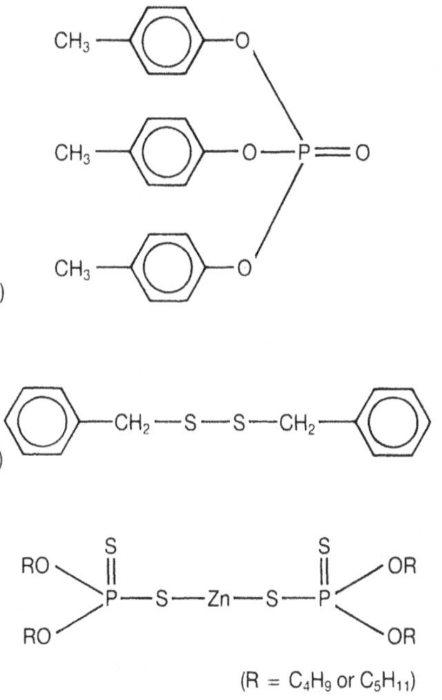

Fig. 4.13 Structures of three EP/antiwear additives: (a) tricresyl phosphate (TCP); (b) dibenzyl disulphide; (c) zinc dialkyl dithiophosphate (ZDP)

Fig. 4.13. The exact mechanisms by which EP additives work are still unclear: there are definite synergistic effects between them, other additives and the base oils, and there is evidence that EHL on a microscopic scale ('micro-EHL') also plays a role.

In general, the more severe the contact conditions in which an oil is required to operate, the more active the EP and anti-wear additives which are needed. Since these compounds function by reacting with the metal surfaces, and some removal of surface material will therefore be inevitable in sliding, the balance is often fine between mechanical damage to the surfaces due to scuffing if insufficient EP protection is provided, and excessive chemical attack by the additives themselves if they are too aggressive.

4.7 SOLID LUBRICATION

Solid materials which exhibit low coefficients of friction may be used as lubricants in preference to liquid or gas films for several reasons. One component of the bearing may be composed of, or coated with, the solid lubricant, or made from a composite containing it, in order to produce a 'self-lubricating' system which needs no external source of lubricant during its lifetime. Such bearings offer obvious savings in maintenance and lubrication costs. Under certain conditions solid lubrication may be the only feasible system, for example in food-processing machinery where contamination of the product by a liquid lubricant must be avoided, or in space where a liquid lubricant would evaporate. At high temperatures, liquid lubricants decompose or oxidize; suitable solid lubricants can extend the operating temperatures of sliding systems beyond 1000 °C, while maintaining relatively low coefficients of friction.

We saw in Sections 3.7 and 3.8 that graphite, molybdenum disulphide (MoS_2) and polytetrafluoroethylene (PTFE) under appropriate conditions all exhibit low coefficients of friction; these three materials are used in the great majority of applications requiring solid lubricants. They can be applied as thin coatings, incorporated into composite bearing materials (see Chapter 9) or, in the case of graphite and PTFE, used as bulk bearing materials. Graphite and MoS_2 can also be applied as replenishable lubricants in the form of dry powders. They are also commonly incorporated into lubricating oils and greases as suspensions of fine particles; they come into play under severe contact conditions and then have boundary lubricating properties.

All three of these materials, however, suffer from limitations. Their use at high temperature is, in each case, limited by decomposition or oxidation in air: graphite to ~500 °C, MoS_2 to ~300 °C, and PTFE to ~250 °C. In addition, the lubricating properties of graphite depend on the presence of condensable vapours (see Section 3.7), while high atmospheric humidity adversely affects the friction and wear rate of MoS_2.

Many other materials have been proposed as solid lubricants, and are used in certain applications. They may be broadly classified as organic polymers, lamellar solids, other inorganic compounds, and soft metal films.

The only polymer used at all widely as a lubricant, in thin films on harder substrates, is PTFE. Several other polymers, notably nylon, polyimides and polyetheretherketone (PEEK), show relatively low friction and good wear resistance, but these tend to be used, as PTFE is also used, in composite bearing materials (see Sections 9.4.2 and 9.4.3).

Many other materials apart from graphite and MoS_2 have lamellar structures, and although this provides no guarantee that the compound will act as a solid lubricant, many do. Transition metal dichalcogenides with structures analogous to MoS_2 have been investigated: NbS_2, TaS_2, WS_2 and the diselenides of the same metals all act as solid lubricants. None is used as widely as MoS_2, however, although some offer lower friction and have other advantages. For example, WS_2 resists oxidation better than MoS_2, and can lubricate at significantly higher temperatures.

Other inorganic compounds with layer structures vary in their lubricating action. Some, such as $CdCl_2$, CdI_2 and $CoCl_2$, typically show $\mu < 0.1$ and are potentially valuable lubricants, while others show high friction and are of no interest for lubrication. Hexagonal boron nitride (BN), which is isoelectronic with carbon and has a structure similar to graphite, has good oxidation resistance, but shows rather high friction in air ($\mu = 0.2$ to 0.4). Like graphite, however, boron nitride is sensitive to the presence of condensable vapours (see Section 3.7), and a relatively low vapour pressure of heptane, for example, is sufficient to halve its coefficient of friction.

Another lamellar solid with potential as a solid lubricant is graphite fluoride. This non-stoichiometric compound has the formula $[CF_x]_n$, where $0.3<x<1.1$. Graphite fluoride has a structure derived from that of graphite by the intercalation of fluorine atoms between the basal planes, increasing the interplanar spacing to 0.75 nm (from 0.34 nm in graphite). The fluorine atoms are covalently bonded to the carbon atoms. The frictional properties of graphite fluoride are excellent, it has good oxidation resistance, and offers promising performance in bonded lubricant films and composite bearing materials. It has not yet, however, become at all widely used.

Certain inorganic solids without lamellar structures also possess low shear strengths, and some of these are useful as lubricants, particularly at high temperatures. A wide range of compounds has been investigated: calcium and barium fluorides (CaF_2 and BaF_2), lead oxide and sulphide (PbO and PbS) and boric oxide (B_2O_3) can be used as high temperature lubricants either in bulk composite form or incorporated into coatings. Ceramic-bonded composite films containing CaF_2 can give a coefficient of friction as low as 0.1 at temperatures up to 1000 °C. It is not easy, however, to find a lubricant which will perform well at low temperatures as well as at high.

Buckminsterfullerene (C_{60}) and other materials with giant molecules of similar near-spherical shape have been suggested as solid lubricants, although their properties are still under investigation.

Thin films of soft metals constitute another class of solid lubricants. As was shown in Section 3.4.3, the coefficient of friction for sliding over a thin soft metal layer on a harder substrate may be simply estimated as

$$\mu = \tau_i / p_o \tag{4.18}$$

where τ_i is the shear strength of the film and p_o is the indentation pressure of the substrate material. The friction force is determined by the shear strength

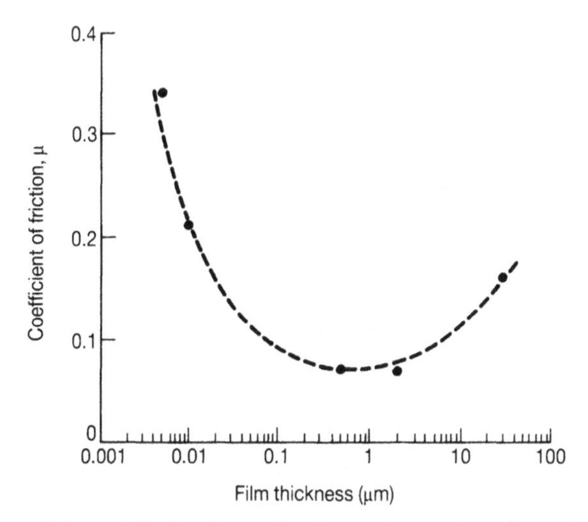

Fig. 4.14 The coefficient of friction for steel sliding against tool steel carrying a thin film of indium metal, plotted against the thickness of the indium film (from Bowden F P and Tabor D, *The Friction and Lubrication of Solids*, Clarendon Press, Oxford, 1950)

of the film material, whereas the normal load is carried by deformation of the substrate; hence the two quantities τ_i and p_o can be varied independently, leading to a low value of μ. Lead, tin, indium, silver and gold have all been widely and successfully used as solid lubricant films, usually deposited by electroplating, physical vapour deposition or sputtering (see Chapter 8). Very thin coatings (as thin as 0.1 μm) will suffice to lubricate suitably smooth hard substrates, and metallic films of this thickness can be used to lubricate rolling element bearings for use in space or ultra-high vacuum, or for high temperature operation. Coefficients of friction can be very low (for example, 0.06 for In, 0.09 for Sn and 0.20 for Ag films on a steel substrate with a steel slider) although the sliding friction for bulk specimens of these soft metals would, in contrast, be extremely high. As the thickness of the metal film is varied, so the coefficient of friction changes. Figure 4.14 shows the behaviour of indium films on hard tool steel. There is an optimum film thickness for minimum friction. Thinner films cannot prevent asperity contacts between the base metals, while for thicker films the frictional behaviour becomes more similar to sliding on a bulk specimen of the soft metal.

Further reading

Arnell R D, Davies P B, Halling J and Whomes T L, *Tribology: Principles and Design Applications*, Macmillan, 1991
Booser E R (Ed.), *CRC Handbook of Lubrication*, Vols. 1 and 2, CRC Press, 1983
Cameron A, *Basic Lubrication Theory*, Ellis Horwood, 1981
Clauss F J, *Solid Lubricants and Self-Lubricating Solids*, Academic Press, 1972
Dowson D and Higginson G R, *Elasto-hydrodynamic Lubrication*, Pergamon, 1977
Jones M H and Scott D, *Industrial Tribology*, Tribology Series no. 8, Elsevier, 1983
Lansdown A R, *Lubrication: A Practical Guide to Lubricant Selection*, Pergamon Press, 1982
Stolarski T A, *Tribology in Machine Design*, Heinemann Newnes, 1990

5

Sliding wear

5.1 INTRODUCTION AND TERMINOLOGY

This chapter is concerned with the wear that occurs when two solid surfaces slide over each other. In most practical applications sliding surfaces are lubricated in some way, and the wear that occurs is then termed *lubricated sliding wear*. In some engineering applications, however, and in many laboratory investigations, surfaces slide in air without a lubricant. The resulting wear is then often called *dry sliding wear*, although it usually takes place in ambient air of appreciable humidity. We shall consider both dry and lubricated wear in this chapter, but we shall deliberately exclude the wear associated with the presence of hard particles, either as a separate component between the sliding surfaces or as an element of the structure of one or both surfaces. This type of wear is termed *abrasive wear*, and is discussed in Chapter 6, together with other kinds of wear caused by hard particles. The distinction between abrasive wear and sliding wear employed here is somewhat artificial, and certainly not respected by nature, since under some conditions sliding wear can generate debris which then causes further wear by abrasion; it must therefore always be borne in mind that the boundary between different types of wear is not a rigid one.

The term *adhesive wear* is sometimes used to describe sliding wear, but its use can be misleading. As we shall see, adhesion plays an important role in sliding wear, but it is only one of the several physical and chemical processes that may be involved. We shall therefore use 'sliding wear' as a general term in preference. The words *scuffing*, *scoring* and *galling* are often associated with severe sliding wear, but are ill-defined, and their usage varies between the two sides of the Atlantic. Scuffing, in UK usage, refers to localized surface damage associated with local solid-state welding between sliding surfaces. The term is frequently used in describing the breakdown of lubrication, usually at high sliding speeds (discussed in Section 5.8). In US usage, the term *scoring* is sometimes used as a synonym for scuffing as described above, and both terms can also imply scratching by abrasive particles (described in Chapter 6). *Galling* represents a more severe form of scuffing, due to local welding, and is associated with gross surface damage. The word often refers to damage resulting from unlubricated sliding at low speeds, characterized by severely roughened surfaces and transfer or displacement of large fragments of material. Galling may occur in nominally lubricated systems when the lubricant film breaks down, and can be followed by seizure of the surfaces and consequent gross failure of the sliding system.

5.2 TESTING METHODS

Many different experimental arrangements have been used to study sliding wear. Laboratory investigations of wear are usually carried out either to examine the mechanisms by which wear occurs, or to simulate practical applications and provide useful design data on wear rates and coefficients of friction. For both purposes, control and measurement of all the variables which may influence wear are very important. It is vital to appreciate that wear rate and friction are often critically dependent on the sliding conditions; apparently minor changes in conditions can lead to radical changes in the dominant mechanism and associated rate of wear. Close control and monitoring are essential if the results of a test are to be useful either as a simulation of a practical application, or for wider scientific purposes.

Figure 5.1 shows the geometrical arrangements employed in several common types of wear testing apparatus. The word *tribometer*, first used in 1774 for an instrument intended to measure friction, is sometimes used for such apparatus; more recently the inelegant term *tribotester* and its associated verb have been coined.

The methods shown in Fig. 5.1 may be divided into two types: those where the sliding surfaces are symmetrically disposed, in which the wear rates of two surfaces of identical material should be the same, and the more common arrangement where the system is inherently asymmetric, in which the two sliding bodies, even of the same material, will almost certainly experience different rates of wear for reasons discussed below.

Symmetrical arrangements are not often used to study wear: examples are the ring-on-ring (or two disc) devices, with contact either along a line (A in Fig. 5.1) or face to face (B). Such devices are only truly symmetrical if both components are rotated.

The most common asymmetric test rigs employ a pin pressed against a disc, either on the flat face (C) or on the rim (D), a block loaded against a ring (E) or a pin on a flat (F). In these cases the contact may initially be over an extended nominal contact area (e.g. with a flat-ended pin on a flat disc, or a conforming block-on-ring), or only at a point or line (e.g. a round-ended pin on a disc, or a plane block-on-ring). Details of these types of *conformal* and *counterformal* (sometimes called *concentrated*) contacts are shown in Fig. 5.2.

In asymmetric arrangements one component of the mating pair, commonly the pin or block, is usually treated as the specimen, and is the component for which the wear rate is measured, while the other, often the disc, flat or ring, is called the *counterface*.

Another test geometry is the *four ball test*, illustrated in Fig. 5.3. The lower three balls are rotated together in a carrier, and move relative to the upper ball which is held stationary and pressed downwards under a fixed normal

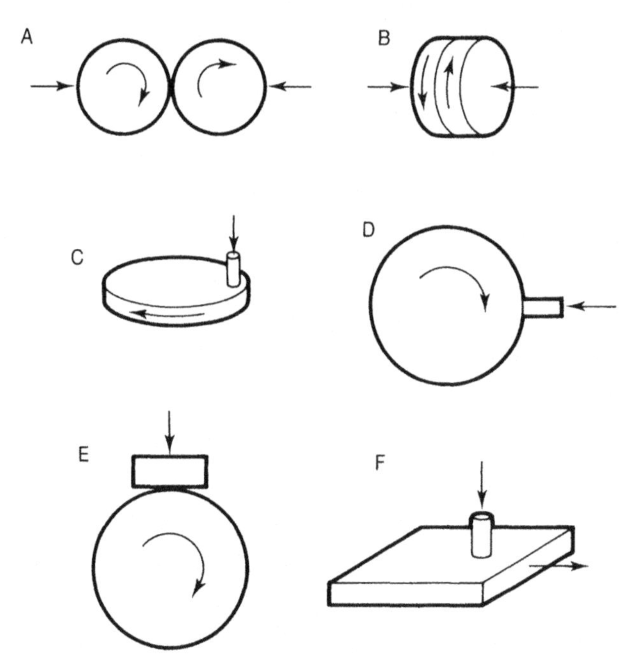

Fig. 5.1 Geometries employed in sliding wear tests (from Bayer R G and Trivedi A K, in Bayer R G (Ed), *Selection and Use of Wear Tests for Metals*, ASTM Sp. Tech. Pub. 615, 1975, pp. 91–101)

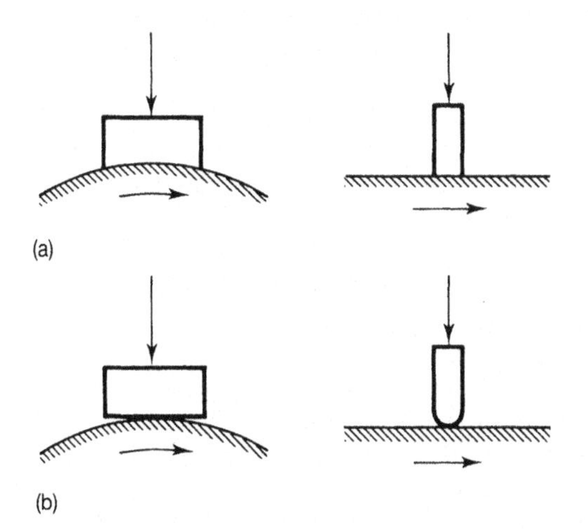

Fig. 5.2 Geometries of (a) conformal and (b) counterformal contacts

load. The balls are usually made from standard rolling bearing steel (see Section 9.2.2), and the test is often used as a method of evaluating lubricant performance rather than to study material behaviour.

The dimensions of specimens in laboratory wear testing usually lie in the range from millimetres to tens of millimetres; in asymmetric tests the pins and

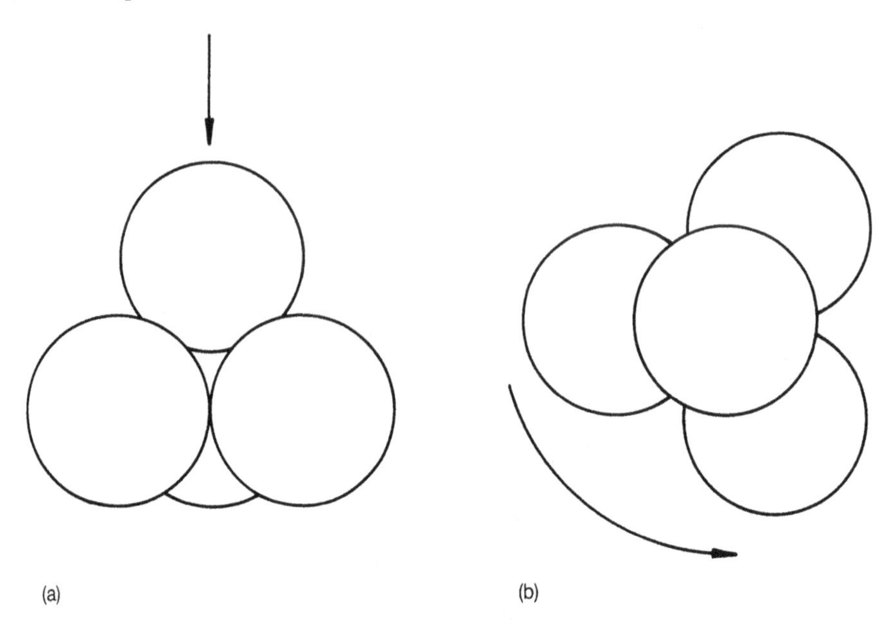

Fig. 5.3 The four ball test, commonly used to evaluate lubricants: (a) perspective view; (b) plan view

blocks are typically smaller than 25 mm in size, while the counterface rings and discs are typically several tens of millimetres in diameter. Loads over a wide range are employed, from fractions of a newton to several kilonewtons. Sliding speeds similarly cover a broad span, from fractions of a millimetre per second to tens or even hundreds of metres per second. The variables which affect friction and wear, and which must therefore be controlled in wear testing, range from the most obvious such as speed and load to the less obvious, such as specimen size and orientation.

Several sliding wear test methods are the subject of national standards. Examples are the block-on-ring (ASTM G77), crossed cylinder (ASTM G83), pin-on-disc (ASTM G99), sphere-on-disc (DIN 50324) and rotating pin-on-flat (ASTM G98) geometries.

As we shall see in the next section, a simple model for wear suggests that the amount of material removed from a sliding body should depend on the distance slid and on the nominal pressure (normal load divided by the nominal contact area) over the contact region. Wear is usually measured either by removing the specimen at intervals and weighing or measuring it, or by continuously measuring its position with an electrical or mechanical transducer and deducing the wear from its change in dimensions. In the case of a counterformal contact (e.g. Fig. 5.2 (b), or Fig. 5.3) wear causes the apparent area of contact to increase, and can therefore be quantified by measuring the size of the wear scar on the specimen; this method of determining wear is commonly used, for example, in the four ball test. Obviously, in any interrupted test care must be taken to replace the specimen in exactly the same position and orientation after measurement or weighing.

It is wise to measure and record the friction force continuously during a wear test. This is usually done by measuring the tangential force on the specimen, or the torque on a rotating counterface. A continuous friction record not only provides numerical values for μ, but also allows changes in sliding behaviour to be monitored (for example, a rise or fall in μ, or a change from a smooth trace to an irregular one). These changes often herald changes in surface nature or topography, or in the mechanism of wear. Running-in, for example, can be explored in this way, as can the breakdown of protective oxide or lubricant films. Knowledge of such changes is sometimes more valuable than measurement of absolute wear rate. For some purposes, continuous measurement of electrical contact resistance between the specimen and counterface is informative, since it provides a means of assessing the thickness of oxide or EHL films. Electrical capacitance measurements provide another source of such information.

Wear under sliding conditions depends on the distance slid, but also to some extent on both the sliding velocity and the duration of the test, independently. The sliding velocity affects the rate of frictional energy dissipation, and hence the temperature at the interface. It certainly cannot be assumed that one wear test will produce the same results as another of half the duration at twice the sliding velocity, since abrupt transitions in wear mechanism and rate may occur as the sliding speed is changed.

Wear also depends on the nominal contact pressure between the sliding surfaces, and, as is shown in the following sections, transitions are commonly induced by changes in contact pressure. The linear dimensions of the specimen are also important, independently of the contact pressure, since wear debris formed near the leading edge of a long specimen will have more influence during its passage through the contact zone than it would have with a shorter specimen.

Apart from the major variables of normal load, contact area, sliding speed and testing time, several other factors must also be considered and monitored in wear testing. The testing temperature is important through its influence on the mechanical properties of the materials and on thermally activated chemical processes, although these may often be dominated by frictionally-generated temperature rises. In lubricated systems it will also be important through its effect on lubricant viscosity. Atmospheric composition is extremely important; reactive components such as water vapour and oxygen strongly influence wear rates and mechanisms in all classes of material. The orientation of the apparatus can also, perhaps surprisingly, affect the results of testing: different behaviour may be seen if wear debris falls readily away from the contact area under gravity, rather than being retained on the counterface.

All the types of apparatus discussed above can be used either unlubricated ('dry') or lubricated, and the lubricant can be introduced in various ways. In a lubricated system the supply of lubricant, and the details of the pressure distribution within the lubricant film and its resulting thickness, are important factors.

The comprehensive list of influences given above may suggest that it is never possible to produce a valid laboratory simulation of a practical application, and that the only valid wear test is a service trial. This is emphatically not the case, but the tribologist must always be aware of the possible consequences of departing too far from the actual conditions which he or she is trying to simulate. Contact stresses, thermal conditions, sliding speeds and chemical environment are all vital ingredients in any wear test, and it is sensible to ensure, by measurement of friction and by close examination of the worn surfaces and wear debris after the test, that the mechanism of wear is the same in the test as in the service application. Only then can the results of a laboratory test be applied with confidence to a practical problem.

5.3 SIMPLE THEORY OF SLIDING WEAR: THE ARCHARD WEAR EQUATION

When two surfaces in contact slide over each other, one or both of the surfaces will suffer wear. A simple theoretical analysis of this type of wear, due originally to Holm and Archard, will be given here. By virtue of its simplicity, it highlights the main variables which influence sliding wear; it also yields a method of describing the severity of wear, by means of the wear coefficient K, which is valuable and is widely used. Although the model was originally developed for metals, it can provide some insight into the wear of other materials as well.

The starting point for the model is the assumption, discussed in detail in Chapter 2, that contact between the two surfaces will occur where asperities touch, and that the true area of contact will be equal to the sum of the individual asperity contact areas. This area will be closely proportional to the normal load, and it can be assumed that under most conditions, for metals at least, the local deformation of the asperities will be plastic (see Section 2.5.3).

Figure 5.4 shows a single asperity contact, which we assume to be circular in plan view and of radius a, at various stages of evolution during sliding. In Fig. 5.4 (c) it has reached maximum size, and the normal load supported by it, δW, will be given by

$$\delta W = P \pi a^2 \tag{5.1}$$

where P is the yield pressure for the plastically deforming asperity (which will be close to its indentation hardness H, see Section 2.5.1).

As sliding proceeds, the two surfaces become displaced as shown in Figs 5.4 (d) and (e), and the load originally borne by the asperity is progressively transferred to other asperity junctions which are in the process of forming elsewhere on the surfaces. Continuous sliding leads to the continuous formation and destruction of individual asperity contacts. Wear is associated with the detachment of fragments of material from the asperities, and the volume of each wear fragment will depend on the size of the asperity junction from which it originated. It is assumed that the volume of material removed by wear, δV, will be proportional to the cube of the contact dimension a, which implies that the shape of the wear particle should be independent of its size.

For argument's sake, the volume can be taken to be that of a hemisphere of

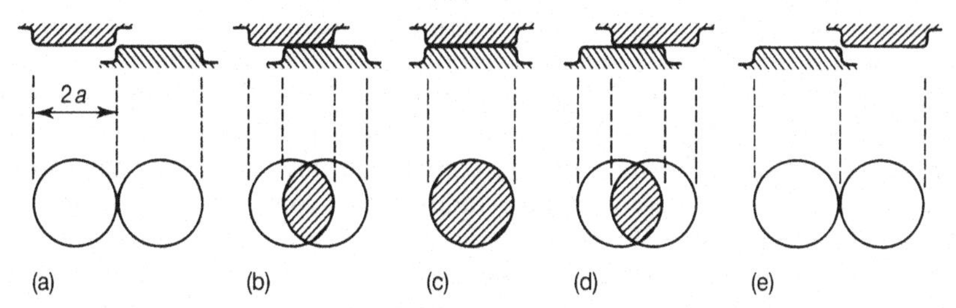

Fig. 5.4 Schematic diagram showing the evolution of a single contact patch as two asperities move over each other (from Archard J F, *J. Appl. Phys.* **24**, 981–988, 1953)

radius a (although this should not be taken to imply that the worn particles are necessarily hemispherical in shape), giving

$$\delta V = 2\pi a^3/3 \tag{5.2}$$

Not all asperity contacts give rise to wear particles. Let us suppose that only a proportion κ does so. The average volume of material δQ worn away per unit sliding distance due to sliding of the one pair of asperities through a distance $2a$ is therefore given by

$$\delta Q = \kappa \delta V/2a = \kappa \pi a^2/3 \tag{5.3}$$

and the overall wear rate Q arising from all the asperity contacts is the sum of the contributions over the whole real area of contact:

$$Q = \sum \delta Q = \frac{\kappa}{3} \sum \pi a^2 \tag{5.4}$$

The total normal load W is given by

$$W = \sum \delta W = P \sum \pi a^2 \tag{5.5}$$

and hence

$$Q = \kappa W/3P \tag{5.6}$$

It is convenient to combine the factor of 1/3 into the constant of proportionality, by putting $K = \kappa/3$, and to assume that $P = H$, the indentation hardness. We can then rewrite the equation in the form:

$$Q = \frac{KW}{H} \tag{5.7}$$

This equation, relating the volume worn per unit sliding distance, Q, to the macroscopic quantities W, the normal load, and H, the hardness of the softer surface, is often called the *Archard wear equation*. The constant K, usually termed the *wear coefficient* or sometimes the *coefficient of wear*, is dimensionless and always less than unity.

The dimensionless wear coefficient K is of fundamental importance, and provides a valuable means of comparing the severity of wear processes in different systems. However, for engineering applications the quantity K/H is often more useful. This is given the symbol k and called the *dimensional wear coefficient*. k is usually quoted in units of $mm^3\ (Nm)^{-1}$, and represents the volume of material removed by wear (in mm^3) per unit distance slid (in metres), per unit normal load on the contact (in newtons). The measure of wear provided by k is particularly helpful for comparing wear rates in different classes of material. In some materials, elastomers for example, there are basic problems with the use of the dimensionless coefficient K since the plastic indentation hardness H cannot be defined.

Although we have so far defined K in terms of the probability of each asperity interaction resulting in the production of a wear particle, it is important to realize that this is not the only possible interpretation. It may also be taken to reflect the number of cycles of deformation required by each asperity before a fragment of material is removed by a fatigue process. Alternatively, it can be correlated with the size of the wear particle produced by every asperity contact.

From a fundamental viewpoint, equation 5.7 follows inevitably from the assumption that the wear rate (volume per unit sliding distance) depends only on the normal load and on the hardness or yield strength of the softer surface; it is the only dimensionally correct relationship which is possible between Q, W and H. Q, the volume worn per unit sliding distance, has the dimensions of area, and the quantity W/H also represents an area of basic importance in contact processes: the real area of contact for fully plastic asperities. K can therefore be interpreted as the ratio between these two areas.

Equation 5.7 implies that if K is a constant for a given sliding system, then the volume (or mass) of material lost by wear should be proportional to the distance slid (i.e. Q should be constant), and if the normal load W is varied then the wear rate should vary in proportion.

It is found by experiment that for many systems the loss of material by wear is indeed proportional to the sliding distance (and so, for sliding at constant velocity, to the time). Transient behaviour is sometimes observed at the start of sliding, until equilibrium surface conditions have become established; the wear rate during this initial *running-in* period may be either higher or lower than the corresponding steady-state wear rate, depending on the nature of the running-in process. Figure 5.5 shows the results of pin-on-ring tests, for a wide range of material combinations under unlubricated conditions in air. In each case, the steady-state wear rate (volume removed per unit sliding distance) is essentially constant for the duration of the test.

Strict proportionality between the wear rate and the normal load is less often found. Although for many systems the wear rate varies directly with load over limited ranges, abrupt transitions from low to high wear rate, and sometimes back again, are often found with increasing load. Figure 5.6 illustrates this behaviour. The wear rate of a brass pin against a tool steel ring is found to increase linearly with load and obeys equation 5.7 well; no transition is seen over this range of load. The ferritic stainless steel pin, however, suffers a rapid increase in wear rate above a critical load, although below that load its behaviour is described well by the Archard equation. The reasons for such transitions will be explored in detail in the following sections;

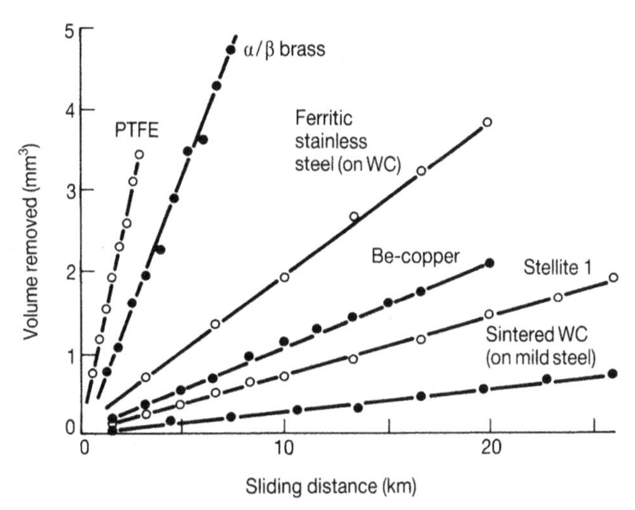

Fig. 5.5 Wear (expressed as the volume removed from the specimen pin sliding against a tool steel ring unless otherwise indicated) plotted against total sliding distance from unlubricated pin-on-ring tests on the materials indicated (from Archard J F and Hirst W, *Proc. Roy. Soc. Lond.* **236A**, 397–410, 1956)

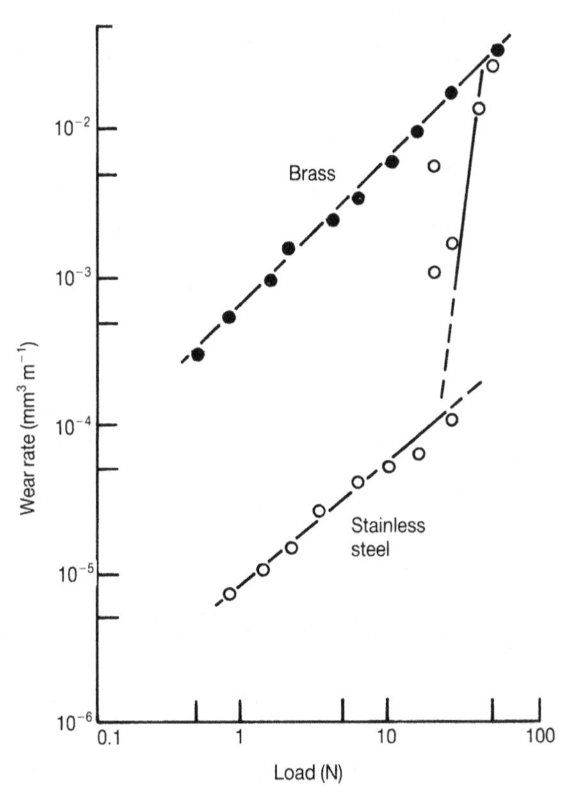

Fig. 5.6 Wear plotted against load (logarithmic scales) for brass and ferritic stainless steel pins sliding against tool steel counterfaces in unlubricated pin-on-ring tests (from Archard J F and Hirst W, *Proc. Roy. Soc. Lond.* **236A**, 397–410, 1956)

Table 5.1 Values of the dimensionless wear coefficient K for various materials sliding against tool steel (unless otherwise stated) in unlubricated pin-on-ring tests in air. It should be noted that these data relate to specific test conditions, and take no account of the mechanism of wear (from Archard J F and Hirst W, *Proc. Roy. Soc. Lond.* **236A**, 397–410, 1956)

Material	Wear coefficient, K
mild steel (against mild steel)	7×10^{-3}
α/β brass	6×10^{-4}
PTFE	2.5×10^{-5}
α brass	1.7×10^{-4}
PMMA	7×10^{-6}
copper–beryllium	3.7×10^{-5}
hard tool steel	1.3×10^{-4}
Stellite 1	5.5×10^{-5}
stainless steel (ferritic)	1.7×10^{-5}
polyethylene	1.3×10^{-7}

they typically correspond to substantial changes, by factors of 100 or more, in the wear coefficient K with normal load.

No mention has so far been made of the influence on wear rate of the apparent area of contact, or of the sliding velocity. According to equation 5.7 the wear rate Q should be independent of these factors. For many purposes this assertion is true, although in some systems, as discussed below, sharp transitions in wear rate are seen with increasing sliding velocity.

Some typical values of K, measured for dry sliding wear in pin-on-ring tests for a range of materials are listed in Table 5.1. It is instructive to contrast these values with the coefficients of friction for the same materials listed in Table 3.1. The values of wear coefficient vary over nearly five orders of magnitude (from ca. 10^{-2} to ca. 10^{-7}), much more than the variation in the coefficient of friction, and no correlation is apparent between the two sets of data.

The Archard wear equation provides a valuable means of describing the severity of wear, via the wear coefficient K, but it must be remembered that its validity cannot be used to support or reject any particular mechanism of material removal. As we have seen, the wear coefficient is open to several interpretations, and it will be shown in Chapter 6 that an equation of identical form can also be derived from completely different starting assumptions, for material removal by abrasive wear. The mechanisms actually responsible for material removal in sliding wear, and for the wide range of values of K observed in practice, will be explored in the following sections.

5.4 WEAR OF BRASS: A PARADIGM FOR MILD AND SEVERE WEAR IN METALS

Several important features commonly observed in the sliding wear of metals can be illustrated by a study of the wear of brass, and we shall examine them in detail in this section.

The behaviour of α/β leaded brass sliding against a smooth hard surface is particularly reproducible and has been studied closely by several investigators. Figure 5.7 shows the wear rate of a leaded α/β brass pin (59% Cu, 39%

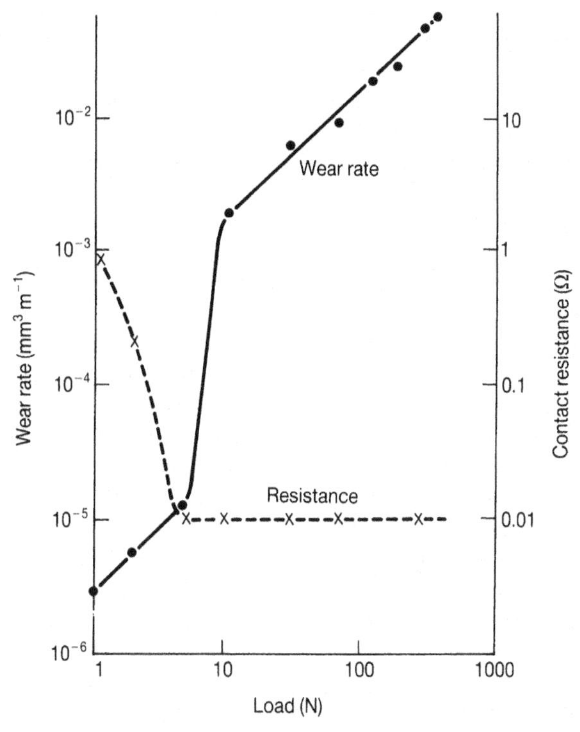

Fig. 5.7 Wear rate and electrical contact resistance for a leaded α/β brass pin sliding against a hard stellite ring, as a function of normal load. Note the sharp transition in wear rate, associated with a change in contact resistance (from Hirst W and Lancaster J K, *J. Appl. Phys.* **27**, 1057–1065, 1956)

Zn, 2% Pb) sliding against a hard stellite ring, as a function of the normal load. Also plotted is the electrical contact resistance between the pin and the ring, which allows the extent of metallic contact to be estimated. Although these particular experiments were carried out in the presence of cetane (*n*-hexadecane), no significant hydrodynamic film was formed and very similar results have been reported under completely dry conditions, against hard steel counterfaces as well as stellite.

At low loads, the wear rate increases with normal load in agreement with equation 5.7, and the wear coefficient K has a value of about 2×10^{-6}. At a load of 5 to 10 N in these experiments, there is a sharp increase in wear rate, by a factor of about 100; as the load is raised further, the behaviour still follows the Archard equation, but with a wear coefficient of $\sim 10^{-4}$. The transition in wear rate is also associated with a drop in the contact resistance, which becomes constant at a much lower value for loads above the transition load. The roughness (R_a) of the worn brass surface also shows a marked change, from less than 0.5 μm below the transition to about 25 μm above the transition. The wear debris formed below the transition point is fine dark-coloured oxide, while that formed above the transition consists of much larger metallic particles.

The regime of wear at low loads, below the transition, is often called *mild wear*, while that leading to the higher rate above the transition is termed *severe wear*. These terms, although apparently vague and subjective, are sufficiently well-defined in the context of the sliding wear of metals to be useful, and they are widely employed.

Mild wear is characterized by finely divided wear debris (typically 0.01 μm to 1 μm in particle size) which is predominantly oxide. The worn surface is relatively smooth. Severe wear, in contrast, results in much larger particles of metallic debris, perhaps 20 to 200 μm in size, which may even be individually visible with the naked eye, and a correspondingly roughened surface. The major distinction between the two regimes lies in the nature of the debris, and in the wear rates which may differ by a factor of 100 or even 1000. Wear rates due to severe wear are nearly always so high as to be completely unacceptable in engineering applications, with dimensionless wear coefficients of up to 10^{-2} to 10^{-3}; even rates of unlubricated mild wear, with K values typically below 10^{-4} to 10^{-5}, may be intolerable under many circumstances.

The transition between mild and severe wear results from a change in the nature of the sliding contact. In the regime of severe wear the wear rate of the harder counterface is insignificantly small, and the wear debris is metallic brass. There is extensive metallic contact over the whole real area of contact, as shown by the low electrical contact resistance. Experiments with radioactive tracers show that severe wear is a two-stage process, in which brass is transferred from the pin to the ring in small discrete particles, and then the transferred material is removed from the ring as composite fragments comprising 50 or so originally transferred particles. The rate-determining step in the process is the transfer from the pin to the ring. Initial wear particles detach from the brass pin by fracture just below the surface where high shear strains are present, as shown in Fig. 5.8.

These fragments tend to be plate-like, small in thickness compared with their other dimensions. Once adhering to the steel surface, these particles

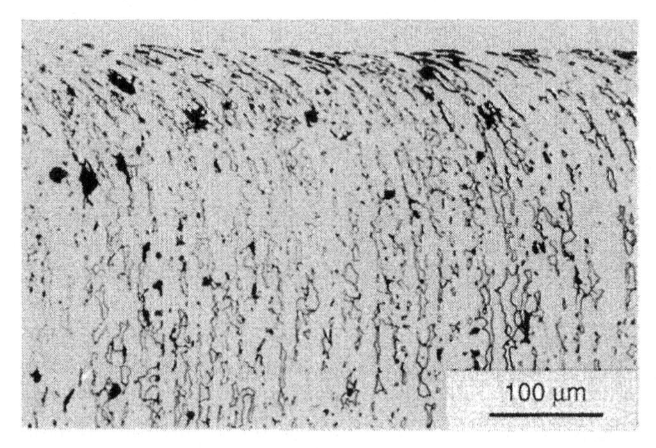

Fig. 5.8 Micrograph showing subsurface deformation in leaded α/β brass after severe sliding wear against tool steel in air (from Lancaster J K, *Proc. Roy. Soc. Lond.* **273A**, 466–483, 1963)

experience further work-hardening so that the brass transfer layer on the steel is stronger than the bulk brass at the surface of the pin; transfer therefore occurs preferentially from pin to ring by failure of the weaker interface. The second stage of wear is detachment of lumps from the transferred layer, probably when the bond between that layer and the underlying steel has been locally weakened by repeated cyclic loading. Figure 5.9 shows a particle of brass debris removed in this way. It is composed of layers of transferred material, the edges of which are clearly visible.

The transition from severe to mild wear occurs when two competing processes balance: the rate of exposure of a fresh metal surface by the severe wear process just discussed, and the rate of oxidation of that surface by the surrounding atmosphere. In mild wear the sliding surfaces are separated by oxide films, with only occasional direct metallic contact; this explains the relatively high contact resistance, and in many sliding systems results in a low coefficient of friction. In the unlubricated sliding of leaded brass on steel, for example, the coefficient of friction is ca. 0.15 in the mild wear regime, compared with 0.25 to 0.3 when severe wear occurs. The debris formed by mild wear consists of mixed oxides of copper, zinc and iron, originating from both the brass slider and the steel ring, and the wear rates of both are comparable despite their different hardnesses.

A detailed study of the development of mild wear in brass sliding on hard steel suggests the following sequence of events. The process starts with a brief period of severe wear, during which a layer of brass is transferred on to the steel counterface. In contrast to the case of continuous severe wear, however,

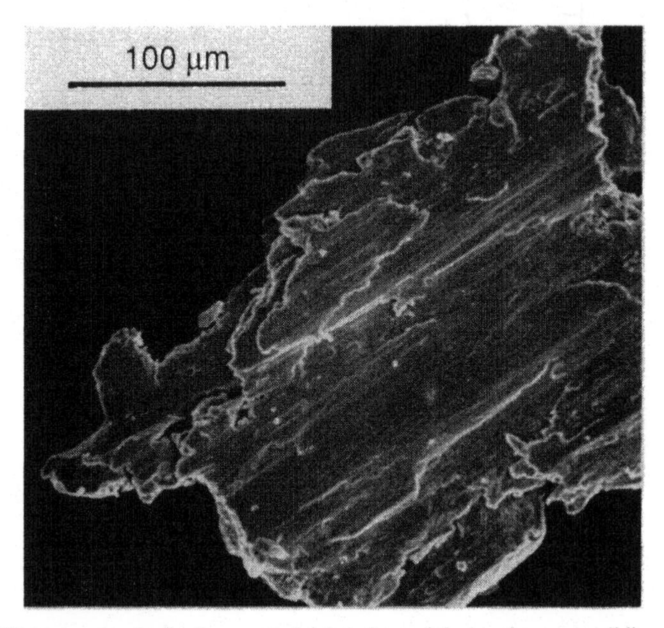

Fig. 5.9 SEM micrograph of a fragment of debris formed during the severe sliding wear of α/β brass (from Kennedy F E and Voss D A, in Ludema K C, Glaeser W A and Rhee S K (Eds), *Wear of Materials 1979*, ASME, 1979, pp. 89–96)

this layer then transfers back to the surface of the pin, probably because that surface becomes hardened by the 'pick-up' of particles of oxide. The interface between the brass transfer layer and the steel becomes weaker than that between the layer and the pin, and back-transfer is therefore favoured. The process of back-transfer continues until all the original transfer layer has been removed from the steel and returned to the pin. The surface of the pin at this point consists of a composite layer of back-transferred brass and oxide; the steel counterface also carries a film of oxide. The oxide incorporated into the surface layer of the pin strengthens it, and also provides a low shear strength at the interface giving low friction.

The structure of the transferred layer is illustrated in Fig. 5.10. For these experimental conditions the overall thickness of the layer is about 40 μm, of which various oxides make up some 25 μm. The enhanced strength of the

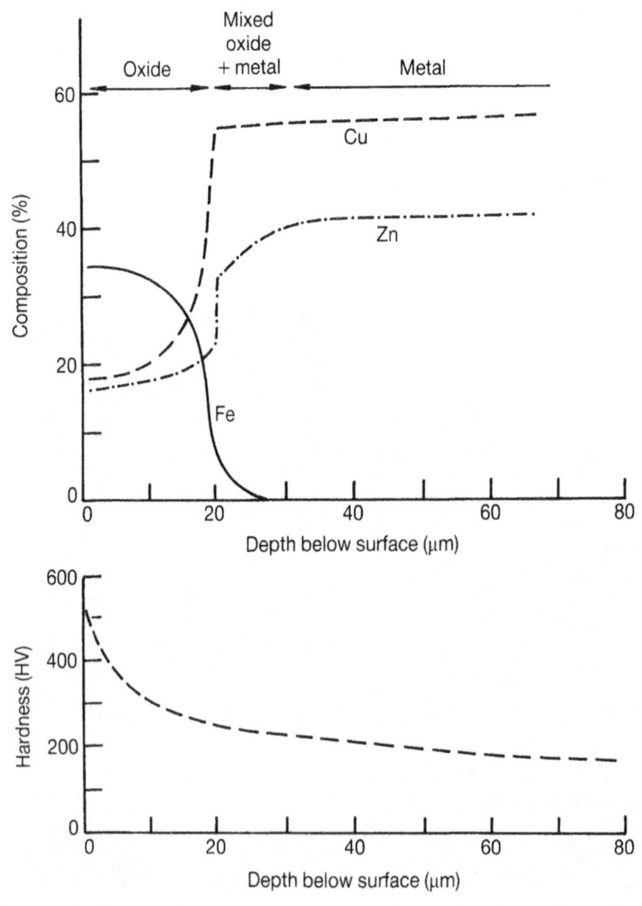

Fig. 5.10 Schematic diagram showing the structure of the surface layers formed on an α/β brass pin after sliding against steel in the mild wear regime, and the corresponding microhardness profile (after Lancaster J K, *Proc. Roy. Soc. Lond.* **273A**, 466–483, 1963)

surface layers compared with the bulk brass is clearly seen from the microhardness profile. The mechanisms of removal of the oxide to form wear debris depend on the precise conditions of load, speed and temperature and can vary between different regions of the contact area. Flakes of oxide may detach as a result of cyclic loading, but abrasion by hard oxide particles derived from the steel can also be important. Abrasion by oxide particles, which is discussed more fully in Chapter 6, is probably also responsible for the wear of the steel, which is, perhaps paradoxically, at least 10^4 times more rapid during the mild wear of the brass than in the severe wear regime.

Since oxide formation plays a crucial role in the process of mild wear, any factor which changes the rate of oxidation of the sliding surfaces will influence the transition between mild and severe wear. Temperature is the most obvious variable affecting oxidation rate: the temperature at the sliding interface will depend on both the ambient temperature and on the frictional power dissipation, which in turn depends on the sliding velocity and load. The effects of both are illustrated in Fig. 5.11.

At 20 °C in air, the severe wear regime extends over a wide range of speeds. At sliding speeds below about 0.2mm s^{-1}, however, the time available for oxidation at any individual point on the surface becomes so long that the mild wear process can occur; a back-transferred film is produced and the wear rate drops dramatically.

At the opposite end of the velocity spectrum, frictional heating raises the interface temperature to a point where oxidation is so rapid that mild wear can dominate again, and at speeds above about 8 m s^{-1} for this geometry the wear rate again falls sharply. The rise in wear rate with sliding speed, which precedes the transition at high speed, is significant, and is also caused by frictional heating; at these speeds thermal softening of the brass occurs,

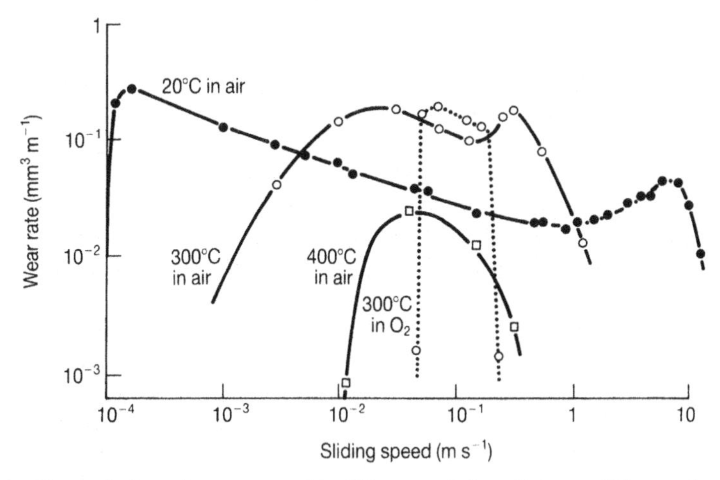

Fig. 5.11 The variation of wear rate with sliding speed for α/β brass sliding against steel at various temperatures in air and in pure oxygen (after Lancaster J K, *Proc. Roy. Soc. Lond.* **273A**, 466–483, 1963)

resulting in an increase in the rate of severe wear. It must be stressed that this effect and the enhanced oxidation rate, which causes the transition to mild wear at higher speed, are due to the mean local temperature at the sliding interface, and not to the appreciably higher 'flash' temperatures which occur at asperity contacts and which are important in the action of EP lubricant additives (see Section 4.6).

At higher ambient temperatures the increased rate of thermal oxidation favours the mild wear regime. Figure 5.11 shows that at 300 °C in air the range of velocities over which severe wear occurs has been considerably reduced. Sliding in pure oxygen also leads to enhanced oxidation, and at 300 °C under these conditions severe wear is seen over only a narrow range of speed.

5.5 UNLUBRICATED WEAR OF METALS

We have seen in the previous section that in the case of brass sliding against a harder counterface, no single, unique wear mechanism operates over a wide range of conditions. Rather, there are several mechanisms which change in relative importance as the sliding conditions are varied. This conclusion is also true for other sliding systems: transitions in dominant wear mechanism, and in the associated rate of wear, are commonly seen with variation of normal load and sliding velocity, and also in some cases with sliding time (or distance). The main factors controlling the importance of the underlying mechanisms are *mechanical stresses*, *temperature* and *oxidation phenomena*. Consideration of all three factors is essential in understanding the sliding wear of metals; it must also be appreciated that the conditions at the interface may be very different, especially in terms of temperature, from those in the surroundings.

The complexity of sliding wear arises from the fact that all three controlling factors are interrelated, and may be influenced by both load and sliding velocity. Figure 5.12 illustrates schematically how the extent of mechanical damage (due to surface stresses), and also the interface temperature, will depend on the load and sliding velocity. Increasing the load leads directly to higher stresses, and these will result in more severe mechanical damage. Both the load and the sliding velocity influence the interface temperature. Together, they control the power dissipated at the interface (since that is the product of the sliding speed and the frictional force).

Additionally, the sliding velocity determines the relative importance of heat conduction away from the interface. At low sliding velocity, because the heat generated will be relatively rapidly conducted away, the interface temperature will remain low; in the limit, the sliding process would be isothermal. At high velocity, only limited heat conduction can occur, the interface temperature is therefore high, and the limiting conditions would be adiabatic. A high interface temperature leads to high chemical reactivity of the surfaces, causing, for example, rapid growth of oxide films in air. It will also reduce the mechanical strength of asperities and near-surface material, and may even in extreme cases cause melting.

We shall discuss the influence of mechanical stresses, temperature and

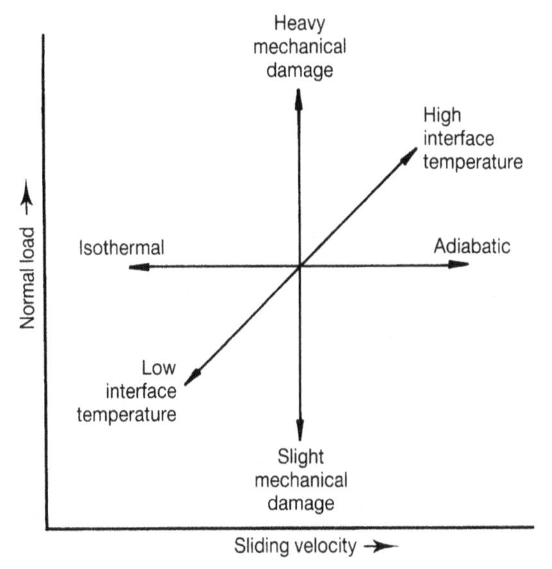

Fig. 5.12 Diagram illustrating the combined influences of load and sliding speed on the sliding wear process in metals

oxidation further in the following paragraphs.

The important mechanical stresses are the normal stress at the surface of each sliding body, and the shear stresses at and below the surface. As we saw in Chapter 2, the normal stress at the asperity contacts will, under conditions of plastic contact, be close to the indentation hardness of the softer body. But if the surfaces are very smooth, closely conforming or lightly loaded then the contacts may be elastic. The normal load applied to the system therefore controls the extent of plastic flow at the asperity contacts; if the load is low enough, or the surfaces conform well enough, wear will proceed only very slowly, perhaps by an elastic (e.g. high-cycle fatigue) process.

The magnitude and position of the maximum shear stress depend on the coefficient of friction. For μ less than ~0.3, the maximum shear stress and associated plastic flow will lie beneath the surface, and the plastic strain accumulated by each sliding pass is small. This condition is typical for a lubricated system, or for one carrying a protective oxide layer. For μ greater than ~0.3, however, the maximum shear stress lies at the surface and large shear strains can be accumulated (as seen, for example, in Fig. 5.8). Several wear mechanisms dominated by plastic flow have been proposed, involving asperity adhesion and shear (as originally envisaged by Archard), or nucleation and growth of subsurface cracks leading to the formation of lamellar wear particles ('delamination wear'), and other models involving fatigue crack propagation. At relatively low sliding speeds and high loads, these 'plasticity-dominated' wear mechanisms prevail: they lead to severe wear, as described in Section 5.4 for the case of brass.

Oxidation phenomena are also important in sliding wear, since nearly all metals form oxide films in air. The rate of film growth depends, as in static

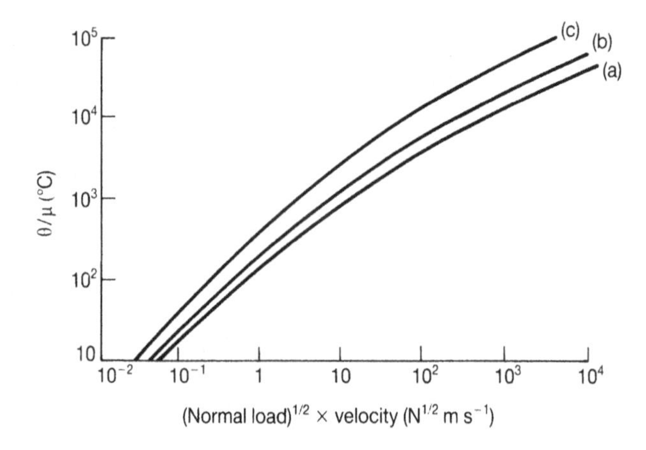

Fig. 5.13 Maximum attainable flash temperature θ for sliding steel contacts. Three different values are assumed for the hardness of the steel: (a) 150 HV; (b) 250 HV; (c) 850 HV (from Archard J F, *Wear* **2**, 438–455, 1958/9)

oxidation, strongly on temperature; however, the local temperature at the sliding interface may be substantially higher than that of the surroundings, and may also be enhanced at the asperity contacts by transient 'flashes' or 'hot-spots'. Estimation and measurement of local flash temperatures are both difficult. Perhaps the simplest approach, which can be used to estimate the highest possible flash temperature, is to suppose that for a brief moment the whole load applied to the system, both normal and tangential, is carried by a single asperity, and to determine the interface temperature resulting from this instantaneous rate of local power dissipation. The temperature rise varies with sliding speed because the power dissipation depends on velocity and also because the motion of the heat source over the counterface must be considered. Figure 5.13 shows the predictions of this model, for steels of different hardness.

Flash temperature rises predicted in this way may exceed 1000 K, and steady interfacial temperature rises due to the mean frictional power dissipation can add a further hundred degrees or so. Temperatures of this order have been confirmed experimentally by thermoelectric and infra-red measurements. Such high temperature transients can lead to local phase transformations: for example, to the formation of austenite, which then quenches to martensite, in carbon and low-alloy steels. They are also responsible for rapid surface oxidation.

5.6 WEAR-REGIME MAPS FOR METALS

A useful method of depicting the various regimes in which different modes of wear dominate is provided by a *wear-regime map* or *wear-mode map*. An example, for steels sliding in air in the common pin-on-disc geometry, is shown in Fig. 5.14.

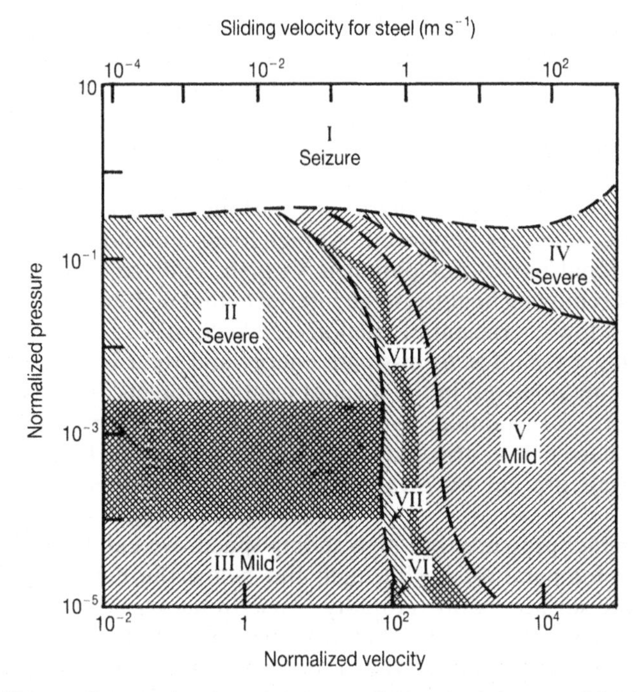

Fig. 5.14 Wear-mode map for the unlubricated sliding of steel on steel in the pin-on-disc configuration (from Lim S C, Ashby M F and Brunton J H, *Acta Metall.* **35**, 1343–1348, 1987)

Although the details of the map are specific for steels, the general form will be similar for the unlubricated sliding of most metals in air. The map is plotted with the same coordinates as the diagram in Fig. 5.12. Dimensionless variables are employed: a normalized contact pressure (defined as the normal load on the specimen pin, divided by the nominal contact area times the indentation hardness of the softer material), and a normalized velocity (which can be thought of as the sliding velocity divided by the velocity of heat flow). Representative values of sliding velocity are also given along the upper abscissa, for typical steels.

The dominant modes of wear, and the resulting regions on the diagram, are determined in two ways: by plotting empirical data from experiments performed under widely differing conditions, and by the use of simple analytical models for wear rates due to the various mechanisms. Although wear-regime maps can, at present, provide only approximate predictions of wear rates, they nevertheless give a valuable illustration of the regimes over which the different wear mechanisms are important.

Eight distinct regimes can be identified in Fig. 5.14. In regime I, associated with very high contact pressure, gross seizure of the surfaces occurs. This regime corresponds to catastrophic growth of asperity junctions, as described in Chapter 3, so that the real area of contact becomes equal to the apparent area. The boundary of this regime can be calculated from the junction growth model discussed in Section 3.4.3.

In regime II, corresponding to high loads and relatively low sliding speeds, the thin surface oxide film on the steel is penetrated at asperity contacts, high surface tractions occur, and metallic wear debris is formed; this is a regime of severe wear. At lower loads, in regime III, the oxide is not penetrated, and a low rate of mild wear results due to removal of particles from the oxide layer itself.

In regimes II and III, thermal effects are negligible, whereas in regimes IV and V they become very important. At high loads and sliding speeds (regime IV) the frictional power dissipation is so high, and thermal conduction so ineffective at removing heat from the interface, that melting occurs. Although this leads to a low coefficient of friction (see Section 3.5.5), viscous forces in the molten layer continue to dissipate energy and wear is rapid and severe, with metal being removed in the form of molten droplets. This extreme form of wear occurs, for example, as the driving band of an artillery shell slides along a gun barrel. It is also put to use in the process of *friction cutting*, in which materials are cut by a rapidly rotating disc of steel with a smooth periphery.

At low contact pressure, but high sliding speed, we enter regime V. Here the interface temperature is still high, but below the melting point of the metal, and the surface oxidizes rapidly. An extreme type of oxidational wear occurs, in which a thick, hot oxide layer deforms plastically and may even melt, and where the wear debris consists of oxide from this layer, rather than metal. Following our earlier definition (see Section 5.4), this is a form of mild wear.

Regimes VI, VII and VIII occur over narrow ranges of sliding velocity, and represent various types of transitional behaviour between the extremes of low speed, isothermal conditions, and high speed, adiabatic conditions. At low loads, all three regimes can be experienced as the sliding speed is increased. In regime VI, thermal effects begin to play a role, not by raising the interface temperature significantly, but through the occurrence of hot-spots at asperity contacts. These local flash temperatures lead to patchy oxide growth, and the main source of debris in this regime of mild wear is the detachment of oxide particles by spalling, typically when they reach a thickness of about 10 μm. At higher loads, metallic contact between asperities occurs, despite the enhanced oxidation at hot-spots, and the severe wear regime VII, with metallic debris, is entered.

Regime VIII, associated with higher flash temperatures, occurs when martensite is formed at the interface. This results from local heating of the asperities, followed by quenching by conduction of heat into the bulk. It is probable that the local temperature must rise above the α—γ allotropic transformation temperature for iron (910 °C), rather than simply above the eutectoid temperature (723 °C), since only then could austenitization occur on the extremely short timescale of a hot-spot transient. Dissolution of cementite in ferrite, which would be necessary below 910 °C, is rather sluggish, whereas dissolution of cementite in austenite is extremely rapid. There is also some evidence that the transformation is facilitated by shear stresses at the contact points. The high strength of the martensite provides local mechanical support for the surface oxide film, and mild wear proceeds by the removal of this oxide, in a regime similar in many ways to regime VI.

It must be stressed that the patterns of behaviour illustrated in Fig. 5.14 are, necessarily, simplified. No account is taken, for example, of possible transitions in wear behaviour with sliding distance or time. Changes in composition and microstructure of the steel, or in ambient temperature, will obviously shift the boundaries between the various regimes, which in any case are not rigidly defined. In the map shown, regimes of mild and severe wear overlap to a marked extent. These areas of overlap represent uncertainty about the precise location of the transition, not only because of variation in behaviour between different steels and experimental conditions but also because, in some cases, the transitions themselves are poorly defined. Different areas of the rubbing surface may simultaneously exhibit different modes of wear.

Wear maps nevertheless show clearly how relatively small changes in sliding velocity or contact pressure can lead to transitions between mild and severe wear, and how more than one transition might be generated by changing only one variable.

Figure 5.15 shows, as an example, transitions in the sliding wear behaviour of a medium carbon steel. The data were obtained from a pin-on-ring test in air, at a constant sliding velocity.

At light loads the wear rate is low, oxide debris is formed, and the regime can be described as mild wear. The sliding surfaces are separated by an oxide film, and the wear mode probably corresponds to regime VI in Fig. 5.14.

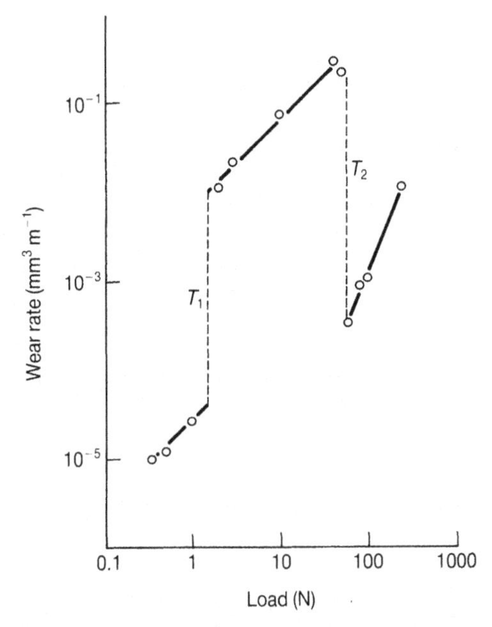

Fig. 5.15 The variation of wear rate with load for a 0.5% plain carbon steel pin sliding at 1 m s^{-1} in air against tool steel in a pin-on-ring test (from Welsh N C, *Phil. Trans. Roy. Soc. Lond.* **257A**, 51–70, 1965)

As the load is increased, an abrupt transition to severe wear occurs, labelled T_1. The wear rate increases by a factor of more than 100, and metallic debris is produced. Regime VII is entered. When tests are made at higher loads the wear rate suddenly drops (at the T_2 transition) and mild wear is encountered once more. At this point, flash temperatures occur which are high enough to transform the surface material locally to martensite, which supports the surface oxide; the system is in regime VIII. However, mild wear under these conditions is established only after an initial period of severe wear, which occurs before sufficient martensite has formed. Furthermore, even when established, this mild wear condition is only marginally stable; removal of the surface oxide or annealing of the martensite layer is sufficient to cause immediate reversion to severe wear.

It is only at an even higher load (above the so-called T_3 transition) that enough martensite and oxide are formed in the first sliding pass to sustain a more stable regime of mild wear.

Transitions between regimes of mild and severe wear, in which mild wear is caused by oxidation phenomena and severe wear involves metallic contact, are commonly observed in many metals. As we have seen, they may result from changes in normal load or sliding speed, but can also occur with time or sliding distance. They are then associated with *running-in* (US *breaking-in*). If a transition occurs during running-in, it is usually from severe to mild wear, and is associated with a smoothing of surface topography, or the formation of an interfacial layer of oxide or a hard phase such as martensite.

Steels of all types, including stainless alloys, have been extensively studied and all exhibit both mild and severe wear regimes. Other transition metals and their alloys sliding in air also show similar behaviour. α/β brass has been discussed in detail in Section 5.4, but other copper alloys perform similarly. So do nickel and its alloys, titanium and tantalum. Aluminium and zinc, in contrast, show only severe wear behaviour when sliding on themselves, although when sliding against each other a regime of mild wear does apparently occur.

The wear mechanisms operating in the various regimes of Fig. 5.14 have been described as either *plasticity-dominated* or *oxidational* in character; these mechanisms of material removal will be discussed in more detail in the next section.

5.7 MECHANISMS OF SLIDING WEAR OF METALS

5.7.1 Plasticity-dominated wear

Many different mechanisms have been proposed for the severe wear of metals. All involve plastic deformation, but differ in the detailed processes by which material becomes removed.

It is never easy to discriminate unequivocally between the possible mechanisms which may be operating in a particular case, but critical examination of both the worn surfaces and the wear debris can provide useful information. The worn surfaces will presumably contain fragments of debris at all stages of their life cycle, but identifying them is not simple. As well as examination of the surface topography and subsurface microstructures by optical microscopy and SEM, use has been made recently of transmission electron microscopy (TEM) to study deformation and structure in material very close to the worn surface. A wide range of analytical techniques, such as X-ray diffractometry, energy- and wavelength-dispersive X-ray microanalysis and Auger electron spectroscopy has also provided compositional and structural information about this material.

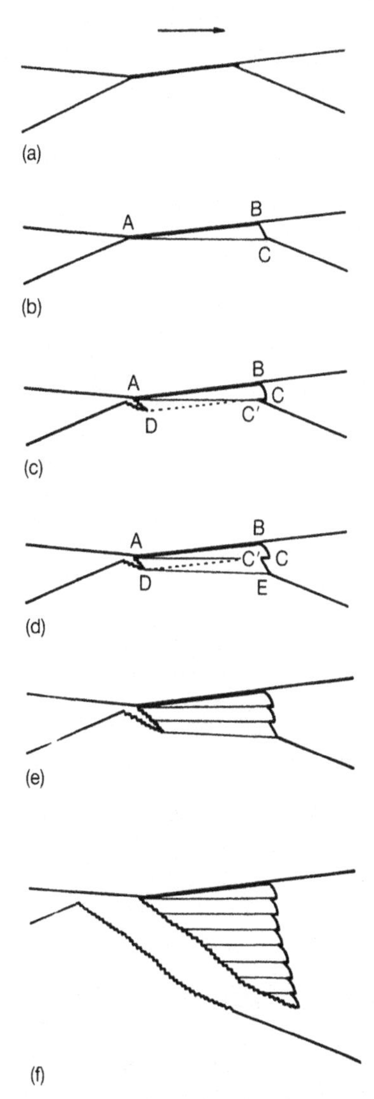

Fig. 5.16 Diagram illustrating one way by which detachment of a fragment of material might result from plastic deformation of an asperity tip (from Kayaba T and Kato K, in Ludema K. C., Glaeser W A and Rhee S K (Eds), *Wear of Materials 1979*, ASME, 1979, pp. 45–56)

Examination of wear debris by these methods can also give valuable clues, although it is important to question whether the debris collected from a wear test is in the same state in which it first became detached from the surface, or whether its appearance and even nature may have been changed subsequently. The rolling of debris particles between sliding surfaces, and the oxidation of metallic debris after detachment, are examples of possible changes.

Early theories of sliding wear suggested that material is removed as lumps or fragments from asperity peaks by adhesive processes, and some debris particles certainly have irregular, blocky shapes of dimensions which seem appropriate to such processes. Several mechanisms have been proposed by which asperity contacts can lead directly to the formation of wear debris. Figure 5.16 shows a representative example, in which plastic flow at an asperity tip is followed by detachment of a wear particle.

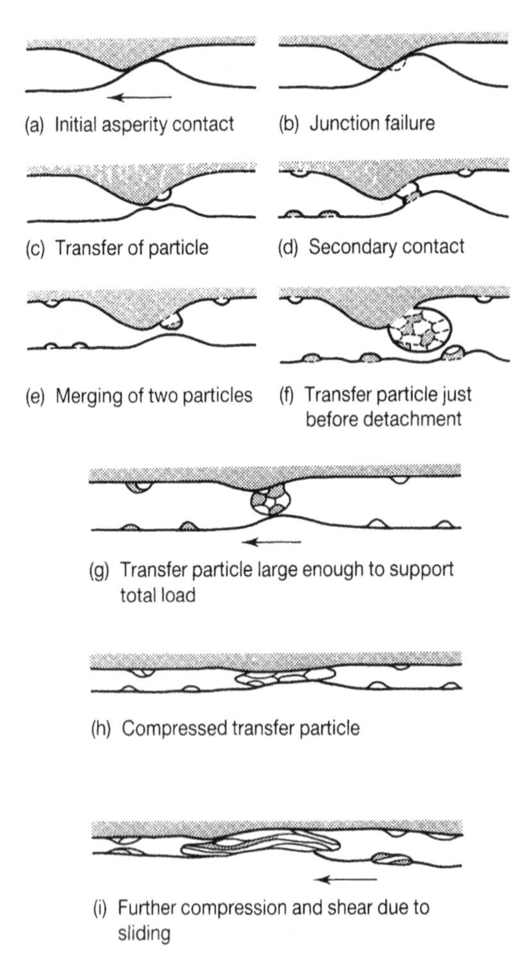

Fig. 5.17 Formation of a transfer particle by asperity rupture and aggregation (from Sasada T, in *Tribology in the 80s*, NASA Conf. Pub. 2300, vol I, 197–218, 1984)

In this case, plastic shearing of successive layers (for example along the plane AC) occurs in conjunction with the propagation of a shear crack (AD), along which the particle detaches. Adhesive forces are not necessary for material to be removed by such a mechanism, which depends only on the mechanical interaction between two asperities, but they must be invoked to explain any subsequent transfer of debris to one of the sliding surfaces.

Another model for wear supposes that the formation of a debris fragment by asperity rupture is followed immediately by its adhesive transfer to the counterface, to form a new asperity on that surface (Fig. 5.17).

Further sliding causes more fragments to be formed, which adhere to the original fragment until a much larger conglomerate wear particle eventually becomes detached. This particle of debris may be roughly equiaxed, as shown at (f) in Fig. 5.17, or it may be flattened and elongated in the direction of sliding by further plastic deformation, forming a plate-like composite wear particle (as at (i)).

Plate-like debris particles are often found in sliding wear (as shown for example in Fig. 5.18), with ratios of thickness to width or length usually in the range from 0.1 to 0.5. Several other mechanisms have been proposed to explain their formation. For example, shear fracture of the asperity summit in Fig. 5.16 along the plane AC would generate a thin flat particle (ABC). Another mechanism has become known as the *delamination theory*. This involves the nucleation of subsurface cracks and their propagation parallel to the surface. The cracks originate within the plastically deformed material beneath the surface, perhaps but not necessarily from voids caused by the shearing of inclusions, and after growing and linking together eventually extend up to the free surface. A plate-like wear particle thus becomes

Fig. 5.18 Flakes of debris formed by sliding copper–1.8% beryllium against a tool steel in dry argon in a block-on-ring test (from Rigney D A, Chen L H, Naylor M G S and Rosenfield A R, *Wear* **100**, 195–219, 1984)

detached. A model for this kind of mechanism was assumed in regime II of the wear map shown in Fig. 5.14.

Subsurface microstructures associated with severe sliding wear in metals certainly show heavily deformed regions. Figure 5.19 illustrates schematically how the magnitude of the shear deformation increases towards the surface. A common nomenclature has been developed by several investigators for the regions observed. In zone 1, bulk material remains undeformed. Zone 2 contains plastically deformed material, with plastic shear strains increasing towards the surface. Grain reorientation and refinement may be found towards the boundary of zones 2 and 3. The material nearest the surface, in zone 3, often has a different structure from that in zone 2, with very fine grains (sub-micrometre, sometimes as fine as 10 nm) and containing components not present in the original bulk material, such as material transferred from the counterface, and oxides. Some investigators have suggested that zone 3 can be further sub-divided, with foreign material being incorporated only into the outermost layer of the zone, while the region nearer zone 2 contains only very fine-grained material of bulk composition. The material in zone 3 of mixed composition, which contains intimately mixed particles of oxide and counterface material, shows many similarities to material produced by the commercial process of mechanical alloying.

Material near the top of zone 2 and in zone 3 often shows a lamellar structure, in the former case resulting from alignment of dislocation cell walls and subgrain boundaries parallel to the sliding surface. The microstructure is similar to that seen in metals subjected to large plastic shear strains in other ways, for example in torsion testing. At large strains there is a tendency for shear to localize, and fracture originating from this concentrated deformation

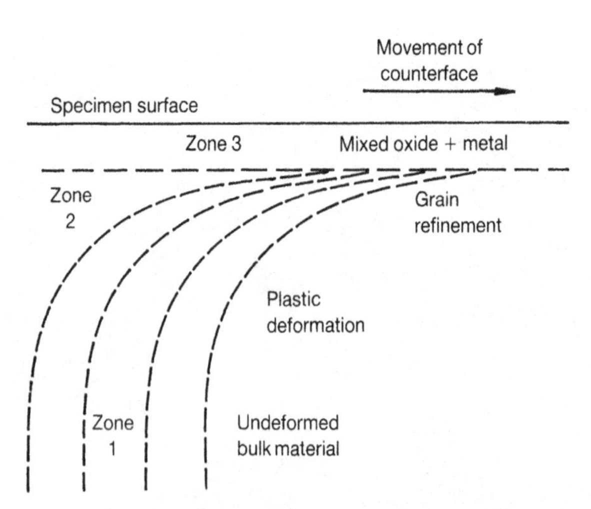

Fig. 5.19 Schematic diagram showing how the severity of plastic deformation is distributed beneath a worn metal surface in the severe wear regime (adapted from Rice S L, Nowotny H and Wayne S F, in Rhee S K, Ruff A W and Ludema K C (Eds), *Wear of Materials 1981*, ASME, 1981, pp. 47–52)

may be responsible for the formation of the initial wear particles which are then further deformed and mixed to form the transfer layer in zone 3.

Mechanisms of wear have been proposed in which cracks propagating parallel to these lamellar features result in the removal of wear debris, which may be composed solely of the mechanically alloyed transfer layer, or contain substantial amounts of heavily sheared substrate material from zone 2. Although the wear debris may be largely metallic, and the mechanism is correctly described as severe wear, the effects of oxidation are nevertheless important, since oxide is incorporated into zone 3, and probably plays a role in stabilizing the fine grain size and in the adhesion of the transfer layer.

Although several attempts have been made to model the sliding wear of metals as a fatigue process, the validity of this approach is not well established. Models have been proposed for the propagation of subsurface fatigue cracks under cyclic (elastic) shear forces, while others have assumed that asperities are removed as a result of low-cycle (macroscopically plastic) fatigue. There is experimental evidence for the existence of subsurface fatigue cracks under some conditions of sliding in some materials, but there is disagreement about whether the rate-controlling step in wear by a high-cycle fatigue mechanism might be the initiation of cracks or their propagation. It is certainly true that no simple correlation is found between the wear rates of metals and their fatigue properties; this may be because the cracks formed in wear are much shorter than those used in establishing conventional fatigue data, or because they are stressed in shear rather than in tension, or because the surface material in which they propagate is very different from the bulk material used in fatigue tests.

The complexity of sliding wear in the severe wear regime is well illustrated by the diversity of the mechanisms discussed above. Experimental evidence has been published for each of them, yet it would be unrealistic to claim that any one represents the 'true' mechanism for sliding wear. Some will produce debris in a single sliding pass, while others require many passes. Transfer layers are necessary for some mechanisms but ignored in other models. Several different mechanisms may well operate in sequence or simultaneously during the lifetime of a sliding system, and details of the surface topography, composition and microstructure of the near-surface material, and the nature of the surface films and adsorbed species will all be very important. As yet, not enough is known about the influence of these factors on the process of severe wear.

5.7.2 Oxidative wear

We have seen (in Section 5.5) that flash temperatures of several hundred degrees can readily be generated in sliding, and that the magnitude of these temperatures depends strongly on velocity. Only moderate sliding speeds are needed with most metals before these hot-spot temperatures become high enough to cause significant surface oxidation. For example, 700 °C is reached at around $1\ \mathrm{m\,s^{-1}}$ for steels, and the associated oxidation results in a transition in wear behaviour as seen in Fig. 5.14. Growth of a substantial oxide film suppresses the plasticity-dominated mechanisms discussed in the previous section, by reducing the shear strength of the interface and hence the development of the large subsurface shear strains necessary for these mechanisms (as discussed in Section 5.5). In oxidative wear, debris is removed from the oxide layer. Under vacuum, or in an inert atmosphere, of course, oxidative wear cannot occur, and plasticity-dominated processes then occur over a much wider range of conditions.

The role of temperature in oxidative wear is clear, and various mechanisms have been suggested by which thermal oxidation can lead to wear. At low ambient temperatures significant oxidation occurs only at asperity contacts, and is associated with the transient flash temperatures. The fact that some FeO is found in the wear debris from steels sliding at low bulk temperature, for example, suggests that it forms at the hot-spots rather than at the lower mean temperature of the surface, since FeO grows under static conditions only at above 570 °C.

At higher ambient temperatures, however, general oxidation of the surface outside the contact zone becomes more important, and may even dominate. In the oxidative wear of steels, the predominant oxide present in the debris changes with sliding speed and ambient temperature: at low speeds and temperatures α-Fe_2O_3 dominates, with FeO forming at high speeds and temperatures and Fe_3O_4 under intermediate conditions.

As we saw in Section 5.5, both the measurement and estimation of temperature rises due to sliding are difficult. Determining the temperature dependence of oxidation rates due to sliding therefore poses problems. It is nevertheless clear that oxide growth during sliding, like thermal oxidation under static conditions, obeys an Arrhenius equation:

$$k_p = A \exp(-Q/RT) \tag{5.8}$$

Here k_p is the parabolic rate constant for growth of the oxide film, T is the absolute temperature and R is the gas constant. Q, the *activation energy*, determines the temperature dependence of the growth rate, while A, known as the *pre-exponential factor* or *Arrhenius constant*, acts as a multiplying factor.

Table 5.2 shows values of A and Q for the oxidation of low alloy steels at various temperatures, under static conditions and also in sliding wear

Table 5.2 Typical values of Arrhenius constant A and activation energy Q for oxidation of steels under static conditions and in sliding contacts (from Quinn T F J, *Proc. Int. Conf. on Tribology – Friction, Wear and Lubrication*, Inst. Mech. Engrs. Conf. Series 1987–5, pp. 253–259, 1987)

Temperature range:	<450 °C	450–600 °C	>600 °C
Arrhenius constant, A ($kg^2\,m^{-4}\,s^{-1}$):			
static oxidation	1.5×10^6	3.2×10^{-2}	1.1×10^5
under sliding conditions	10^{16}	10^3	10^8
Activation energy, Q ($kJ\,mol^{-1}$):	208	96	210

experiments. The values of Q are those for static oxidation, since it is generally assumed that the activation energy will not vary substantially between static and sliding conditions. This view is open to question, however, and there is some evidence that rather lower values of Q might be applicable to sliding wear.

The data in Table 5.2 indicate clearly that oxidation under sliding conditions is much more rapid than would be expected from static oxidation models, especially at lower temperatures, and this conclusion is confirmed by many studies. Oxide thicknesses which would take a year to grow under static conditions can develop in hours or even minutes during sliding wear. This strong enhancement of oxidation by sliding may result from increased diffusion rates of ions through a growing oxide layer which has a high defect content (e.g. voids, dislocations and vacancies) due to mechanical perturbation.

There have been many proposals for detailed mechanisms by which oxide particles become detached as wear debris. Growth of oxide on an asperity tip immediately after it has been scraped clean by contact with the counterface may be followed by removal of that oxide in the next contact event. This process has been termed 'oxidation-scrape-reoxidation'. The fine particles of oxide removed in this way might agglomerate into larger transfer particles before finally being released as debris, in a similar way to that shown for metallic particles in Fig. 5.17.

Experimental evidence suggests that under many circumstances an appreciable thickness of oxide (perhaps several micrometres) builds up on the metal surface before it becomes detached as debris particles, and mechanisms have been proposed in which 'islands' of oxide grow until they reach a critical thickness, when they spall off. These islands may be homogeneous oxide layers growing by tribochemically enhanced oxidation of the surface, as has been assumed in modelling oxidative wear in regime VI in Fig. 5.14. Alternatively, the oxide islands may be aggregates of fine particles generated at asperity contacts by oxidation-scrape-reoxidation, or perhaps aggregates of fine metallic particles which have oxidized after detachment from the asperity peaks. The possibilities of transfer and back-transfer of oxide particles, and of abrasive wear caused by hard oxide particles, add further complexities to the range of mechanisms which have been put forward. As with plasticity-dominated wear, there is undoubtedly no single mechanism which applies to all cases, and the conditions under which the various possible mechanisms become important are still only poorly understood.

5.8 LUBRICATED WEAR OF METALS

In a fluid-lubricated system, the ratio λ between the thickness of the lubricant film and the root-mean-square asperity height (see equation 4.16) determines the regime of lubrication. As we saw in Chapter 4, increasing the normal load on the contact, decreasing the sliding speed or decreasing the lubricant viscosity all tend to reduce the film thickness. Under full-film hydrodynamic or elastohydrodynamic lubrication (EHL) conditions, where the asperities on the opposing surfaces do not come into contact, then the wear rate will be very low.

If λ falls below about 3, however, some asperity contact occurs, and we enter the regime of partial (or mixed) EHL. The wear rate will then inevitably be higher than under full-film conditions. For λ less than ~1, only the presence of boundary or solid lubricants can prevent the wear rate from rising to the level found in an unlubricated system; such lubricants are also invaluable in mitigating the wear rate in the partial EHL regime.

Figure 5.20 illustrates these regimes of lubrication and wear schematically. It also shows the variation of coefficient of friction μ, which follows the Stribeck curve familiar from Fig. 4.6. Typical values of the dimensionless wear coefficient K for metals are indicated. Although the absolute values of K vary considerably between different systems, the relative changes between the various regimes represent widely observed behaviour.

A valuable alternative method of representing data on lubricated wear is provided by the *IRG transition diagram*, developed by the International Research Group on Wear of Engineering Materials sponsored by the OECD. The IRG diagram displays regimes of lubricated wear on a map with coordinates representing normal load and sliding speed. The boundaries between the regimes are determined from experiments with concentrated

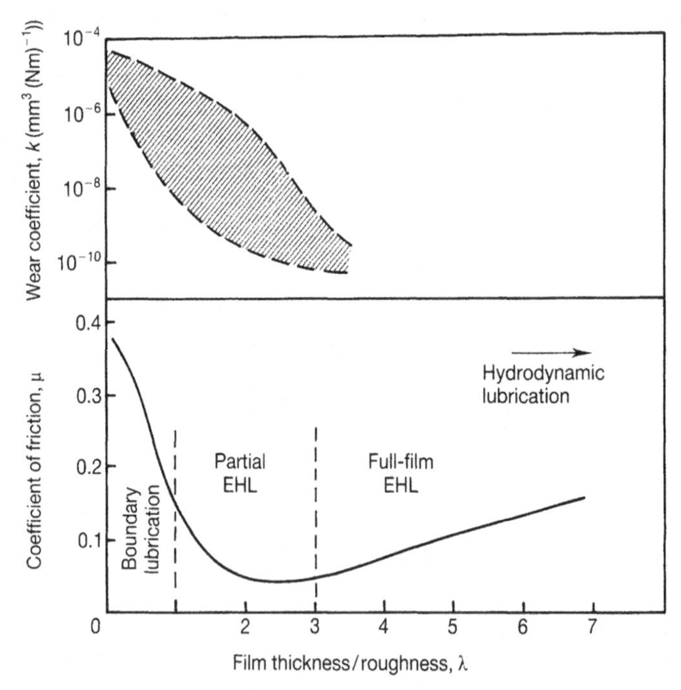

Fig. 5.20 Regimes of lubrication and wear in the lubricated sliding of metals, as a function of the ratio of film thickness to surface roughness, λ (from Czichos H and Habig K-H, in Dowson D, Taylor C M, Godet M and Berthe D (Eds), *Mixed Lubrication and Lubricated Wear*, Proc. 11th Leeds-Lyon Symposium on Tribology, Butterworths, 1985, pp. 135–147)

(counterformal) contact geometries, such as the four-ball, sphere on flat or crossed cylinder methods (see Section 5.2).

Figure 5.21 shows a typical IRG transition diagram for steel sliding on steel, completely submerged in an oil bath at constant temperature. Boundaries separate three regions, conventionally labelled I, II and III.

Regime I corresponds to partial EHL. The coefficient of friction is typically 0.02 to 0.1, and remains steady at this value for long periods of running after a very brief transient higher value. The wear rate during this initial period may be quite high but drops rapidly and continues to fall with further sliding, eventually becoming very small (with dimensional wear coefficient k less than $\sim 10^{-9}$ mm^3 (Nm)$^{-1}$). This behaviour may be explained in the following way. On initial sliding, frictional interactions lead to local heating and oxidation of asperities, and material is removed by oxidational wear. The regime may be described as one of mild wear. As high spots are removed from the surfaces by wear, the effective value of λ increases, and the conditions change from partial to full-film EHL. Friction and wear thus both fall to low and stable levels.

In regime II the initial coefficient of friction is higher than in regime I, typically 0.3 to 0.4 for steels. The wear rate is also higher, although still tolerable for some engineering applications ($k \approx 10^{-6} - 10^{-8}$ mm^3 (Nm)$^{-1}$). In this regime, boundary lubrication dominates; fluid film effects are negligible because of the high contact pressures and consequent low values of λ. After a longer running-in period than in regime I, μ falls (to about 0.1); this drop is associated primarily with general oxidation of the sliding surfaces. As wear continues and the surfaces become smoother, λ increases, fluid film lubrication can become important and stable sliding under partial or even full-film EHL conditions can be maintained. Wear under steady-state condi-

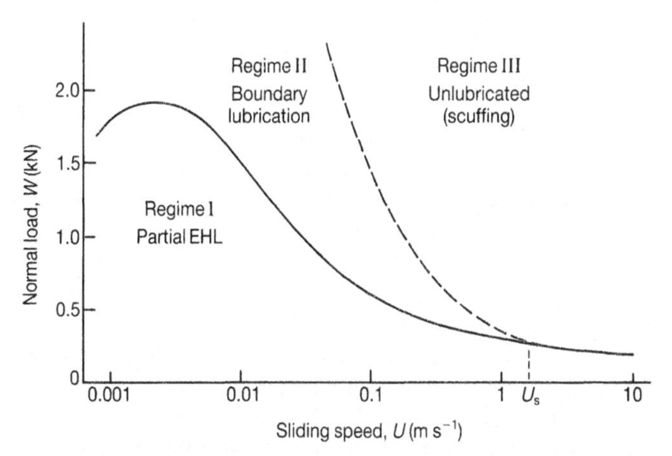

Fig. 5.21 Example of an IRG transition diagram for counterformal contact between steel components, fully submerged in an oil bath (from de Gee A W J, Begelinger A and Salomon G, in Dowson D, Taylor C M, Godet M and Berthe D (Eds), *Mixed Lubrication and Lubricated Wear*, Proc. 11th Leeds-Lyon Symposium on Tribology, Butterworths, 1985, pp. 105–116)

tions in regime II is oxidative in character, although the initial higher wear rate during running-in is associated with adhesive mechanisms involving unprotected asperity contacts, and can be viewed as incipient scuffing (see below).

In regime III conditions are so severe that no lubricant film (boundary or partial EHL) can prevent metallic contact. Rapid wear occurs between the effectively unlubricated surfaces. The coefficient of friction is typically 0.4 to 0.5, with k greater than $\sim 10^{-5}$ mm^3 (Nm)$^{-1}$. The term *scuffing* is often used to describe this regime of wear in concentrated contacts, which is associated with the breakdown of lubrication.

The breakdown of partial EHL, leading either to boundary lubrication (i.e. a transition between regimes I and II) or, at higher sliding speeds, directly to unlubricated sliding and scuffing (I to III) is indicated by the lower transition curve (solid curve) in Fig. 5.21. This transition can be understood by reference to equation 4.14, which shows that the film thickness in EHL depends, among other factors, on the viscosity of the lubricant and on the sliding velocity. Increasing either of these leads to a thicker film, and hence raises the load-carrying capacity of the contact. At very low sliding speeds (below about 1 mm s^{-1} in Fig. 5.21) the transition load does increase with sliding speed, since the viscosity of the oil remains effectively constant. At higher speeds, however, thermal effects become important, and largely determine the shape and location of the lower transition curve. As the sliding speed is increased, frictional heating in the contact zone leads to a drop in the local viscosity of the lubricant; this effect outweighs that of the increased sliding velocity in equation 4.14, and the film thickness drops. The load-carrying capacity of the contact therefore falls with increasing speed. To a good approximation the falling part of the lower transition curve fits the equation

$$WU^n = \text{constant} \tag{5.9}$$

The value of the exponent n varies with the temperature-dependence of viscosity of the lubricant, and lies in the range 0.3 to 0.8 for a wide range of mineral and synthetic oils.

The lower transition is largely associated with mechanical effects, being controlled primarily by the roughness of the surfaces, the lubricant viscosity and its dependence on temperature and pressure. A subsidiary role is also played by chemical effects, as indicated by the fact that transition loads in the absence of oxygen may be significantly lower than in air; the composition of the lubricant, particularly the presence of boundary lubricating or EP agents (see Section 4.6) is also important.

The upper transition (the broken curve in Fig. 5.21), from boundary lubrication (regime II) to unlubricated sliding and scuffing wear (regime III) depends, in contrast, mainly on chemical effects. It is independent of the viscosity of the lubricant and of the initial roughness of the surfaces, but strongly influenced by the lubricant chemistry and by the microstructure and composition of the surfaces.

The phenomenon of scuffing in lubricated systems is of great practical importance, since it leads to unacceptably high friction and rates of wear.

There is evidence that scuffing first occurs when the surface temperature reaches a critical value, possibly because this causes some desorption of the boundary lubricant, resulting in exposure of a critical fraction of bare, unprotected surface. Scuffing can therefore be avoided by ensuring a low rate of frictional energy dissipation, through low friction, low sliding speed or low load. As can be seen from the transition diagram (Fig. 5.21), at low sliding speeds some warning of impending scuffing may be given, since the system must pass through the boundary lubricated regime (II) before full scuffing occurs. At higher speeds, however, above the bifurcation speed U_s, the system will immediately move with increasing load from the desirable state of partial EHL (regime I) to catastrophic scuffing (regime III).

A tenacious boundary lubricant film, together with suitable EP additives, is very valuable in increasing the scuffing load. Oxidation of the surface can also be valuable, both in aiding adhesion of the boundary lubricant, and through the lubricating action of the oxide itself. Fe_3O_4, for example, provides a ductile film of low shear strength on steel asperities during lubricated sliding in air. Fe_2O_3, in contrast, forms on steels during sliding in the presence of water, and is abrasive; it enhances rather than reduces the propensity for scuffing.

The process of running-in under lubricated conditions usually involves the formation of beneficial oxide films, typically tens of nanometres thick, by local thermal oxidation of high spots, as well as the smoothing of the surfaces by wear of the higher asperities. Solid lubricant films, and other surface coatings (see Section 8.3), can also raise the upper transition curve very substantially.

The other important factor which controls the onset of scuffing is the nature of the sliding surfaces. Dissimilar metals show less tendency for adhesion at asperity contacts than similar metals (see Sections 3.5 and 7.6.2), and should therefore show higher scuffing loads. However, in many engineering applications the use of steels is dictated by reason of strength and economy. Alloying elements in steels have been shown to influence the scuffing transition, both by changing the chemical composition of the surface and therefore the strength of adhesion of boundary lubricants, and also by altering the microstructure of the steel; for example, an increase in the proportion of retained austenite in a steel tends to produce a downward shift in the upper transition curve.

5.9 FRETTING WEAR OF METALS

The term *fretting* denotes a small oscillatory movement between two solid surfaces in contact. The direction of the motion is usually, but not necessarily, tangential to the surfaces. When the amplitude of the motion lies in a range typically from 1 to 100 μm, surface degradation occurs which is called *fretting damage* or *fretting wear*.

Although fretting wear can be regarded formally as reciprocating sliding wear with very small displacements, there are enough differences in both wear rates and mechanisms to merit the use of a distinct term.

Whereas sliding wear usually results from deliberate movement of the surfaces, fretting often arises between surfaces which are intended to be fixed in relation to each other, but which nevertheless experience a small oscillatory relative movement. These small displacements often originate from vibration. Typical examples of locations where fretting may occur are in hubs and discs press-fitted to rotating shafts, in riveted or bolted joints, between the strands of wire ropes, and between the rolling elements and their tracks in stationary ball and roller races. It may also occur between items packed inadequately for transport, where vibration can lead to fretting damage at the points of contact between the items. Fretting wear can lead to loosening of joints, resulting in increased vibration and a consequently accelerated rate of further wear. Since the debris formed by fretting wear is predominantly oxide, which in most metals occupies a larger volume than the metal from which it originates, fretting wear can also lead to seizure of parts which are designed to slide or rotate with a small clearance. Whether fretting leads to increased clearance or to seizure depends on the ease with which the wear debris can escape from the contact region.

A further important phenomenon associated with fretting damage is the development of fatigue cracks in the damaged region, leading to large reductions in the fatigue strength of a cyclically-loaded component. The greatest reduction in fatigue strength is seen when the fretting process and cyclic stressing occur together; the cyclic displacement responsible for fretting damage is often directly linked to the cyclic load which causes the fatigue cracks to grow. Fatigue failure originating from fretting damage is given the name *fretting-fatigue*.

Fretting wear is usually studied in the laboratory in systems of simple geometry: either sphere-on-flat or crossed cylinders. Both geometries lead to circular areas of contact and, at least for elastic displacement, the conditions in the contact zone can be analysed theoretically. We saw in Section 2.5.1 that for elastic contact of a sphere on a plane surface under a normal load, the contact pressure reaches a maximum at the centre of the contact circle, and falls to zero at the edges. Figure 5.22 illustrates this pressure distribution, and also shows a plan view of the area of contact.

If a small cyclic tangential force is superimposed on the normal force, some displacement may occur between the surfaces around the edges of the contact zone, where the normal pressure is lowest and the frictional stress opposing movement is therefore least (Fig. 5.22(c)). The contact zone can then be divided into two regions: a central area where there is no relative tangential movement, and an annular zone in which *microslip* occurs. Similar behaviour is seen if a cyclic torque (about a vertical axis) is applied to the sphere, rather than a tangential force. Fretting damage occurs in the microslip region. As the amplitude of the cyclic tangential force is increased, the central area within which no slip occurs shrinks, until eventually slip occurs over the whole contact area (Fig. 5.22(d)).

The tangential force may be translated into an equivalent macroscopic tangential displacement, which may be a more convenient measure of the conditions of fretting. Figure 5.23 shows examples of the fretting damage developed on a stainless steel flat in laboratory sphere-on-flat experiments

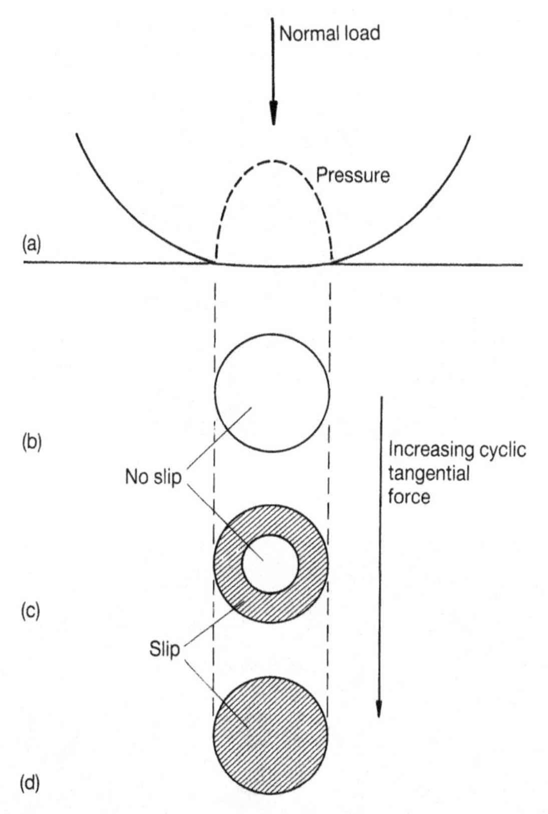

Fig. 5.22 (a) The distribution of elastic normal stress beneath a sphere pressed against a flat. (b) to (d) show plan views of the area of contact with increasing values of applied cyclic tangential force. The shaded regions represent areas over which local slip occurs between the surfaces

where, in one case (a), the displacement amplitude caused microslip over only part of the contact circle, and in the other (b), gross slip occurred over the whole area.

The regimes of normal load and displacement amplitude corresponding to complete sticking, mixed sliding and microslip, and gross slip in such experiments are plotted in Fig. 5.24. At a fixed normal load, increasing the amplitude of the cyclic tangential displacement (or load) leads to an increase in the extent of slip; a similar effect results from reducing the normal load at a fixed amplitude.

Fretting wear occurs most severely in the area of the contact zone undergoing slip, and is usually quantified in terms of the mass or volume loss from the surface. During a laboratory test there is often a brief initial period of rapid wear (Fig. 5.25), followed by either a steady (curve B) or decreasing (curve D) rate of wear. In a few materials, however, a subsequent increase in wear rate may be observed (curves A or C).

In materials showing linear wear (curve B), the wear rate is found to be approximately proportional to the normal load, and it is useful then to

(a)

(b)

Fig. 5.23 Surface fretting damage caused by small-scale cyclic displacement of a stainless steel sphere against a flat of the same material: (a) partial sticking leads to less damage in the centre of the contact area, while (b) gross slip causes damage over the whole region. The elliptical shape of the contact areas results from the tilt employed in the SEM imaging (from Bryggman U and Söderberg S, *Wear* **110**, 1–17, 1986)

express it as a *specific wear rate*, equivalent to the dimensional wear coefficient k introduced in Section 5.3. The dimensional wear coefficient is the volume removed by wear per unit sliding distance per unit normal load; the sliding distance is simply related to the duration of the test, the frequency of the vibration and the cyclic displacement amplitude. The dimensional wear coefficient k provides a measure of the severity of wear, and allows the results of tests at different normal loads and amplitudes to be compared.

Figure 5.26 shows schematically how the wear rates of steels, defined in this

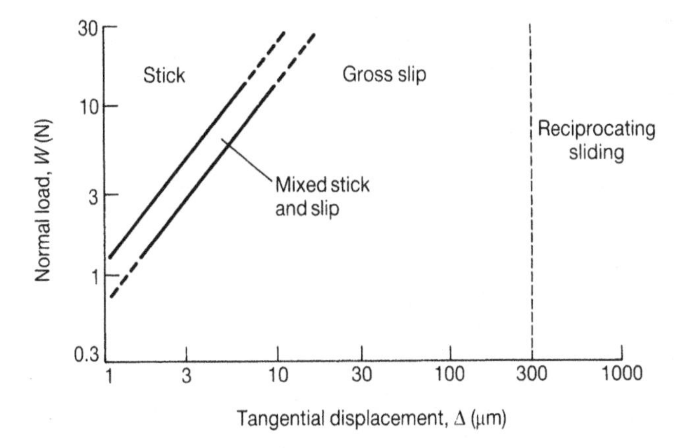

Fig. 5.24 Schematic illustration of fretting regimes for counterformal contact between stainless steel surfaces (sphere-on-flat geometry). Normal load W is plotted against tangential displacement Δ (from Vingsbo O and Söderberg S, in Ludema K C, *Wear of Materials 1987*, ASME, 1987, pp. 885–894)

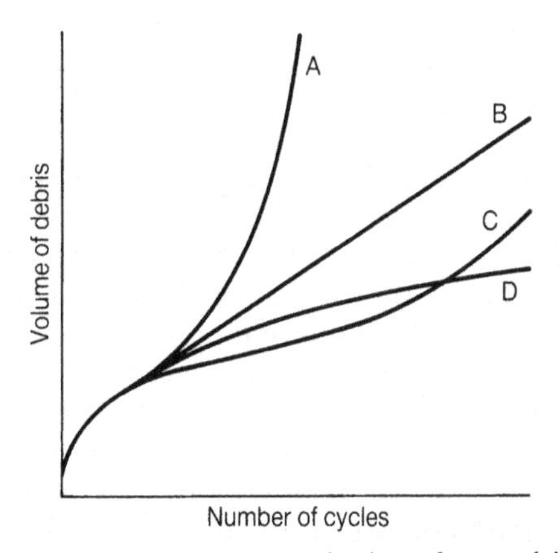

Fig. 5.25 Schematic representation of the production of wear debris during fretting with increasing number of displacement cycles (from Waterhouse R B, in Scott D (Ed.), *Wear, Treatise on Materials Science and Technology*, vol. 13, Academic Press, 1979, pp. 259–286)

way, depend on the amplitude of the fretting motion. At very small amplitudes (less than ~1 μm) the wear rate is negligibly small. Under these conditions no microslip occurs in the contact and there is consequently very little damage. As the amplitude is increased, significant microslip develops and the wear rate rises slowly. Once gross slip occurs (at amplitudes greater than ~10 μm) a rapid rise in k occurs, which eventually levels off to the constant value expected for reciprocating sliding wear at amplitudes greater than ~300 μm. The fretting amplitudes at which, on the one hand, damage becomes negligible and, on the other, the process becomes indistinguishable from reciprocating sliding, are the subject of debate. They certainly depend on the mechanical properties of the material. Various values have been

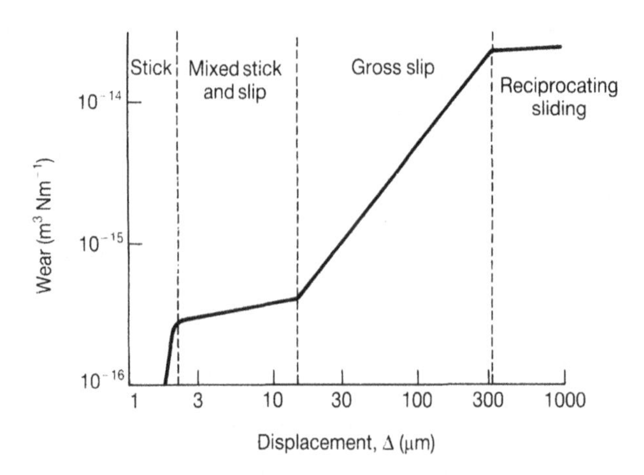

Fig. 5.26 The variation of fretting wear rate with amplitude of displacement Δ for a wide range of steels (from Vingsbo O and Söderberg S, in Ludema K C, *Wear of Materials 1987*, ASME, 1987, pp. 885–894)

reported; the threshold for damage in some systems is less than 0.5 μm, perhaps as low as 0.1 μm, while the transition to sliding wear (identified by a value of k which is independent of amplitude) has been observed at amplitudes as small as 50 μm.

The fatigue life of a cyclically-loaded specimen also varies with the amplitude of fretting displacement at its surface. A minimum in the fatigue life occurs at the transition from mixed microslip and sticking in the contact zone to gross slip. It can be shown that in the mixed stick-slip regime (as shown in Fig. 5.22(c)), the highest alternating stresses occur in the surface at the boundary between the two regions, and it is therefore at that point that fatigue cracks would be expected to initiate. It is important to note that the conditions giving the shortest fatigue life do not correspond to those giving the highest wear rate; in practice, failure by fretting fatigue is most likely at loads and fretting amplitudes which are responsible for only low rates of wear.

The mean sliding velocity during fretting is very low; a displacement amplitude of 20 μm at a frequency of 50 Hz provides a mean velocity of only 2 mm s^{-1}. Reference to Fig. 5.14 suggests that unidirectional sliding wear in steels at such low speeds and at moderately high normal pressure should result in severe wear, with metallic wear debris. This indeed occurs during the early stages of fretting wear, and gives rise to low electrical contact resistance and to initial wear rates which are comparable with those in unidirectional sliding.

However, fretting differs from continuous sliding, even at an equivalent velocity, in that the wear debris can escape from the contact zone only with difficulty. Further fretting in an oxidizing environment causes plastic smearing of the metallic debris fragments formed during the earliest stages, combined with their oxidation. In the steady-state fretting wear of steels the wear debris consists of very fine (0.01–0.1 μm) plate-like particles, mainly of

α-Fe_2O_3 with some metallic iron, of a characteristic red-brown colour, redder than the normal corrosion product; it is sometimes colloquially known as 'cocoa' or 'red rust'. The wear debris appears to form by delamination from a very thin composite surface layer of mixed oxide and metal, and the associated wear rate is substantially lower than that in the initial period of severe wear.

The coefficient of friction in steady-state fretting wear of steels may be very low, typically 0.02 to 0.1, and is probably accounted for by the 'lubricating' action of the fine debris particles moving between the sliding surfaces. The transition from initial severe wear by plastic deformation to steady-state mild wear by an oxidative mechanism is responsible for the decrease in the wear rate with increasing number of cycles seen in Fig. 5.25, curves B and D.

Other metals wear in a similar way, although the duration of the initial stage depends on their mechanical properties and rate of oxidation: titanium, for example, forms oxide debris almost immediately from the start of fretting wear. In some materials the fretting debris may be hard enough once formed to cause further wear by abrasion (see Section 6.3). The debris formed by aluminium, for example, is mainly fine γ-Al_2O_3, black in colour and very hard compared with the aluminium itself. For this reason, aluminium shows an accelerating fretting wear rate (e.g. Fig. 5.25, curve A or C).

The importance of oxidation in fretting wear, and the fact that the debris after the initial stages is predominantly oxide, has led to the use of the term *fretting corrosion* as a synonym for fretting wear, although as we have seen, the earliest stages of fretting wear do not involve appreciable chemical attack. It is preferable to use the more general term *fretting wear* to denote all types of wear due to fretting motion, and to restrict the term *fretting corrosion* to cases, as in the steady-state fretting wear of most metals in air, where the debris is predominantly the product of a chemical reaction.

Removal of oxygen or other chemically reactive species from the atmosphere leads to a marked reduction in the fretting wear rates of metals; experiments in nitrogen or in vacuum, for example, yield steady-state wear rates for mild steel which are only about 15% of that measured in dry air. Fretting in an inert atmosphere leads to the transfer of metallic debris from one surface to the other, but little escapes from the contact zone. The volume changes associated with the formation of oxide are absent, and the surface damage is considerably less than in air. As we shall see in Section 7.4, exclusion of oxygen can provide one practical method of reducing fretting wear.

Water vapour also influences the rate of fretting wear in air, although there is a negligible effect of adding water vapour to an inert atmosphere such as nitrogen. Steady-state wear rates of steels in air decrease linearly with increasing humidity, possibly because the hydrated oxide debris formed in humid air is softer and therefore causes less abrasion than the α-Fe_2O_3 formed in dry air.

5.10 WEAR OF CERAMICS

5.10.1 Introduction

We have seen in Section 3.6 that ceramic materials differ from metals in the nature of their interatomic bonding, and that this leads to very limited capacity for plastic flow at room temperature. In comparison with metals they are therefore much more inclined to respond to stress by brittle fracture.

In some ceramics, notably oxides, crack growth is sensitive to environmental influences. In alumina, for example, the presence of water increases the crack growth rate dramatically. Environmental factors can also influence plastic flow in many ceramics, by affecting the mobility of near-surface dislocations. This example of a *chemomechanical effect* is also known as the *Rebinder effect*. Surface chemical reactions, such as oxidation in susceptible materials, occur in ceramics just as in metals, and the rates of these reactions may be greatly enhanced by sliding.

We therefore see similarities in sliding wear between ceramics and metals, in that some local plastic flow can occur in both, and tribochemical reactions can modify the sliding interface and also lead to the formation of transfer films. However, plastic flow is much less important in ceramics, and tribochemical effects are correspondingly more significant than in metals. Brittle fracture, essentially absent in metals, plays a major role. Environmental factors, through their influence on both flow and fracture, are very important.

5.10.2 Brittle fracture

As we saw in Chapter 3, the coefficient of friction for the unlubricated sliding of ceramics in air lies typically in the range 0.25 to 0.8. Sliding under these conditions results in significant tangential forces, which in ceramic materials may lead to fracture rather than plastic flow. In conformal contacts, where the mean contact pressure is low, the scale of the fracture when it occurs will usually be small. Typically, it takes place along grain boundaries, leading to the removal of individual grains and a rough surface as seen in Fig. 5.27.

The relative ease of plastic flow in ceramic materials is usually strongly anisotropic, depending on crystallographic orientation. For example, sliding on single crystals of Al_2O_3 parallel to prismatic planes in the *c*-axis direction leads to extensive fracture, whereas in other directions on these planes, or on the basal plane, it is accompanied by plastic flow. It is therefore quite possible for the surface of a polycrystalline ceramic to exhibit regions of intergranular fracture alongside areas of predominantly plastic flow, associated with differences in the crystallographic orientation of the grains.

Brittle fracture due to a concentrated (counterformal) contact occurs on a much larger scale, and can be understood in terms of the Hertz elastic stress distribution. (For contact by a sharp hard indenter the stress field and the resulting fracture are different, and discussed in Section 6.3.2) For a sphere pressing against a plane surface we saw in Section 2.5.1 that the radius of the contact circle, a, was proportional to the cube root of the normal load, w (equation 2.7). The tensile stress σ_r in the plane surface outside the contact

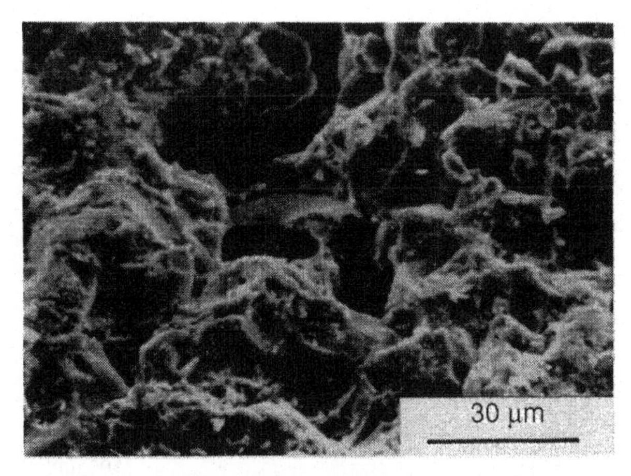

Fig. 5.27 The surface of an alumina specimen after sliding against alumina in a pin-on-disc test, showing evidence of severe intergranular fracture (courtesy of M G Gee)

area reaches a maximum value σ_{rmax} just at the edge of the circle of contact, given by

$$\sigma_{rmax} = (1 - 2\nu)\, p_{mean} \tag{5.10}$$

Here p_{mean} is the mean normal stress exerted over the contact area ($p_{mean} = w/\pi a^2$), and ν is Poisson's ratio. It is this tensile stress component σ_{rmax} which gives rise to fracture. When the normal load reaches a critical value, a crack initiates just outside the contact circle, and rapidly propagates to form a Hertzian cone crack, as shown in Fig. 5.28.

If a tangential force is applied to the contact, as in sliding, then the stress distribution is modified and the normal load necessary to initiate fracture is greatly reduced. For example, in experiments with a TiC sphere on a flat of the same material, the critical normal load was reduced by a factor of 10 under sliding conditions in air (with $\mu \approx 0.2$), compared with its value with no tangential force. Sliding in vacuum resulted in a greater tangential traction

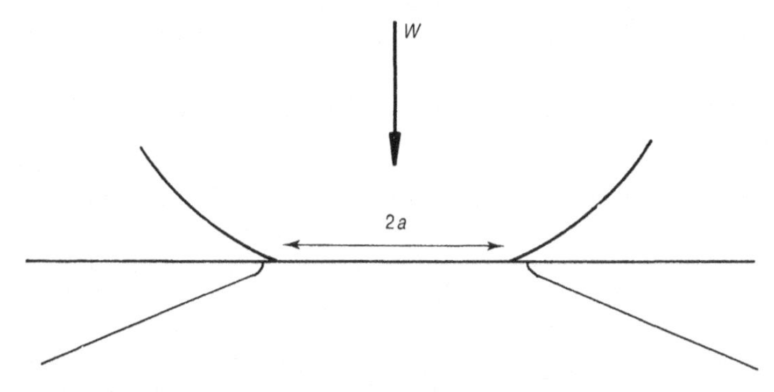

Fig. 5.28 The geometry of a Hertzian cone crack formed by a sphere loaded normally on to the plane surface of a brittle material

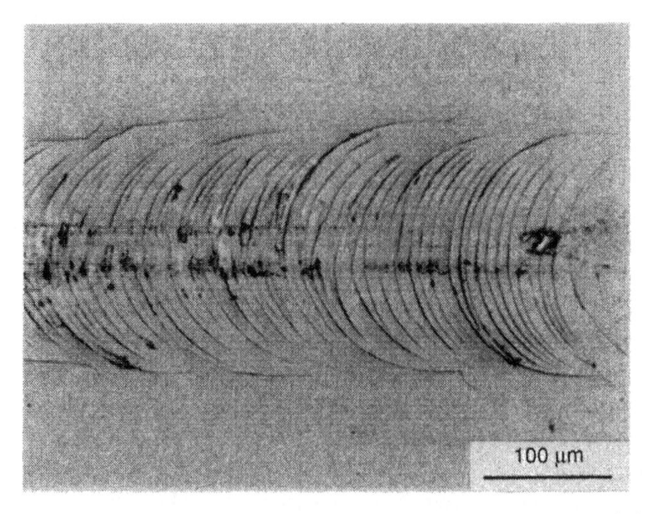

Fig. 5.29 Series of arc-shaped fractures caused by sliding a sphere over a brittle solid under normal load. This example shows the damage due to a tungsten carbide sphere on a soda-lime glass plane. The sphere slid from left to right (courtesy of P H Shipway)

($\mu \approx 0.9$) and a corresponding further reduction, by another factor of 50, in the normal force needed for cracking. Under sliding conditions, the cracks which form no longer intersect the surface in complete circles, but in a series of arcs as shown in Fig. 5.29.

Ceramics with low fracture toughness form such cracks more readily than tougher materials, and cracking is also favoured by high contact stresses (i.e. high loads and counterformal contact) and by large surface tractions (i.e. a high coefficient of friction).

5.10.3 Wear transitions: mild and severe wear

Sliding wear mechanisms in ceramics can involve fracture, tribochemical effects and plastic flow. Transitions between regimes dominated by each of these commonly lead to sharp changes in wear rate with load, sliding speed or environmental conditions (e.g. humidity or oxygen content). The terms *mild wear* and *severe wear* are often used, as for metals, to describe conditions on either side of a transition.

Mild wear in ceramics is associated with a low wear rate, smooth surfaces, a steady friction trace and mechanisms of wear dominated by plastic flow or tribochemical reactions. The wear debris is often finely divided, and may be chemically different from the bulk sliding material, for example through oxidation or hydration.

Severe wear, in contrast, causes a higher wear rate together with a rougher surface, a fluctuating friction trace and mechanisms of wear dominated by brittle fracture. The wear debris is often angular and not chemically different from the substrate. Ceramic materials obey the Archard equation (equation

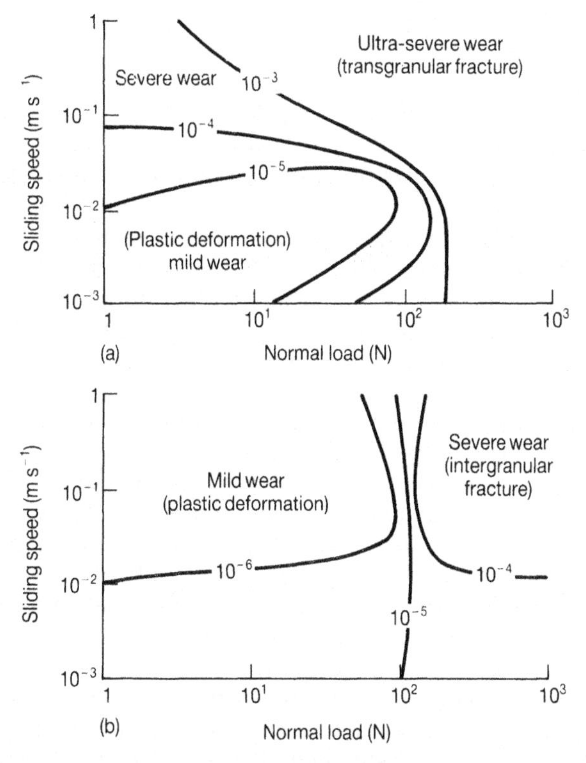

Fig. 5.30 Wear regime maps for polycrystalline alumina specimens, derived from experiments in a modified four-ball test (ball on three flats): (a) under unlubricated conditions; (b) lubricated with paraffin oil. The contours represent wear rates in mm^3 per metre sliding distance (from Lee S W, Hsu S M and Munro R G, in Rohatgi P K, Blau P J and Yust C S (Eds), *Tribology of Composite Materials*, ASM, 1990, pp. 35–41)

5.7) well enough, by showing linear dependence of wear on sliding distance and load, to render the wear coefficient (either dimensionless, K or dimensional, k) a useful quantity. Typical values of K for engineering ceramics undergoing severe wear are 10^{-4} to 10^{-2}, with K for mild wear being lower by factors of 10 to 100 or even more.

The development of wear mechanism maps for ceramics is less advanced than for metals, but it is already clear that the approach outlined in Section 5.6 can also be valuable for these materials. As an example, Fig. 5.30 shows maps for the sliding wear of polycrystalline alumina specimens, derived from experiments with a sphere-on-flat geometry under both dry and lubricated conditions. It should be noted that the axes, representing load and sliding speed, are interchanged relative to those in Figs. 5.12 and 5.14. At low loads and sliding speeds, both systems exhibit mild wear involving plastic deformation. Transitions to severe wear, involving intergranular fracture, can be induced by changes in the sliding conditions, and in the unlubricated case a further 'ultra-severe' regime involving transgranular fracture can also be identified.

5.10.4 Chemical effects

Figure 5.31 shows an example of a transition in wear rate in Si_3N_4, caused by an increase in sliding speed. At low speeds in air containing water vapour, Si_3N_4 forms a surface layer of hydrated silicon oxide by the reactions outlined in Section 3.6, which provides the source of the wear debris. Having a lower shear strength than the bulk ceramic, this layer also lowers the coefficient of friction. It is possible that adsorbed water on the surface also increases the ease of plastic flow, by the Rebinder effect (see Section 5.10.1). As the sliding speed is raised, increasing the interface temperature, the tribochemically reacted layer ceases to provide protection and the coefficient of friction rises. The increased surface shear stress causes cracking, and a transition ensues from mild wear (largely tribochemical in origin) to severe wear (with extensive brittle fracture). The resulting increase in surface roughness is responsible for the rise in the coefficient of friction.

The effect of water vapour and liquid water on the wear of Si_3N_4 provides an example of the environmental sensitivity commonly observed in the sliding wear of ceramics. Figure 5.32 shows how the same transition in Si_3N_4 can be induced in a different way: by changing the humidity of the surrounding air. A similar tribochemical reaction occurs in SiC, also leading to the formation of silicon oxide, but in this case the presence of water leads to an increase in wear rate.

Alumina and zirconia also show strong sensitivity to water which, in both ceramics, also causes the wear rate to increase, typically tenfold in comparison with dry sliding. This phenomenon may reflect the increased surface

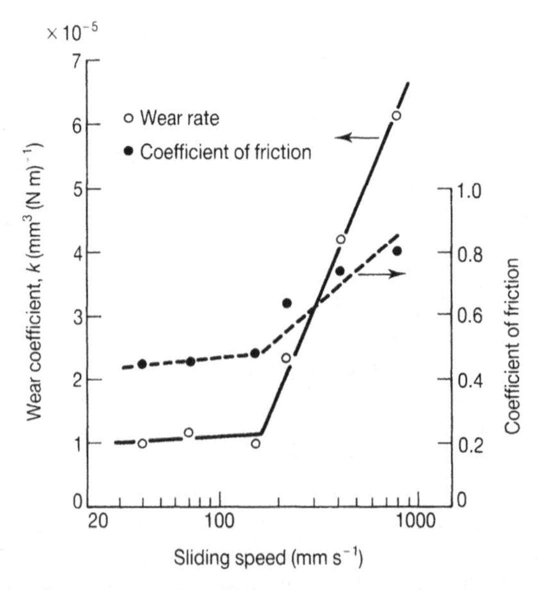

Fig. 5.31 Variation of wear rate and coefficient of friction with sliding speed for hot-pressed silicon nitride sliding against itself in pin-on-disc tests in air (from Ishigaki H, Kawaguchi I, Iwasa M and Toibana Y, in Ludema K C (Ed.), *Wear of Materials 1985*, ASME, 1985, pp. 13–21)

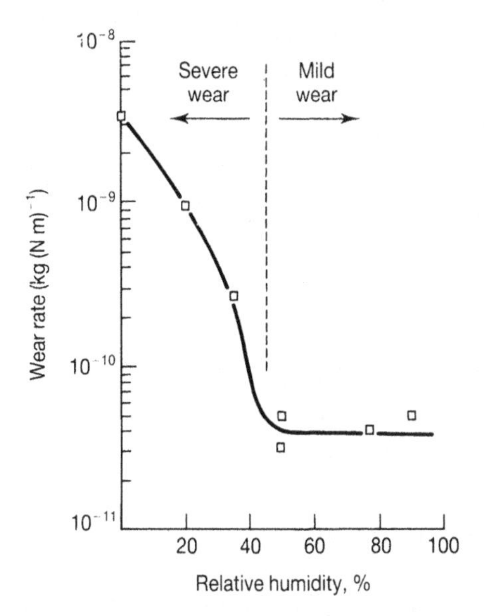

Fig. 5.32 Variation of wear rate with humidity for silicon nitride sliding against itself in a pin-on-disc test in air (the wear rate plotted relates to the disc) (from Fischer T E and Tomizawa H, in Ludema K C (Ed.), *Wear of Materials 1985*, ASME, 1985, pp. 22–32)

plasticity of these ceramics in the presence of water (the Rebinder effect) which can also be detected from microhardness measurements; on the other hand, it may be associated with enhanced crack growth (stress-corrosion cracking) which is common in oxide ceramics exposed to water.

Chemical effects can also be important in sliding at high speeds, where surface heating leads to increased rates of reaction. Oxidation, for example, is the major mechanism of wear in carbon/carbon composite materials used in high performance (e.g. aircraft) brakes. Many other non-oxide ceramics oxidize at high interfacial temperatures. Phase transformations may also result from the high temperatures; for example α-Si_3N_4 is formed on the sliding surface of β-Si_3N_4 in this way, and the rapid wear of diamond if it is used to machine or grind steel is caused by local transformation of the diamond to graphite.

5.10.5 Lubricated wear

Ceramic materials respond to conventional lubricants in a similar way to metals. Effective lubrication lowers the incidence of asperity contact, reduces the surface shear forces, and decreases the wear rate, as seen in Fig. 5.30. However, even under lubricated conditions chemical effects can be very important. Surface oxidation of Si_3N_4 and SiC occurs even in mineral oils, and the mechanism of lubricated wear of these ceramics is probably tribochemical. Boundary lubricants such as octadecanoic acid (stearic acid) function on ionic materials to which their polar end groups can bond; for example, Al_2O_3 is lubricated effectively by stearic acid, whereas SiC (with covalent bonding) is not. However, on some ionic ceramics, notably Al_2O_3 and ZrO_2, such lubricants lead to sharply enhanced wear rates despite the lowered coefficient of friction. This effect appears to be due to stress-corrosion cracking, which leads to intergranular fracture.

A phenomenon which can occur in the lubricated sliding of materials with low thermal conductivity and high thermal expansion coefficients is *thermo-elastic instability*. Zirconia is particularly susceptible to this. Contact between asperities in thin-film lubricated sliding causes intense local power dissipation and consequent heating. In a material which conducts heat poorly and shows high thermal expansion, the local temperature rise can significantly distort the topography of the surface and cause further asperity contact to be concentrated in the same area. The instability results in very high flash temperatures and rapid wear, and is responsible for a poor performance of zirconia under some lubricated sliding conditions.

5.11 WEAR OF POLYMERS

5.11.1 Introduction: interfacial and cohesive wear

In contrast to metals and ceramics, polymers exhibit lower coefficients of friction, with values typically between 0.1 and 0.5, whether self-mated or sliding against other materials (see Section 3.8). They are therefore often used unlubricated in tribological applications, usually sliding against harder counterfaces.

Polymers are much more compliant than metals or ceramics, with values of elastic modulus typically one tenth or even less. Their strengths are also much lower, and it is therefore reasonable to consider metallic or ceramic counterfaces when sliding against polymers to act as rigid bodies. Nearly all the deformation due to contact or sliding takes place within the polymer, and the surface finish of the hard counterface has a strong influence on the mechanism of the resulting wear.

If the counterface is smooth, then wear may result from adhesion between the surfaces, and involve deformation only in the surface layers of the polymer. On the other hand, if the counterface is rough then its asperities will cause deformation in the polymer to a significant depth; wear then results either from abrasion associated with plastic deformation of the polymer, or from fatigue crack growth in the deformed region. These two classes of wear mechanism, involving surface and subsurface deformation respectively, have been termed *interfacial* and *cohesive* wear processes.

As with metals and ceramics, chemical effects can also play important roles in the wear of polymers, for example through environmental effects on fatigue crack growth, or through surface degradation which constitutes a further example of an interfacial wear process. We shall examine the various cohesive and interfacial wear processes in the following sections.

The level of counterface roughness at which the transition from interfacial to cohesive wear mechanisms occurs depends on the nature of the polymer, but corresponds typically to R_a values between 0.01 and 1 μm. In general, susceptible polymers sliding on highly polished hard counterfaces will experience adhesive wear, while turned or ground surfaces (see Table 2.1) promote cohesive mechanisms. The transition between the two regimes can lead to a pronounced minimum in wear rate at a certain surface roughness, as illustrated in Fig. 5.33 for ultra-high molecular weight polyethylene sliding against stainless steel counterfaces of different roughnesses. The roughness at which the transition occurs can also be influenced by environmental factors and especially by the presence of species which reduce adhesion at the interface.

5.11.2 Cohesive wear mechanisms

Cohesive wear results from the deformation of surface and subsurface material, caused by the passage of a protuberance on the counterface over the polymer surface. The protuberance may be an asperity on a hard surface arising from its topography, or a particle of harder material partially embedded into a softer counterface, or possibly a lump of polymeric debris transferred to the counterface. The resulting deformation in the polymer may be either plastic (permanent) or elastic (recoverable). In the first case the mechanism of wear can be termed *abrasion*, while in the second it is associated with *fatigue*.

The distinction between these two types of mechanism in the wear of polymers is not sharp. It is fairer to envisage a progressive change, rather than an abrupt transition from one mechanism to the other as the mechanical properties of the polymer are changed. Figure 5.34 shows schematically how

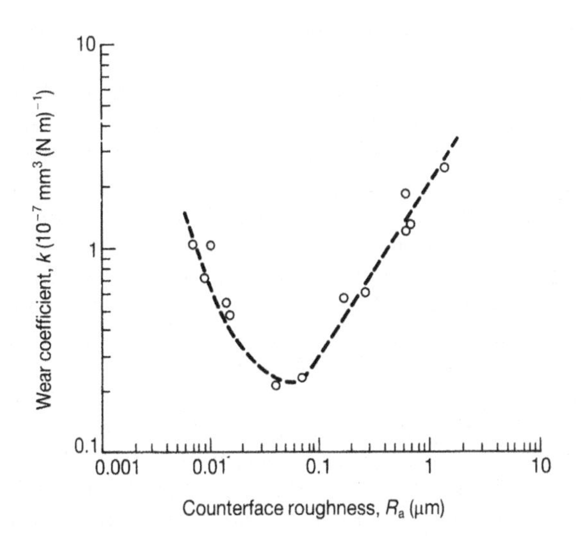

Fig. 5.33 Wear rate of ultra-high molecular weight polyethylene (UHMWPE) sliding against a steel counterface, as a function of the roughness of the steel surface (from Dowson D, Challen J M, Holmes K and Atkinson J R, in Dowson D, Godet M and Taylor C M (Eds), *The Wear of Non-metallic Materials*, Proc. 3rd Leeds-Lyon Symposium on Tribology, Mechanical Engineering Publications, 1976, pp. 99–102)

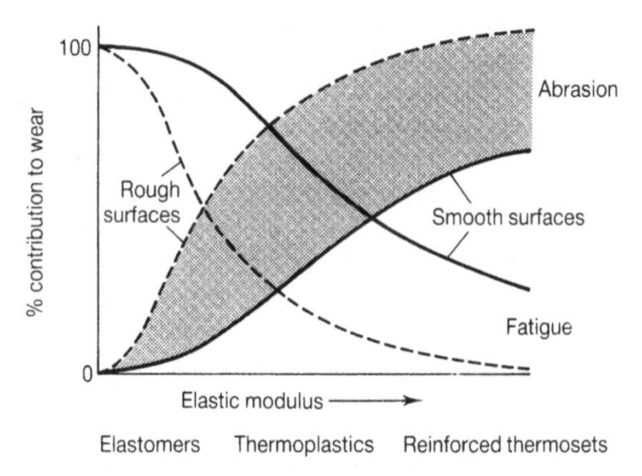

Fig. 5.34 Schematic diagram showing the variation in importance of the abrasion and fatigue mechanisms in polymers, as defined in the text, with the elastic modulus of the polymer and the roughness of the counterface (from Evans D C and Lancaster J K, in Scott D (Ed.), *Wear, Treatise on Materials Science and Technology*, Academic Press, **13**, 85–139, 1979)

the relative proportions of abrasion and fatigue depend on the surface roughness and the nature of the polymer. In elastomers, for example, with low elastic modulus, the contact deformation will be almost totally elastic and fatigue mechanisms will dominate. High modulus polymers such as thermosets, on the other hand, show appreciable plastic deformation due to asperity contact and suffer wear by abrasion. We shall now examine the processes of wear by abrasion and fatigue.

Abrasive wear is discussed in detail in Chapter 6. The mechanisms of interest here result from plastic displacement (ploughing or cutting) of the polymer by the rigid protuberances on the counterface. For this to occur, the contact must be plastic, and the plasticity index discussed in Section 2.5.3 can be used to establish whether this will be so. The term E/H in the expression for the plasticity index ψ (see equation 2.16) is approximately inversely proportional to the yield strain of the material, and is much smaller (one tenth or less) for polymers than it is for metals. The other term in the plasticity index expression depends on the surface roughness; the roughness at which the contact becomes predominantly plastic is therefore much greater for polymers than for metals. Behaviour similar to that shown in Fig. 2.13 will occur, but with the plastic regime starting at a counterface roughness (R_a value) typically greater than a few micrometres.

The simple theory of abrasive wear by plastic deformation presented in Section 6.3.1 predicts that the wear rate should be proportional to the normal load. This behaviour is indeed seen in polymers for wear by this mechanism. Equation 6.4 also predicts that the wear rate should be inversely proportional to tan α, where $(\pi/2-\alpha)$ is the mean surface slope and depends on the mean roughness R_a. Abrasive wear rates of polymers do increase with surface roughness, but considerably more rapidly than equation 6.4 suggests.

The theory also predicts that the wear rate should be inversely proportional to the hardness, H. In practice, H correlates only with the general trend of behaviour for polymers in this regime of wear, as can be seen from Fig. 6.19. The correlation of abrasive wear rate with hardness is much worse than for metals. The reason for this disparity is twofold. Hardness, defined as the load per unit area of the residual impression in an indentation test, provides a measure of rather different properties for metals and for polymers. In metals it accurately reflects the ease of plastic flow and correlates closely with the yield stress, whereas in polymers much of the material displaced around the indentation is accommodated elastically and the measured hardness incorporates contributions from both elastic and plastic sources. The second reason for the poor correlation is that even in the harder polymers sliding on rough counterfaces appreciable elastic deformation occurs; this causes damage by fatigue processes which are not controlled by plastic properties.

A much better correlation is found between abrasive wear rates and values of $1/\sigma_u \varepsilon_u$, where σ_u and ε_u are the ultimate tensile stress and elongation for the polymer, measured in a conventional tensile test. Although the conditions under which σ_u and ε_u are measured differ considerably, especially in terms of strain rate, from those at a sliding contact, the validity of this correlation, sometimes called the *Ratner–Lancaster correlation*, is widely recognised. Figure 5.35 shows the good agreement found for several thermoplastics.

The product $\sigma_u \varepsilon_u$ is roughly proportional to the area under the stress–strain curve for the polymer to the point of tensile failure, and thus provides a measure of the work done in producing tensile rupture.

Figure 5.34 shows that for low modulus polymers, such as the softer thermoplastics and elastomers, fatigue processes occur rather than abrasion. Fatigue is also important in harder polymers sliding against smooth counterfaces. Wear due to fatigue results from the formation of cracks associated with predominantly elastic deformation. Damage is cumulative, and develops over a number of contact cycles. Particles of wear debris become removed by the growth and intersection of cracks.

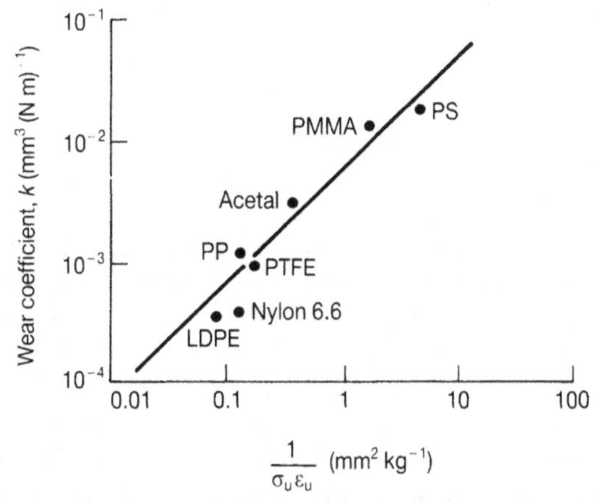

Fig. 5.35 The Ratner–Lancaster correlation between the wear rates of polymers under predominantly abrasive conditions and the reciprocal of the product of the stress and strain at tensile rupture (from Briscoe B J, *Tribology International*, August 1981, 231–243)

In a simple model of fatigue wear, the wear rate would be expected to correlate with the rate of fatigue crack growth. This usually follows the Paris equation:

$$\frac{da}{dN} = A\,(\Delta K)^n \qquad (5.11)$$

where da/dN is the increase in crack length a per stress cycle, ΔK is the range of stress intensity to which the growing crack is exposed during each cycle, and A and n are empirical constants. The use and validity of this equation for fatigue crack growth in many materials are discussed in basic texts on materials science and engineering. The value of the exponent n lies typically in the range from 1.5 to 3.5 for elastomers and 3 to 10 for rigid thermoplastics and thermosets.

ΔK is proportional to the range of stress $\Delta\sigma$ experienced by elements of the polymer during sliding. If it is assumed that the crack length at which material becomes detached is much greater than the size of the defect from which it grows, then integration of equation 5.11 shows that the number of cycles needed to remove a wear fragment, N_f, is proportional to $1/\Delta\sigma^n$. The wear rate will be inversely proportional to N_f. $\Delta\sigma$ can be assumed for simplicity to be the mean elastic contact stress due to a single spherical asperity of radius r under a load w, and depends on these quantities as follows (see Section 2.5.1):

$$\Delta\sigma \propto w^{1/3} r^{-2/3} \qquad (5.12)$$

The radius a of the contact area is given by:

$$a \propto w^{1/3} r^{1/3} \qquad (5.13)$$

If the breadth and width of the wear particle removed after N_f cycles are both proportional to a, and its depth is independent of a (being just the crack length at which detachment occurs), then the volume of material removed per unit sliding distance, q, will be given by:

$$q \propto \frac{a^2}{N_f} \qquad (5.14)$$

Hence, combining the preceding equations,

$$q \propto r^{2(1-n)/3} w^{(2+n)/3} \qquad (5.15)$$

We would therefore expect the overall wear rate due to fatigue to depend strongly on the surface roughness (since this controls r and also the number of asperity contacts per unit area). This is indeed observed: a tenfold increase in the roughness of the counterface (e.g. from 0.1 to 1 μm R_a) can lead to an increase in wear rate of a polymer by as much as one hundred times.

Strong dependence of wear rate on normal load is also observed, often following the power law suggested by equation 5.15. The exponent of load depends on both the nature of the polymer and the topography of the counterface.

Despite the success of fatigue models in explaining the dependence of wear on load and roughness, direct microscopic evidence for the presence of fatigue cracks during dry sliding wear is sparse. This may be because they are obscured by other changes in the surface layers, of chemical or thermal origin. However, strong evidence for a link between wear and fatigue is seen from experiments carried out in liquid environments. Figure 5.36 shows how the wear rate of polyethersulphone (PES) sliding against a stainless steel counterface depends on the nature of the liquid present. Under these sliding conditions fluid film lubrication did not occur, nor was abrasion an important mechanism.

Also plotted in the figure are the values of A and n from the Paris equation (equation 5.11), derived from conventional fatigue experiments carried out in the same liquids. On the horizontal axis are plotted values of δ_s, the solubility parameter for the solvent. The value of δ_s provides a measure of the chemical affinity between the liquid and the polymer; when δ_s is equal to δ_p, the solubility parameter for the polymer, this affinity is a maximum. Two important points arise from Fig. 5.36. First, the wear rate correlates strongly with the mutual affinity of the solvent and the polymer; similar results are found for many, though by no means all, other polymers. Second, the wear rate correlates even more strikingly with the crack growth rates measured in conventional fatigue tests.

Elastomers are distinguished from other engineering polymers by their low elastic modulus and large values of ultimate tensile elongation. True abrasive wear by a cutting mechanism occurs only on sliding against counterfaces with extremely sharp, needle-like asperities. More usually, wear occurs by prog-

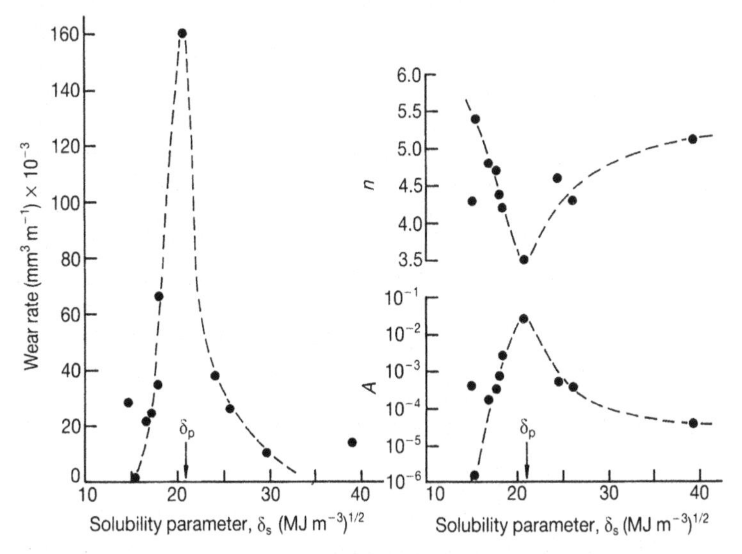

Fig. 5.36 The wear rate of polyethersulphone (PES) sliding against a smooth stainless steel counterface in a variety of organic solvents under boundary lubrication conditions. The liquids are described by the values of the solvent solubility parameter δ_s. Also shown are the values of A and n from the Paris equation (5.11) (from Atkins A G, Omar M K and Lancaster J K, *J. Mat. Sci. Lett.*, **3**, 779–782, 1984)

Fig. 5.37 Surface of a natural rubber specimen after abrasion against a rough hard counterface, showing a characteristic ridged 'abrasion pattern'. The direction of sliding of the counterface was from right to left (from Muhr A H and Roberts A D, *Friction and Wear*, ch. 6 in Roberts A D (Ed.), *Natural Rubber Science and Technology*, Oxford University Press, 1988)

ressive crack growth, with cracks propagating from the surface under cyclic loading. A rough counterface is not essential for the development of these cracks, since they can arise directly from surface tractions associated with sliding on a smooth counterface with a high coefficient of friction. This crack growth process in elastomers often leads to the formation of a characteristic 'abrasion pattern' on the surface.

The pattern, as shown in Fig. 5.37, consists of regular parallel ridges lying perpendicular to the direction of sliding. In cross-section the ridges have a saw-tooth profile; each ridge is associated with the progressive growth of a single crack into the rubber. Excellent agreement is found between wear rates measured for this type of wear and the predictions of a model based on incremental crack growth. The wear rate depends strongly on the frictional traction at the surface. With a high coefficient of friction, and for elastomers of low strength, the crack may propagate far enough in one contact cycle to generate a wear particle; this debris can be deformed by rolling between the sliding surfaces into a characteristic elongated shape, leading to this wear process being sometimes termed 'roll formation'.

5.11.3 Interfacial wear mechanisms

In interfacial wear, material removal results from processes occurring close to or in the surface of the polymer. The most important such process is wear by adhesion, which is directly related to the adhesive component of sliding friction discussed in Section 3.8.3.

Adhesive wear occurs only when the counterface is smooth, and involves the transfer of polymer to the harder counterface and its subsequent removal as wear debris. In this respect the process is similar to that seen in metals. A running-in period often occurs before steady-state conditions are reached; during steady-state wear, the wear rate is often directly proportional to the normal load over quite a large range (see Fig. 5.38), in conformity with the Archard equation (equation 5.7). Corresponding values of the dimensional wear coefficient k lie typically in the range 10^{-6} to 10^{-3} mm^3 $(Nm)^{-1}$.

Not all polymers show adhesive wear of this type. Thermosets, for example, do not form transfer films, but wear instead by fatigue processes or abrasion, or at high interfacial temperatures by thermal degradation (pyrolysis) of the surface — another form of interfacial wear. In amorphous polymers such as polystyrene, polyvinylchloride (PVC) and polymethylmethacrylate (PMMA) below their glass transition temperatures (T_g), the interface between polymer and counterface is weaker than the bulk polymer, and wear occurs mainly by fatigue or abrasion as discussed in the previous section. Once the interface temperature rises above T_g, however, amorphous thermoplastics wear by adhesive transfer.

Wear results from the adhesion of polymer material to the counterface by electrostatic forces (including van der Waals forces). The junction between the polymer and the counterface is stronger than the polymer itself, and failure occurs within its bulk, leaving a transferred fragment. Repeated sliding over the counterface leads to the progressive build-up of a transferred layer which eventually becomes detached. Under these conditions, sometimes known as 'normal' transfer, polymer is transferred to the counterface without significant chain scission or chemical degradation, in irregular lumps or patches typically 0.1 to 10 μm thick. The polymer chains may show some orientation in the direction of sliding. Back-transfer to the polymer may also occur, as may incorporation of material from the counterface into the transfer layer. The rate of wear appears to be dictated by the rate of removal of the

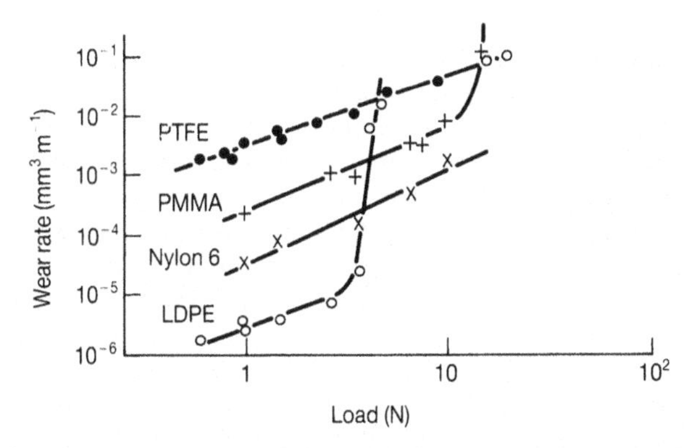

Fig. 5.38 The variation of steady-state wear rate with normal load for polymers sliding against a smooth mild steel conterface ($R_a = 0.15$ μm) under unlubricated conditions (from Evans D C and Lancaster J K, in Scott D (Ed.), *Wear, Treatise on Materials Science and Technology*, Academic Press, **13**, 85–139, 1979)

transfer film from the counterface, rather than by the rate of polymer transfer into the film. There is evidence that polymers are reluctant to transfer further on to their own transfer films, and if the transfer film adheres strongly to the counterface then the wear rate is low.

A small number of polymers, notably polytetrafluoroethylene (PTFE), high density polyethylene (HDPE) and ultra-high molecular weight polyethylene (UHMWPE), form transfer layers of a rather different type under certain conditions, which are associated with low friction and wear rates. These were discussed in Section 3.8.3. At low sliding speeds (typically $<0.01\ m\ s^{-1}$) a very thin transfer layer is laid down, perhaps 5 to 10 nm thick. In this layer the polymer chains are highly oriented in the direction of sliding. The layer adheres well to the counterface, and further sliding occurs between the surface of the bulk polymer, which contains similarly oriented molecules, and the transfer film on the counterface.

The conditions under which this regime of low friction and wear is maintained are precarious; a return to 'normal' transfer occurs with rougher counterfaces or higher sliding velocities. However, the thin oriented transfer layer can be stabilized by measures which improve its adhesion to the counterface. Suitable fillers in the polymer will do this and lower the wear rate by a factor of 10^2 to 10^3, without significant change in the coefficient of friction. The incorporation of lead and copper oxides into HDPE, for example, provides an effective enhancement of its wear resistance against steel, although details are still unclear of the chemical mechanisms by which the adhesion of the transfer film is improved. Metallic fillers (e.g. bronze) also appear to contribute to a chemical enhancement of film adhesion. Other fillers, for example carbon particles, are thought to reduce wear rates by abrading the counterface slightly; this process will reduce the surface roughness and also provide a cleaner surface, both tending to increase the adhesion of the transfer layer. Fillers of these types are widely used in composite polymeric bearing materials (see Section 9.4.2), which may also contain solid or dispersed liquid lubricants, as well as strong filler particles which carry a proportion of the normal load.

Thermoplastic polymers generally have low thermal conductivities and low softening or melting temperatures compared with metals; both factors lead to marked thermal softening under much milder sliding conditions than would be needed for metals. The phenomenon is, however, exactly the same as the transition from effectively isothermal to almost adiabatic sliding conditions discussed in Section 5.5. The same method of analysis used to generate the curves in Fig. 5.13 indicates that the rapid increases in wear rate of low density polyethylene (LDPE) and polymethylmethacrylate (PMMA) at the higher loads in Fig. 5.38 arise from thermal softening. These transitions occur when the flash temperature at the interface reaches the softening point of the polymer. Above this thermal transition the rate of wear is high, with material being removed from a molten layer at the surface of the polymer. The transition has been described as one from 'mild to 'severe' wear, although these terms have different meanings from those used for metals or ceramics.

The same effect is also responsible for a sharp increase in wear rate for thermoplastics above a certain sliding speed, as shown for nylon 6.6 in Fig.

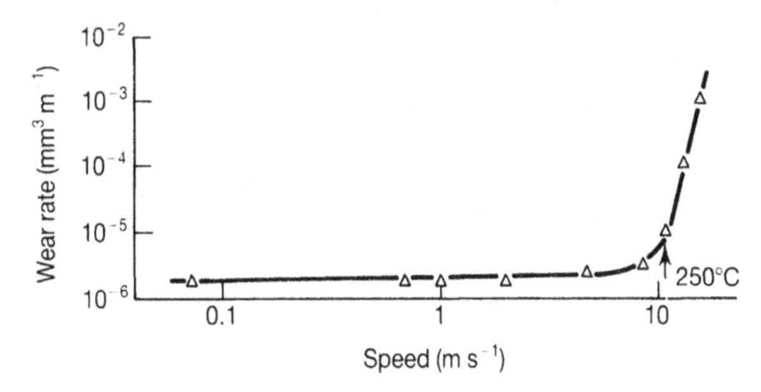

Fig. 5.39 The variation of steady-state wear rate with sliding speed for nylon 6.6 sliding against a smooth mild steel counterface ($R_a = 0.15$ μm) under unlubricated conditions (from Evans D C and Lancaster J K, in Scott D (Ed.), *Wear, Treatise on Materials Science and Technology*, Academic Press, **13**, 85–139, 1979)

5.39; the arrow indicates the speed at which, under these experimental conditions, the calculated flash temperature reaches 250 °C, the softening point of the polymer.

The nature of the counterface also influences the conditions at which thermal softening occurs. For a metallic counterface, a moderately good conductor of heat, the critical speed may be some hundred times higher than for a polymeric counterface with much lower conductivity. Part of the beneficial effect of metallic fillers in reducing the wear rates of polymer composites in bearing applications (see Section 9.4) is due to the increase in thermal conductivity caused by the filler.

5.11.4 Lubricated wear of polymers

Hydrodynamic lubricant films or even EHL films (see Chapter 4) will separate the sliding surfaces and reduce the contact stresses, lowering the wear rates due to all the mechanisms discussed above. Polymers, with their low values of elastic modulus, remain in the EHL regime at much lower sliding speeds and higher normal loads than metals. In the regime of boundary lubrication, however, when the load is carried predominantly by solid contact, the mechanical effects of lubrication are often of secondary importance compared with the chemical and physical effects of the lubricant on the surface of the polymer. Although lubricants always reduce the coefficient of friction, they may either increase or decrease the rate of wear.

Boundary lubricants, reviewed in Section 4.6, act by forming adsorbed layers on the sliding surfaces. They bond poorly to the low energy surfaces of polymers, and so have little effect on the friction and wear of self-mated polymers, although they may still play a useful role when they can chemisorb on to a metallic counterface.

Liquid lubricants interact with polymers in several ways. In some cases, in glassy thermoplastics for example, they can cause increased plasticity of the surface which leads to reduced wear. However, the same polymers can also experience stress-crazing or cracking in other liquids, resulting in greatly enhanced wear. We have already seen in Fig. 5.36 how the relative affinity of the polymer and the liquid can influence the rate of fatigue crack growth, and hence the rate of wear by fatigue mechanisms.

Polymers which show low wear rates because they form thin oriented transfer films tend to suffer greater wear when lubricated, because the lubricant interferes with the adhesion of the transfer film to the counterface. Water has a particularly marked influence in this respect; for example, the wear rates of PTFE composites containing glass fibres or bronze particles, sliding against metal counterfaces, may be 100 to 1000 times greater when wet than when dry.

Further reading: references

Briscoe B J, Wear of polymers: an essay on fundamental aspects, *Tribology International* **14,** 231–243, 1981

Briscoe B J and Tabor D, Friction and wear of polymers, in Clark D T and Feast W J (Eds), *Polymer Surfaces*, John Wiley, pp 1–23, 1978

Buckley D H and Miyoshi K, Tribological properties of structural ceramics, in Wachtman J B (Ed.), *Structural Ceramics, Treatise on Materials Science and Technology* **29**, pp 293–365, 1989

Fischer T E, Tribochemistry, *Ann. Rev. Mater. Sci.* **18**, 303–323, 1988

Friedrich K (Ed.), *Friction and Wear of Polymer Composites*, Composite Materials Series, Vol. 1, Elsevier, 1986

Lim S C and Ashby M F, Wear-mechanism maps, *Acta Metall.* **35**, 1–24, 1987

Ling F F and Pan C H T (Eds), *Approaches to Modeling of Friction and Wear*, Springer-Verlag, 1988

Merchant H D and Bhansali K J (Eds), *Metal Transfer and Galling in Metallic Systems*, TMS, 1987

Peterson M B and Winer W O (Eds), *Wear Control Handbook*, ASME, 1980

Rigney D A (Ed.), *Fundamentals of Friction and Wear of Materials*, ASM, 1981

Rigney D A, Sliding wear of metals, *Ann. Rev. Mater. Sci.* **18**, 141–163, 1988

Scott D (Ed.), *Wear, Treatise on Materials Science and Technology*, Vol. 13, Academic Press, 1979

Sullivan J E, The role of oxides in the protection of tribological surfaces, *Inst. Mech. Engrs. Conf. Series* 1987–5, Vol. 1, 283–302

Viewpoint set on materials aspects of wear (various authors), *Scripta Met. et Mat.*, **24**, 799–844, 1990

Waterhouse R B, *Fretting Corrosion*, Pergamon, 1972

6

Wear by hard particles

6.1 INTRODUCTION AND TERMINOLOGY

In this chapter, we shall concentrate on *abrasive wear* and *erosion*. In abrasive wear, material is removed or displaced from a surface by hard particles, or sometimes by hard protuberances on a counterface, forced against and sliding along the surface. Several qualifying terms can be used in describing abrasion. A distinction is often made between *two-body abrasive wear*, illustrated in Fig. 6.1(a), and *three-body abrasive wear*, shown in Fig. 6.1(b). Two-body wear is caused by hard protuberances on the counterface, while in three-body wear hard particles are free to roll and slide between two, perhaps dissimilar, sliding surfaces. A drill bit cutting rock might experience two-body wear, while grit particles entrained between sliding surfaces, perhaps present as

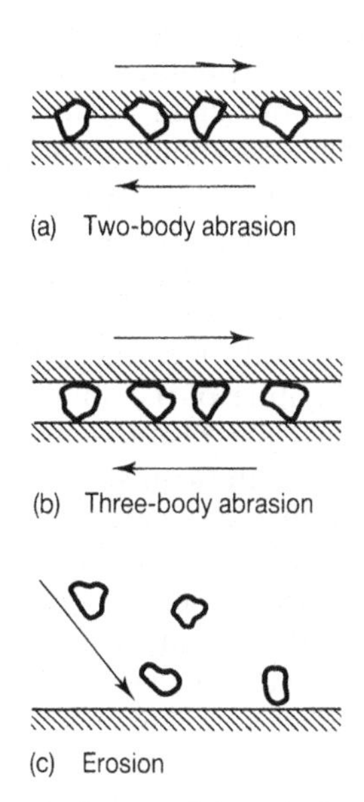

Fig. 6.1 Illustration of the differences between (a) two-body abrasion; (b) three-body abrasion; and (c) erosion

contaminant in a lubricating oil, would cause three-body wear. Wear rates due to three-body abrasion are generally lower than those due to two-body abrasion, although the various mechanisms of material removal in the two cases differ only in relative importance rather than in nature.

Other terms sometimes used to describe abrasive wear are *high-stress* and *low-stress* abrasion. In high-stress abrasion, the crushing strength of the abrasive particles is exceeded, so that they are broken up during the wear process, while in low-stress abrasion the particles remain unbroken. The term *gouging abrasion* is often used to describe high-stress abrasion by large lumps of hard abrasive material, for example in rock crushing machinery; gouging results in the removal of material from the worn surface in large fragments. The scale of surface deformation above which the term 'gouging' is appropriate is, however, ill-defined.

In some situations, wear is caused by hard particles striking the surface, either carried by a gas stream or entrained in a flowing liquid (Fig. 6.1(c)). This type of wear is called *erosion*, often qualified as *solid particle erosion* or *solid impingement erosion* to distinguish it from the damage caused by the impact of liquid jets or drops. Particle velocities in erosive wear are commonly between 5 and $500\,\mathrm{m\,s^{-1}}$, although the term can still be applied outside this range. If the hard particles are carried by a liquid, the wear may be termed *slurry erosion*.

The particles responsible for erosion or abrasion may be intrinsic to a given application; for example, tools used to cultivate soils must inevitably be exposed to hard particles, as must pipelines carrying sand/water slurries in mining operations. In other situations the particles may be contaminants which are difficult to avoid, such as fine airborne grit particles finding their way into lubricating oil. Hard particles may be generated locally by oxidation or wear from components of the tribological system; Fe_2O_3 wear debris, for example, may be produced in the sliding wear of steels, and can then cause further damage by abrasion. Two-body abrasion may result simply from differences in hardness and roughness between two sliding surfaces, incompletely lubricated: a relatively rough alumina ceramic sliding against a steel counterface would cause wear of this kind.

Abrasion and erosion can also be employed usefully, in grinding and polishing processes and in methods of cutting and shaping materials.

Although the sources and nature of the abrasive and erosive particles may differ among these examples, the resulting wear processes have much in common, which we shall explore in the rest of this chapter.

We have so far assumed that the particles responsible for abrasion or erosion are 'hard'. In the next section we shall define more closely what is implied by this term, and discuss the properties of common abrasive and erosive particles which are important in wear. We shall then discuss in detail the phenomena of abrasive and erosive wear of all types of material.

6.2 PARTICLE PROPERTIES: HARDNESS, SHAPE AND SIZE

6.2.1 Particle hardness

The hardness of the particles involved in abrasion or erosion influences the rate of wear: particles with lower hardness than that of the surface cause much less wear than harder particles. For particles significantly harder than the surface, then the exact value of their hardness matters much less. This behaviour is illustrated in Fig. 6.2, which shows the relative wear rates by two-body abrasion of a wide range of metals and ceramics, abraded by various types of grit particle. The wear rate becomes much more sensitive to the ratio of abrasive hardness H_a to the surface hardness H_s when H_a/H_s is less than ~1.

The reason for this behaviour can be understood by examining the

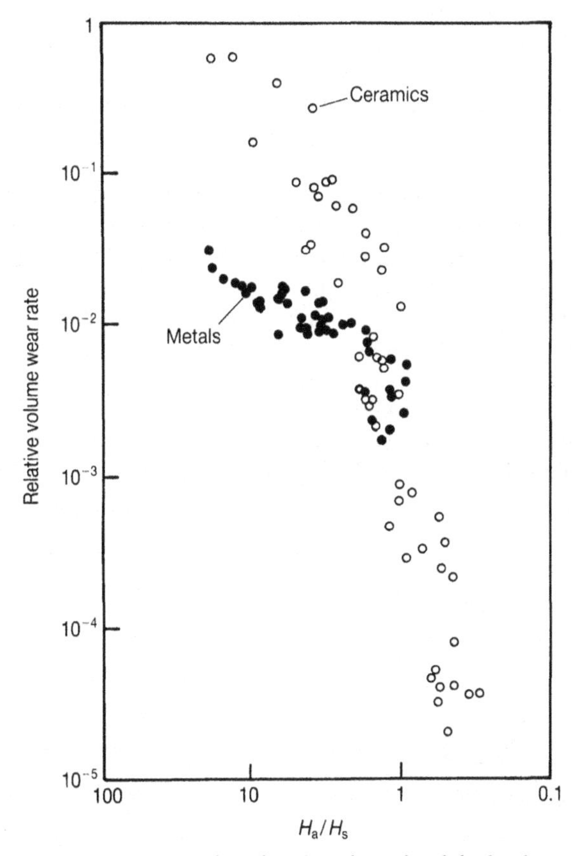

Fig. 6.2 Relative volume wear rate plotted against the ratio of the hardness of the abrasive to that of the surface (H_a/H_s) for a range of metallic and ceramic materials and abrasive particles, for two-body abrasion (data from Moore M A, *Materials in Engineering Applications* **1**, 97–111, 1978)

mechanics of contact between a discrete grit particle and a plane surface. If the surface material flows plastically once its yield point is exceeded, significant plastic flow will occur in the surface when the mean contact pressure reaches about three times its uniaxial yield stress Y, as discussed in Section 2.5.1. This contact pressure is the indentation hardness of the surface, and depends little on the detailed shape of the particle. Plastic indentation of the surface will occur as the normal load on the particle is increased only if the particle can sustain this contact pressure without deforming (Fig. 6.3(a)). If the particle fails by flow or fracture before the pressure on the surface reaches $\sim 3Y$, then insignificant plastic deformation will occur in the surface (Fig. 6.3(b)).

For a spherical particle pressed against a flat surface, the maximum contact pressure is about 0.8 times the indentation hardness of the particle material. We would therefore expect a sphere of hardness H_a to cause plastic indentation in a surface of hardness H_s if H_s is less than $\sim 0.8\,H_a$; that is, if $H_a/H_s > 1.25$. A similar result can be derived from slip-line field theory for other contact geometries, and it is observed experimentally that abrasive grit particles of any shape will cause plastic scratching only if $H_a/H_s > 1.2$. Abrasion under conditions where $H_a/H_s < 1.2$ is sometimes termed *soft abrasion*, in contrast to *hard abrasion* when $H_a/H_s > 1.2$.

The observation that a certain minimum ratio of hardness is needed for one material to be able to scratch another provides the physical basis for the scale of hardness devised by the Austrian mineralogist Mohs in 1824. Mohs assigned integer hardness numbers to a sequence of ten minerals, each of which would scratch all those, but only those, below it in the scale. As Fig. 6.4 shows, the ratio of indentation hardness between neighbouring standard minerals (except for diamond) is nearly constant. For the Mohs scale it is about 1.6, rather higher than the minimum necessary to cause scratching, but allowing a wide range of hardness to be spanned with only ten standards.

Knowing that erosion or abrasion will lead to much greater wear rates when the particles have more than about 1.2 times the hardness of the surface, we can now examine the hardnesses of common abrasive particles and structural materials. Table 6.1 lists values for a range of materials. Silica (quartz) is the

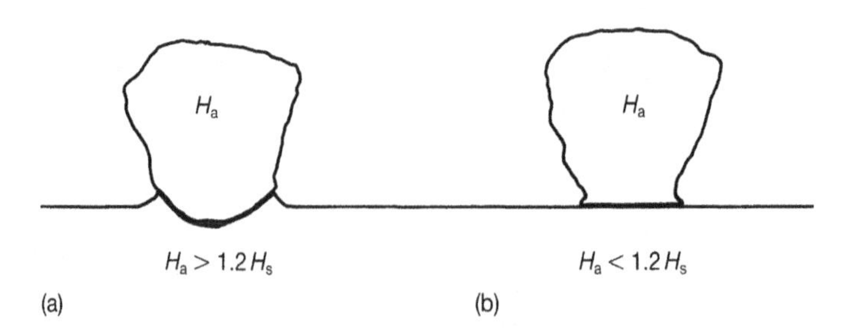

Fig. 6.3 Illustration of contact between a grit particle under normal load and a plane surface. (a) If H_a is greater than $\sim 1.2\,H_s$, the particle will indent the surface; (b) if H_a is less than $\sim 1.2\,H_s$, plastic flow will occur in the particle, which will be blunted

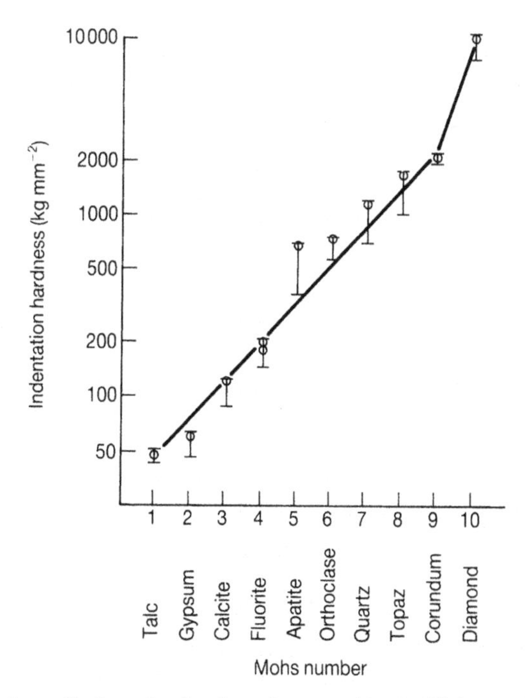

Fig. 6.4 Comparison of indentation hardness (measured by the Vickers or Knoop method) with Mohs hardness number for ten standard minerals (from Bowden F P and Tabor D, *The Friction and Lubrication of Solids, Part II*, Clarendon Press, Oxford, 1964)

Table 6.1 Hardness values for common abrasive particles, bulk structural materials, hard coating materials and alloy phases. It should be noted that the precise value of hardness will depend on the composition and microstructure of the material: for example on grain size, level of cold work and state of heat treatment. In many materials hardness also depends significantly on the indenter load used to measure the hardness. The values given here, from a wide variety of sources, must therefore be taken only as a rough guide to relative values of hardness

1. Typical abrasive materials	*Material Hardness (HV)*
Diamond	6000–10000
Boron carbide, B_4C	2700–3700
Silicon carbide	2100–2600
Alumina (corundum)	1800–2000
Quartz (silica)	750–1200
Garnet	600–1000
Magnetite, Fe_3O_4	370–600
Soda-lime glass	~500
Fluorite, CaF_2	180–190
2. Phases or constituents of steels and cast irons	
Ferrite, α-Fe	70–200
Pearlite (plain C)	250–320
Pearlite (alloyed)	300–460
Austenite (12% Mn)	170–230
Austenite (low alloy)	250–350
Austenite (high Cr)	300–600
Martensite	500–1000
Cementite, Fe_3C	840–1100
Chromium/iron carbide, $(Fe,Cr)_7C_3$	1200–1600

Table 6.1–*cont.*

Material	*Material Hardness (HV)*
3. Other metals (pure unless indicated)	
Aluminium (commercial purity)	25–45
Aluminium alloy (age-hardened)	100–170
Brass (α or β)	80–140
Chromium (cast)	100–170
Chromium (electroplated)	500–1250
Copper (commercial purity)	40–130
Copper–beryllium	150–400
Gold	30–70
Indium	1
Iron	70–200
Lead	4
Molybdenum	160–180
Nickel	70–230
Nickel (electroplated)	200–400
Rhodium (electroplated)	800
Silver	25–80
Tin	5–6
Tungsten	260–1000
Zinc	30–35
4. Ceramic materials (bulk or as coatings) (see also section 1 above)	
Carbides:	
Chromium carbide, Cr_7C_3	1600
Chromium carbide, Cr_3C_2	1300
Hafnium carbide	2270–2650
Molybdenum carbide, Mo_2C	1500
Niobium carbide	2400–2850
Tantalum carbide	1800–2450
Titanium carbide	2000–3200
Tungsten carbide, WC	2000–2400
Vanadium carbide	2460–3150
Zirconium carbide	2360–2600
Nitrides:	
Chromium nitride	2200
Hafnium nitride	1640
Niobium nitride, NbN	1400
Niobium nitride, Nb_2N	1720
Tantalum nitride	1220
Titanium nitride	1200–2000
Vanadium nitride	1520–1900
Zirconium nitride	1150
Borides:	
Chromium diboride	1800
Hafnium diboride	2250–2900
Molybdenum diboride	2350
Tantalum diboride	2450–2910
Titanium diboride	2200–3500
Tungsten diboride	2400–2660
Vanadium diboride	2070–2800
Zirconium diboride	2250–2600

most commonly occurring natural abrasive contaminant, constituting about 60% of the Earth's crust, and has a hardness of about 800 kgf mm^{-2} (i.e. 800 HV). Even a martensitic steel will have a hardness lower than 1.2 times this, and it is therefore clear that steels and non-ferrous metals will be especially vulnerable to abrasive wear and erosion by quartz particles. Materials containing harder phases, such as cermets and 'wear-resistant' alloys, will often also have softer constituents. Ceramic materials may be sufficiently hard and homogeneous to suffer little plastic deformation from common abrasive or erosive particles and, as we shall see below and in Chapter 7, they can confer valuable wear resistance when used either as bulk materials or as coatings.

6.2.2 Particle shape

Most particles responsible for abrasive or erosive wear are roughly equiaxed, but there can be considerable variation in their angularity depending on their

Fig. 6.5 SEM micrographs of silica particles: (a) rounded and (b) angular (courtesy of A J Sparks)

origins. Wear rates depend strongly on the shapes of the particles, with angular particles causing greater wear than rounded particles. The reasons for this dependence are discussed below, in Sections 6.3 and 6.4.

Angularity is difficult to define. Figure 6.5 shows two shapes of quartz particle: rounded and angular. Differences in particle shape of this magnitude may result in differences in wear rate by a factor of ten or more, yet the angularity of abrasive particles is seldom measured quantitatively. This is largely because of the difficulty of identifying and quantifying the features of a complex three-dimensional shape that are responsible for its abrasivity.

One of the simplest descriptions of shape is based on measurements of the perimeter and area of a two-dimensional projection of the particle, usually generated by optical microscopy. A *roundness factor* F can then be defined as the ratio between the actual area A of the projection, and the area of a circle with the same perimeter P as the projection. In terms of these quantities,

$$F = \frac{4\pi A}{P^2} \tag{6.1}$$

If $F = 1$ the projection is a circle; the more the outline of the particle departs from circular, the smaller the value of F. By averaging the values of F derived from the two-dimensional outlines of many particles, oriented randomly, an indication can be gained of the departure from sphericity of three-dimensional particles.

Some success has been achieved in correlating abrasive wear rates with values of the roundness factor F, but it provides only a crude measure of the deviation of the particle from a perfect sphere. Many other, more sensitive measures of angularity have been proposed, and there is considerable potential for future research into their applicability.

6.2.3 Particle size

The sizes of abrasive particles cover a wide range. Those responsible for most abrasive and erosive wear are between 5 and 500 μm in size, although polishing may employ submicrometre particles, while gouging wear may involve hard objects some tens or even hundreds of millimetres across.

A consistent pattern of behaviour is found in laboratory studies of both abrasion and erosion of metals, as shown in Fig. 6.6. The units of wear plotted there are explained further in Sections 6.3 and 6.4; for the present it is sufficient to note that if particles of different sizes were equally effective at removing material, the wear rates expressed in these units would be constant. In fact, as Fig. 6.6 shows, wear rates for particles smaller than about 100 μm drop markedly with decreasing particle size. Similar behaviour is seen for quite different particle materials and different metals.

Several explanations have been proposed for this effect of particle size on the abrasion and erosion of metals, but only one is convincingly applicable over the wide range of conditions for which the effect is observed. The behaviour is thought to reflect a true size effect in the strength of the metal

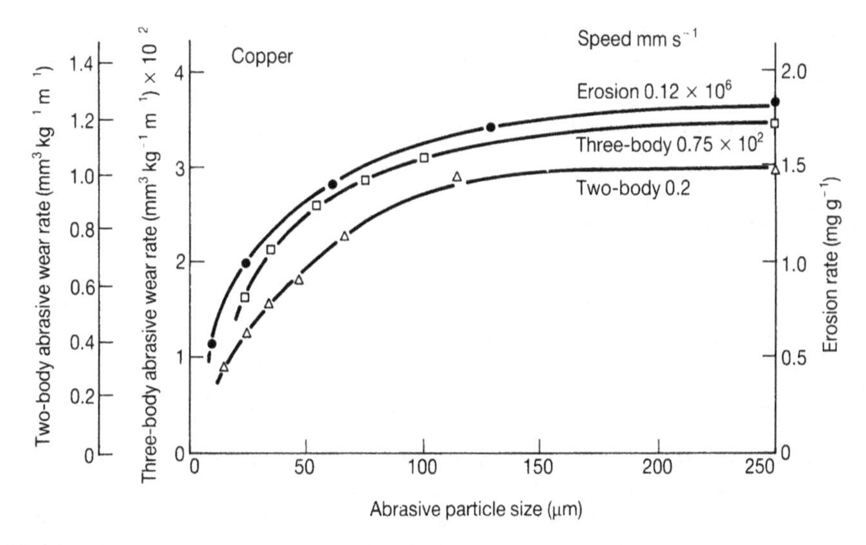

Fig. 6.6 Wear rates of copper under conditions of two-body and three-body abrasion and erosion, due to silicon carbide particles of different sizes (from Misra A and Finnie I, *Wear* **65**, 359–373, 1981)

itself, which is also found in indentation and scratching experiments. The flow stress of a very small volume of material is higher than that of a larger volume, perhaps because of the difficulty of nucleating or moving dislocations in a small volume. This increase in local flow stress as the scale of the deformation is diminished leads to a reduction in wear rates by plastic processes (further discussed in Sections 6.3.1 and 6.4.1), and hence to the observed particle size effect.

Behaviour like that illustrated in Fig. 6.6 is seen in metals and other materials where wear involves plastic flow. As we shall see below, wear of brittle materials may involve fracture, and they then exhibit an even stronger dependence of wear rate on particle size. Whatever the dominant mechanism, the fact that small particles cause proportionately less wear than larger ones is fortunate. It means that because methods of removing contaminant particles from a system, such as filtration or centrifugal methods, eliminate large particles more readily than smaller ones, they can reduce wear rates by abrasion and erosion very effectively (see Section 7.4).

6.3 ABRASIVE WEAR

Mechanisms of abrasive wear can involve both plastic flow and brittle fracture. Under some circumstances plastic flow may occur alone, but both often occur together, even in materials conventionally thought of as ideally brittle. Models for abrasive wear by each type of mechanism in isolation have been developed, but the models usually ignore the possibility of the other mechanism. In order to understand abrasive wear in simple terms, we shall also separate the two groups of mechanisms, and examine in the following sections models for abrasive wear by plastic deformation and by brittle fracture. We shall then discuss the behaviour of engineering materials in the light of these somewhat idealized models.

6.3.1 Abrasive wear by plastic deformation

A simple model for abrasive wear involves the removal of material by plastic deformation. It is closely related to the model for the ploughing (deformation) component of friction examined in Section 3.4.2.

Figure 6.7 shows an abrasive particle, idealized as a cone of semiangle α, being dragged across the surface of a ductile material which flows under an indentation pressure P. It forms a groove in the material, and wear is assumed to occur by the removal of some proportion of the material which is displaced by the particle from the groove. As we saw in Section 3.4.2, the normal load w carried by the particle is supported by plastic flow beneath the particle, which causes a pressure P to act over the area of contact between the particle and the surface. Since the cone is moving and therefore in contact only over its front surface,

$$w = P\frac{\pi a^2}{2} = \frac{1}{2}P\pi x^2 \tan^2 \alpha \tag{6.2}$$

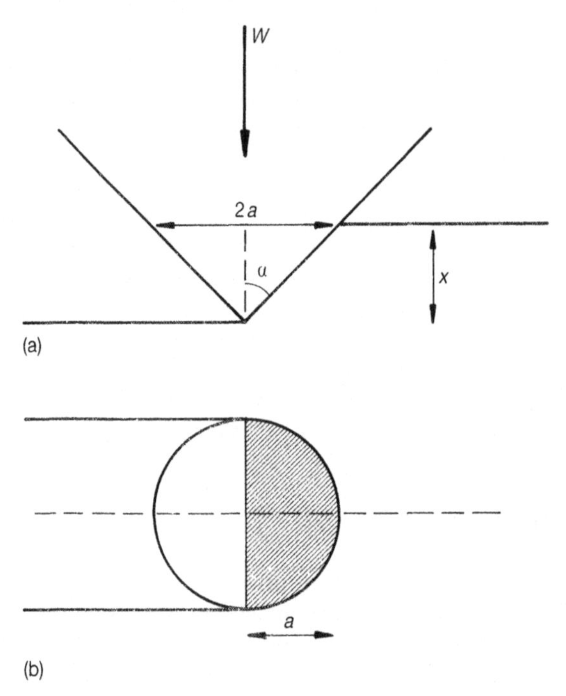

Fig. 6.7 Geometry of contact between an idealized conical abrasive particle and a surface: (a) in elevation; (b) in plan view

The volume of material displaced from the groove by the cone in sliding a distance l along the surface is $l\,a\,x$, or $l\,x^2 \tan\alpha$. The quantities a and x are defined in Fig. 6.7. So if a fraction η of the material displaced from the groove is actually removed as wear debris, then the volume of wear debris produced by this one particle per unit sliding distance, q, will be given by

$$q = \eta x^2 \tan\alpha \tag{6.3}$$

We can substitute for x^2 from equation 6.2, to find

$$q = \frac{2\eta w}{\pi P \tan\alpha} \tag{6.4}$$

Summing over many abrasive particles, and assuming that $P \approx H$, the indentation hardness of the material, we can then show that the total volume removed per unit sliding distance Q is given by

$$Q = \frac{KW}{H} \tag{6.5}$$

where W is the total applied normal load and the constant K depends on the fraction η of displaced material actually removed, and on the geometry of the abrasive particles (i.e. on α).

Equation 6.5 is exactly the same as the Archard equation for sliding wear (equation 5.7), although derived from completely different initial assumptions. K, the dimensionless wear coefficient, can be used as a measure of the severity of wear. Typical values of K in the two-body abrasive wear of metals lie between $\sim 5 \times 10^{-3}$ and $\sim 50 \times 10^{-3}$. For three-body abrasion, K is lower, typically between 0.5×10^{-3} and 5×10^{-3}. As in the case of sliding wear, it is sometimes more useful to employ the dimensional wear coefficient k ($= K/H$), with usual units $\mathrm{mm^3\,(Nm)^{-1}}$.

Another quantity sometimes used to express the severity of abrasive wear, which allows particularly revealing comparisons between different abrasion conditions and with erosive wear, is the specific energy for material removal, U. This is defined as the frictional work expended per unit volume of material removed. In terms of the quantities discussed above,

$$U = \frac{\mu W}{Q} \tag{6.6}$$

where μ is the coefficient of friction (ratio of tangential to normal force) between the sliding bodies. U can also be expressed in terms of μ and k:

$$U = \frac{\mu}{k} \tag{6.7}$$

For two-body abrasion of metals in air, μ lies typically between 0.4 and 1, while for three-body conditions it is often lower: 0.2 to 0.5. A rough estimate of U can therefore readily be made from equation 6.7 and the value of the dimensional wear coefficient k.

The simple model for wear by plastic deformation which leads to equation

6.5 predicts that the volume of material removed by two-body abrasion should be directly proportional to the sliding distance, and also to the normal load. This behaviour is usually observed in practice, as illustrated by Fig. 6.8 for three ductile metals subjected to two-body abrasion by silicon carbide particles.

Equation 6.5 also suggests that the wear rate should vary inversely with the hardness of the material, H. Many pure metals do behave in this way, although alloys often exhibit more complex behaviour. Figure 6.9 shows the results of two-body abrasion tests on a range of annealed pure metals (open points) and heat-treated and work-hardened steels (solid points). The relative wear resistance, which is proportional to the reciprocal of the volume wear rate Q, is plotted against the indentation hardness of the bulk metals. The experimental points for the pure metals lie close to the straight line through the origin expected from equation 6.5. However, the results for the steels

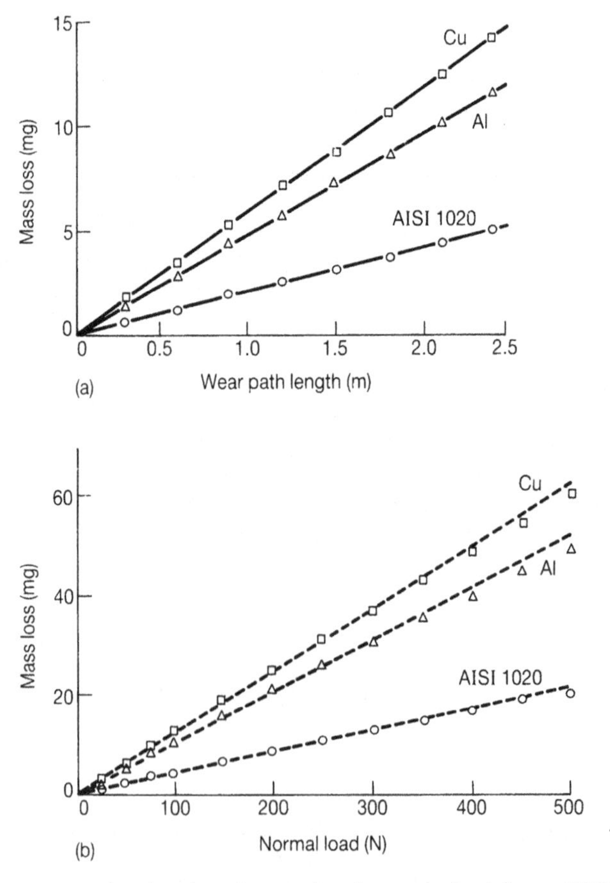

Fig. 6.8 Wear, measured as mass loss, for samples of copper, aluminium and 0.2% carbon steel (AISI 1020) subjected to two body abrasion by 115 μm silicon carbide particles: (a) variation with sliding distance; (b) variation with normal load (from Misra A and Finnie I, *Wear* **68**, 41–56, 1981)

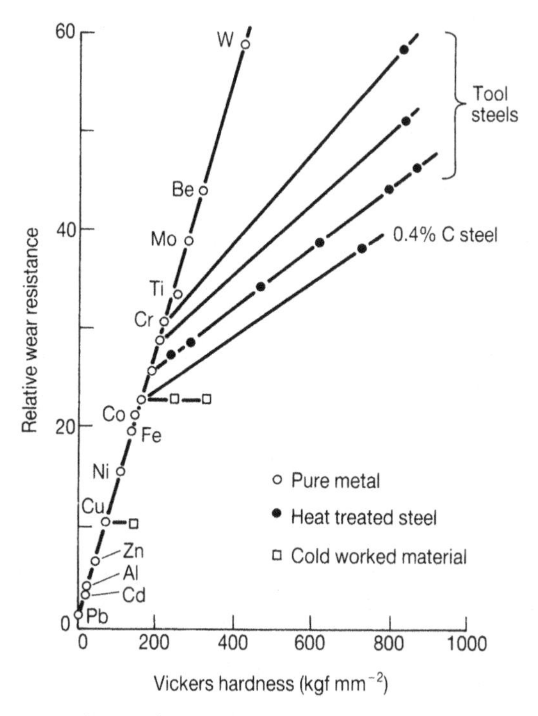

Fig. 6.9 Relative wear resistance (proportional to 1/wear rate) for pure metals (open circles) and heat-treated and work-hardened steels (solid circles) under conditions of two-body abrasion, plotted against indentation hardness (data from Khrushchov M M, *Proc. Conf. on Lubrication and Wear*, Inst. of Mechanical Engineers, London, 1957, pp. 655–659)

(solid points) do not lie on this line, but on other lines with different slopes. In order to understand the reasons for this, we must explore the factors which influence the value of the quantity K in equation 6.5.

In plotting wear resistance against the bulk hardness of the metals in Fig. 6.9, we ignored the fact that the material at the worn surface will have been strain-hardened by plastic flow, and that its hardness will generally be greater than that of the bulk. Abrasion introduces very high shear strains into the surface material; in copper and brass, for example, true strains of up to 8 have been recorded. The shear strain decreases with depth into the bulk as shown in Fig. 6.10; the depth of deformation and the strain at a given depth are both proportional to the depth of indentation of the abrasive particles. The energy of frictional work expended in abrasion is largely accounted for by the plastic work done in this deformed subsurface region. As a result of this strain hardening, the flow stress of the deformed surface material may be two or even three times that of the bulk, although its precise value depends on the response of the particular metal to the very high shear strains, and possible high temperatures and strain rates, to which it is subjected during abrasion. Better correlation is therefore often found between wear resistance and the hardness of the worn surface than with the hardness of the bulk material.

One consequence of the high strains imposed by abrasion is that any cold

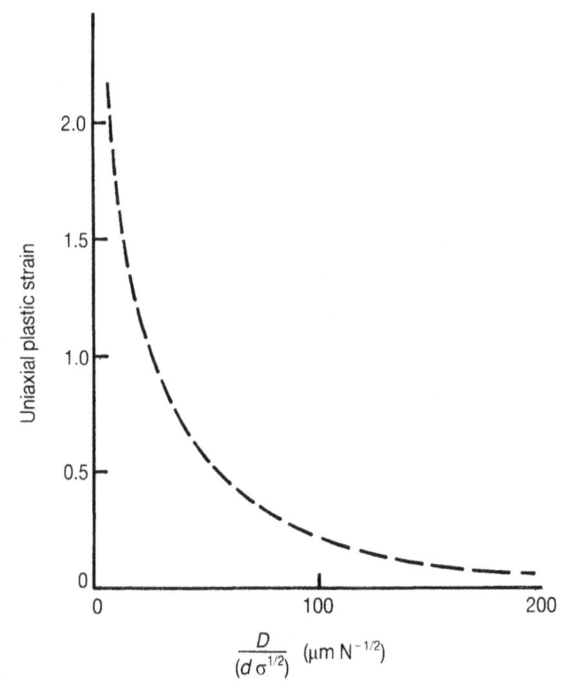

Fig. 6.10 Distribution of plastic strain with depth beneath abraded metal surfaces (here, a copper/silver laminate). In this graph the depth D is scaled by dividing by the product of the particle size d and the square root of the contact pressure σ (the normal load per unit area) (from Moore M A in Rigney D A (Ed.), *Fundamentals of Friction and Wear of Materials*, ASM, 1981, pp. 73–118)

work imposed on the metal before wear will have a negligible effect on its abrasive wear resistance. Although it may raise the bulk hardness, any effect on the wear rate will be swamped by the much greater strains introduced by the wear process itself. This behaviour is seen in Fig. 6.9; cold working of the 0.4% carbon steel, despite causing a significant increase in bulk hardness, has no effect on its wear resistance.

A second important consequence is that alloying will increase the abrasion resistance of a metal only if the strengthening mechanism leads to increased strength at high strains. Fine carbides in steels do cause such an increase, but precipitation hardening by particles which are relatively soft, for example in age-hardened aluminium–copper alloys, does not.

In the model used to derive equation 6.5, we assumed that all the abrasive particles would have the same geometry and remove material from the grooves they form in the same way. In practice, however, particles will have irregular shapes, and will deform the surface in different ways. In particular, abrasive particles can deform the material in ways which lead to the removal of only part of the material displaced from the groove, or even to the removal of no material at all. Figure 6.11 illustrates slip-line fields for three distinct modes of deformation due to a rigid two-dimensional wedge (an idealized

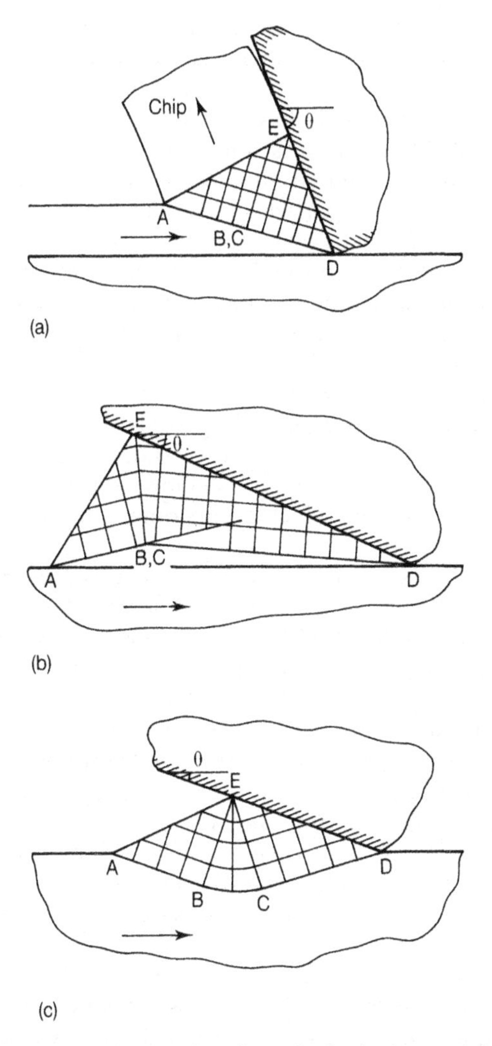

Fig. 6.11 Slip line fields for the deformation of a perfectly plastic material caused by the sliding of a rigid two-dimensional wedge from right to left. Three distinct modes can be identified, as discussed in the text: (a) cutting; (b) wedge formation; (c) ploughing. The angle θ is termed the 'attack angle' (from Challen J M and Oxley P L B, *Wear* **53**, 229–243, 1979)

abrasive particle) sliding over a rigid-plastic material from right to left. Figure 6.11(a) shows the *cutting* mode, in which material is deflected through a shear zone and flows up the front face of the particle to form a chip. This mode of deformation is exactly the same as that caused by a single-point tool in an orthogonal machining process, for example in lathe turning. In this mode, all the material displaced by the particle is removed in the chip. Figure 6.11(c) depicts the other extreme case, the *ploughing* mode, in which a ridge of deformed material is pushed along ahead of the particle, in much the same way that a wrinkle can be pushed along a piece of cloth lying on a table. In the ploughing mode, no material is removed from the surface; material from the raised ridge flows beneath the particle. A simple distinction between these two modes is that in the case of cutting, material flows up the front face of the particle, whereas in ploughing it flows down.

A third mode of deformation is illustrated in Fig. 6.11(b), representing intermediate behaviour. Limited slip, or even complete adhesion, occurs between the front face of the particle and a raised 'prow' of material. It is not possible to construct a steady-state slip-line field for this mode, and Fig. 6.11(b) represents one stage in the development of the prow. The deformation consists of the growth and eventual detachment (by shear along the line AD) of the raised prow, a sequence which is repeated continuously. This mode has been termed *wedge formation* and leads, like the cutting mode, to removal of material from the surface.

The modes of deformation shown in Fig. 6.11(b) and (c) have already been discussed in Section 3.4.4 in the context of sliding friction; but in that case, since asperities on surfaces have very shallow slopes, the possibility of the cutting mode of deformation, which occurs only at high angles of attack, was ignored.

Analysis of the forces involved in each of the three modes of deformation shown in Fig. 6.11 allows the operative mode (that which requires the lowest tangential force) to be identified for any sliding conditions. The controlling factors are θ, the *attack angle* of the particle, and the shear strength of the interface between the particle and the surface. This can be expressed as the ratio f between the shear stress at the interface and the shear yield stress of the plastically deforming material. Perfect lubrication would imply $f = 0$, while complete adhesion would lead to $f = 1$. For $f < 0.5$, only two modes of deformation are possible: cutting and ploughing. Low values of θ favour ploughing, while values greater than a critical value θ_c lead to cutting. For $f > 0.5$, all three modes can operate, with transitions occurring from ploughing to wedge formation and from wedge formation to cutting as the attack angle is increased.

The idealized modes of deformation shown in Fig. 6.11 are two-dimensional. In practice, abrasive grit particles will cause three-dimensional flow patterns which are much more difficult to analyse, but which are nevertheless analogous to the three discussed above. In the ploughing mode, material is displaced both below the particle and to the sides of the groove, and little or none is directly removed. Cutting and wedge formation both lead to wear. The transition from ploughing to the other, debris-forming, modes occurs at a critical attack angle θ_c. Figure 6.12 illustrates these regimes of deformation for the case of a hard spherical indenter sliding over a rigid-plastic material. The depth of penetration of the sphere determines the effective attack angle, which is also plotted in the figure. Although in the rigid-ideal plastic model θ_c depends only on f, for a real material it depends also on the work-hardening rate and on its elastic properties: specifically, on the ratio E/H between the Young's modulus and the surface hardness. θ_c lies typically between 30° and 90° for most metals; a high value of E/H leads to a high value of θ_c.

The distribution of attack angles of the abrasive particles in two-body wear

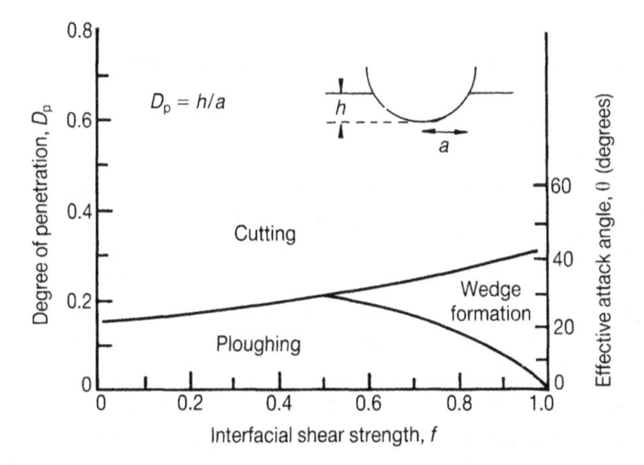

Fig. 6.12 Modes of deformation observed in the sliding of a hard spherical indenter on α-brass, a plain carbon steel (0.45% C) and an austenitic stainless steel (AISI 304). The depth of penetration of the sphere D_p determines the effective attack angle θ which is also plotted (from Hokkirigawa K and Kato K, *Tribology International* **21**, 51–57, 1988)

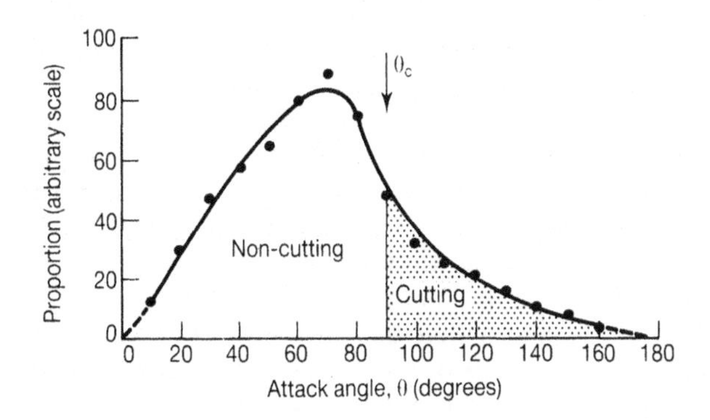

Fig. 6.13 Frequency distribution of attack angle of the contacting abrasive particles in unused 220 mesh size (70 μm) silicon carbide abrasive paper. From curves like this, the proportion of particles for which $\theta > \theta_c$ can be calculated. In this case, $\theta_c = 90°$ has been assumed (from Mulhearn T O and Samuels L E, *Wear* **5**, 478–498, 1962)

depends on the geometry of the abrasive counterface; an example, for silicon carbide coated abrasive paper, is shown in Fig. 6.13. The figure shows how the value of θ_c determines the proportion of the abrasive particles which deform the surface by cutting or wedge formation rather than ploughing, and hence influences the value of K.

We can readily explain the effect of lubrication on abrasive wear by plastic deformation (discussed further in Section 6.3.4), since by lowering the friction between the particle and the surface, lubrication leads to more particles cutting, and hence to a higher rate of wear. The effect of particle shape discussed in Section 6.2.2 can also be understood, since the angularity of the particles controls the distribution of attack angle θ. Angular particles will tend to present higher attack angles—leading to a greater proportion of particles which cut, and thus to a higher wear rate—than more rounded particles. The lower wear rate associated with three-body wear can also be explained, since if the abrasive particles are free to roll between the surfaces then indentation and ploughing will occur more often, and cutting less often, than if the particles are fixed to the counterface as in two-body wear.

When cutting or wedge formation occurs, by no means all the material displaced by the particle is removed as wear debris. Some material will be displaced without being removed, either into side ridges or accommodated by elastic deformation in the bulk away from the groove. The proportion actually removed during cutting or wedge formation varies between different metals. It tends to increase with the hardness of the metal, and so falls with an increase in the ratio E/H defined above. It also decreases with increasing ductility of the metal (defined as the tensile fracture strain or alternatively as the strain at which cracking occurs in rolling experiments).

We have now discussed the influence on K of three factors: the extent of the strain hardening of surface material during abrasion (which causes the surface hardness to become higher than that of the bulk), the fraction of the particles which deform the material by cutting or wedge formation rather than ploughing, and the proportion of the groove volume which is actually removed in cutting or wedge formation. All depend to some extent on the properties of the particles as well as of the metal. A further factor which we incorporated into K in deriving equation 6.5 depends solely on the geometry of the particle, and is equal to $2/(\pi \tan \alpha)$ for a cone. We have seen that experimentally determined values of K for two-body abrasion of metals lie typically between 0.05 and 0.005. Values of this order can be readily understood in terms of the factors we have just enumerated. For angular abrasive particles, α may be around 60°, while for more rounded particles it will be greater. The factor $2/(\pi \tan \alpha)$ will therefore be less than 0.4. Experimental studies suggest that the proportion of particles which remove material by cutting or wedge formation may be between 0.1 and 0.6, and that for those the proportion of the groove volume actually removed may be from 0.2 to 0.9. The combination of these three factors yields values of K from $\sim$0.2 to less than $\sim 8 \times 10^{-3}$, in good agreement with the range actually observed in the two body abrasion of metals, when work-hardening is taken into account.

6.3.2 Abrasive wear by brittle fracture

The second idealized picture of abrasive wear which we shall examine is one in which material removal takes place by brittle fracture, with a negligible contribution from the mechanisms associated with plastic flow, and discussed in the previous section.

As we saw in Section 5.10.2, if a brittle material is indented at a sufficiently high load by a blunt (e.g. spherical) body, and the contact stresses remain elastic, then a Hertzian cone crack will form, as illustrated in Fig. 5.28. If the indenter slides over the surface a series of incomplete conical cracks forms, intersecting the surface in a row of circular arcs (Fig. 5.29). Neither a complete conical crack, nor a row of partial cone cracks, leads readily to material removal. If the slider is a hard angular abrasive particle, however, local plastic deformation can occur at the point of contact and cracks of a different geometry form which can lead immediately to wear.

Figure 6.14 shows how cracks form in a brittle solid subjected to a point load. Cracks like these grow beneath a sharp rigid indenter which generates an elastic-plastic stress field, such as a cone, pyramid or hard irregular grit particle. They are entirely different from the Hertzian cracks which form in the elastic stress field under a blunt indenter.

At the point of initial contact, very high stresses occur. Indeed, if the tip of the indenter were perfectly sharp (i.e. with zero radius of curvature) there would be a stress singularity at this point. These intense stresses (shear and hydrostatic compression) are relieved by local plastic flow or densification around the tip of the indenter; the zone of deformed material is indicated by the letter D in Fig. 6.14. When the load on the indenter increases to a critical value, tensile stresses across the vertical mid-plane initiate a *median vent*

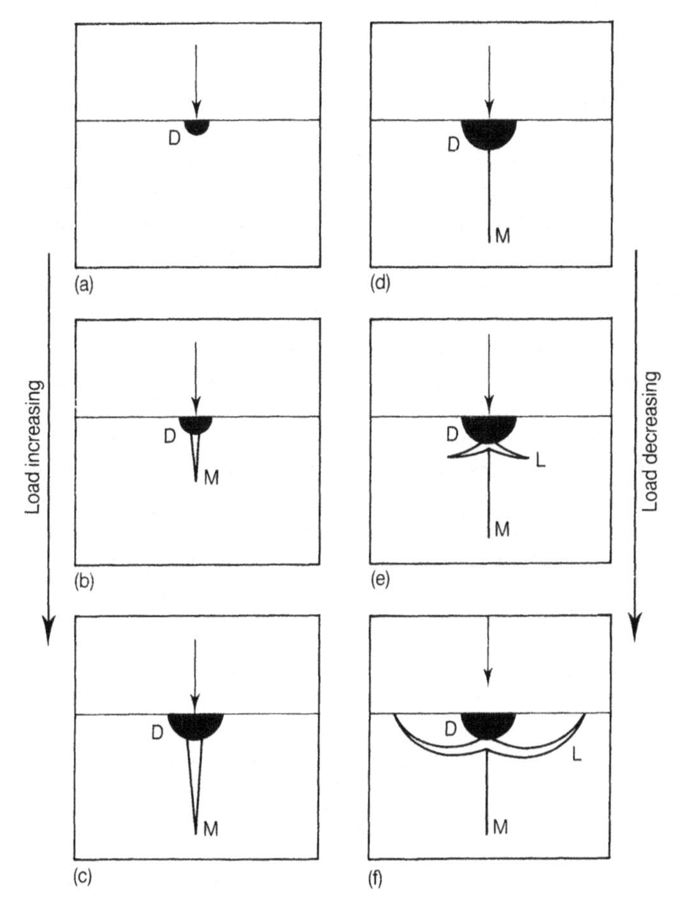

Fig. 6.14 Diagram showing crack formation in a brittle material due to point indentation. The normal load increases from (a) to (c), and is then progressively reduced from (d) to (f) (from Lawn B R and Swain M V, *J. Mat. Sci.* **10**, 113–122, 1975)

crack, indicated by M in Fig. 6.14. Further increase in load is accompanied by progressive extension of the median crack. On reducing the load the median crack closes (Fig. 6.14(d)). Further unloading (Fig. 6.14(e) and (f)) is accompanied by the formation and growth of *lateral vent cracks* (labelled L). The formation of these lateral cracks is driven by residual elastic stresses, caused by the relaxation of the deformed material around the region of contact. As unloading is completed (Fig. 6.14(f)), the lateral cracks curve upwards, terminating at the free surface.

The median cracks, like the Hertzian cone cracks due to a blunt indenter, propagate down into the bulk of the solid with increasing load on the indenter, and do not grow further on unloading. They are not associated in the first instance with the removal of material. Lateral cracks, in contrast, can lead directly to wear.

Lateral cracks form only when the normal load on the indenter has exceeded a critical value, w^*. The value of w^* depends on the fracture toughness of the material, K_c, and on its hardness, H. According to one theory,

$$w^* \propto \left(\frac{K_c}{H}\right)^3 K_c \tag{6.8}$$

K_c is usually taken to be the mode I plane strain fracture toughness K_{Ic}, as measured in conventional notched tensile or bending tests, although the value of toughness applicable to indentation fracture, which occurs under a different stress distribution and at a different size scale, is likely to be significantly different. The ratio H/K_c for a material provides a useful measure of its brittleness: a low value of this *brittleness index* corresponds to a high value of w^*, and therefore indicates a material which is reluctant to fracture on indentation. The quantity $(K_c/H)^2$ has dimensions of length, and can be related to a critical scaling dimension above which fracture will occur during contact by a sharp indenter.

One model for the abrasive wear of brittle materials is based on the removal of material by lateral cracking. As a sharp particle slides over the surface forming a plastic groove, lateral cracks grow upwards to the free surface from the base of the subsurface deformed region (Fig. 6.15), driven

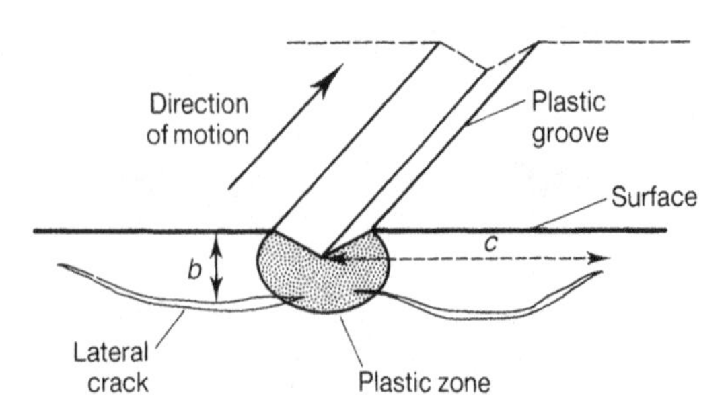

Fig. 6.15 Schematic illustration of material removal in a brittle material by the extension of lateral cracks from beneath a plastic groove (from Evans A G, in *The Science of Ceramic Machining and Surface Finishing II*, Hockey B J and Rice R W (Eds), National Bureau of Standards Sp. Pub. 562, US Govt. Printing Office, 1979, pp. 1–14)

by the residual stresses associated with the deformed material. It is assumed that material is removed as chips from the region bounded by the lateral cracks and the free surface, and the volume wear rate is then estimated from the volume of this region. The sideways spread of the cracks, c, is given by

$$c = \alpha_1 \frac{(E/H)^{3/5} w^{5/8}}{K_c^{1/2} H^{1/8}} \tag{6.9}$$

where w is the normal load on the particle, E and H are the Young's modulus and indentation hardness of the material respectively, and K_c is its fracture toughness. α_1 is a constant which depends only on the particle shape.

The depth of the lateral cracks, b, is assumed to be proportional to the radius of the plastic zone, and given by

$$b = \alpha_2 \left(\frac{E}{H}\right)^{2/5} \left(\frac{w}{H}\right)^{1/2} \tag{6.10}$$

where α_2 is another geometrical constant. An upper limit for the volume removed by lateral fracture, per particle, per unit sliding distance is $2bc$. If N particles are in contact with the surface, each carrying the same load w, then the volume wear rate per unit sliding distance due to all the particles, Q, will be

$$Q = \alpha_3 N \frac{(E/H)\, w^{9/8}}{K_c^{1/2} H^{5/8}} \tag{6.11}$$

where α_3 is a material-independent constant.

Other analyses of the same model lead to slightly different results, since they use alternative methods of calculating c and b. One such approach leads to:

$$Q = \alpha_4 N \frac{w^{5/4}}{K_c^{3/4} H^{1/2}} \tag{6.12}$$

where α_4 is a constant. In this case, Q does not depend on E.

The practical consequences of equations 6.11 and 6.12 are similar, since the ratio E/H does not vary greatly between different hard brittle solids. Both equations predict wear rates that are inversely proportional to both hardness and toughness, raised to powers of about 1/2. Experimental measurements of the abrasive wear rates of some ceramic materials show fair agreement with the predictions of the lateral fracture models. Figure 6.16 shows the correlation between the material removal rate by abrasive machining (i.e. two-body abrasive wear involving brittle fracture) and $H^{-5/8} K_c^{-1/2}$ for several ceramic materials; a very similar correlation is seen with $H^{-1/2} K_c^{-3/4}$.

An important feature of equations 6.11 and 6.12 is that in each case the exponent of w, the normal load on each particle, is greater than 1. This means that the wear rate by lateral fracture is not directly proportional to the normal load, marking an important difference between this mechanism and those mechanisms involving plastic deformation discussed in the previous section. This dependence on load leads to an apparent influence of particle size on wear rate which is quite distinct from the effect on local flow stress discussed

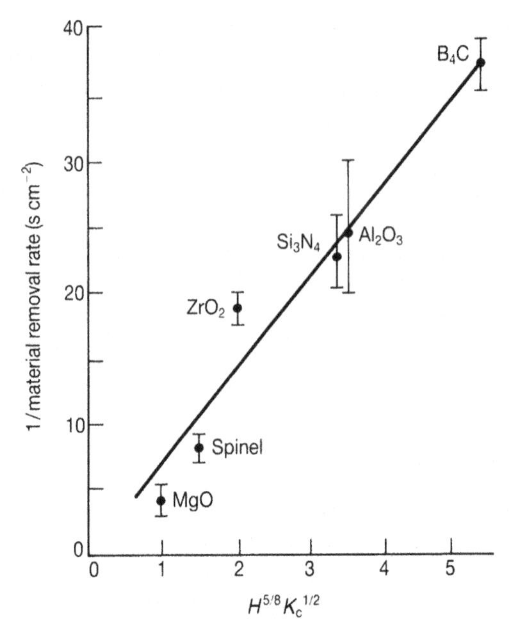

Fig. 6.16 Correlation between the reciprocal of material removal rate in abrasive machining (i.e. two-body abrasion) and the quantity $H^{5/8}\,K_c^{1/2}$, for several ceramic materials (from Evans A G and Marshall D B, in Rigney D. A. (Ed.), *Fundamentals of Friction and Wear of Materials*, ASM, 1981, pp. 439–452)

in Section 6.2.3. If we assume there to be N abrasive particles spread over an area A of the surface, each carrying the same normal load w, then the total applied load W is just Nw. If the particles are each of linear dimension d then N will be proportional to Ad^{-2}. (This supposes that the pattern in which the particles are distributed over the surface scales linearly with d—a reasonable supposition for many cases of abrasive wear.) Taking equation 6.12 to represent the wear rate, it follows that the volume removed per unit sliding distance by all the particles will be given by

$$Q = \alpha_5 \frac{W^{5/4}\,d^{1/2}}{A^{1/4}K_c^{3/4}\,H^{1/2}} \tag{6.13}$$

For a fixed load W and apparent contact area A, we would therefore expect the wear rate to increase with the square root of the particle size.

Several other models for abrasive wear involving fracture have been proposed, based on various assumptions about the geometry and origin of the cracks. Some are applicable only to particular classes of material. For example, Fig. 6.17 shows how wear can occur in grey cast iron, which contains flakes of brittle graphite. The frictional traction imposed on the surface as an abrasive particle slides across it causes cracks to open on the planes of the graphite lamellæ lying normal (or nearly normal) to the sliding direction. Material is then detached by the propagation of shear cracks parallel to the surface, at a depth equal to the depth of penetration of the abrasive particle.

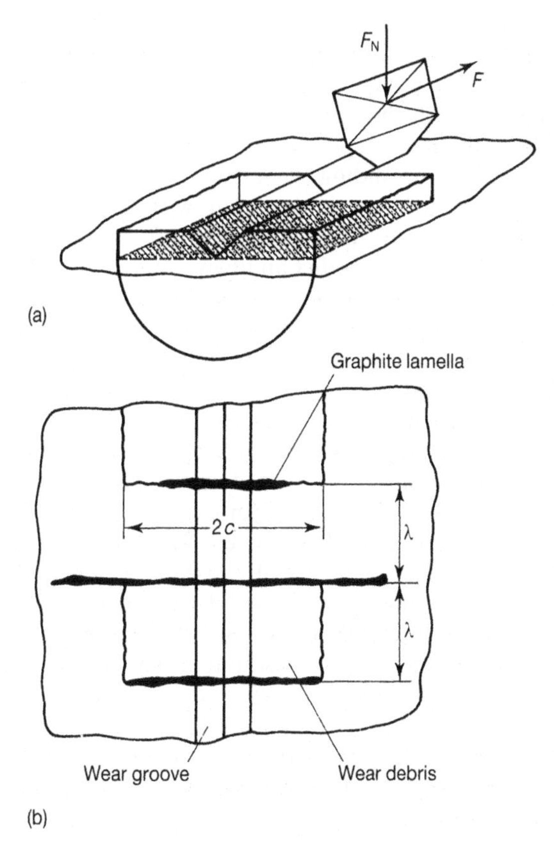

Fig. 6.17 (a) Perspective and (b) plan views showing crack formation modified by the presence of weak graphite lamellæ during the abrasive wear of grey cast iron (from Zum Gahr K-H, *Microstructure and Wear of Materials*, Elsevier, 1987)

The size of the debris fragment is therefore defined by this depth, by the width of the cracks in the graphite flakes ($2c$ in Fig. 6.17(b)), and by the mean separation of the graphite flakes λ. This model leads to a wear rate due to N particles given by

$$Q = \alpha_6 N \frac{w^{3/2} H^{1/2}}{K_c^2} \tag{6.14}$$

where α_6 is a factor which incorporates the coefficient of friction for the system, the geometry of the particles and the spatial distribution of the graphite lamellae. Exactly the same functional dependence of Q on w, H and K_c is found for a similar analysis of wear by fracture in a material containing microcracks or brittle grain boundaries. These analyses are, of course, only valid when the load on each particle is sufficient to cause fracture in the cast iron; this condition is determined by a relationship similar to equation 6.8.

The models for wear by brittle fracture described here have several important features in common. They predict wear rates considerably higher than those due to plastic mechanisms; in Figs 6.15 and 6.17, for example, the volume encompassed by the cracks is much greater than that displaced from the plastic groove. They suggest that the wear rate should increase more rapidly than linearly with the applied load. In this respect the models depart significantly from the linear dependence assumed in the plastic models and in the Archard equation for sliding wear (see Section 5.3). As we have seen above, this leads to an increase in wear rate with abrasive particle size which is quite separate in origin from the size effect associated with plastic deformation. The models all predict an inverse correlation between wear rate and some power of the material's fracture toughness: the dependence on toughness is often stronger than the dependence on hardness. Finally, they all suggest that wear by fracture will occur only when a critical load on each abrasive particle is exceeded.

This last point is important, for it means that if the abrasive particles are sufficiently small and numerous in the contact region, the load carried by each may be below the threshold w^* needed to cause cracking (equation 6.8). This is indeed observed. In soda-lime glass, for example, a load of some 0.1 to 1 N on a single sharp particle is needed to cause indentation fracture. Below this threshold a hard abrasive particle will cause only plastic deformation, and wear will occur by the plastic processes discussed in Section 6.3.1. Plastic scratches up to about 1 μm wide can be formed in glass, conventionally thought of as extremely brittle, without fracture. We therefore see a transition in wear mechanism in brittle materials with increasing load or particle size, which originates from this effect. At low loads or with small particles, fracture may be suppressed and abrasive wear may occur by plastic processes. For higher loads, or larger particles, brittle fracture occurs, leading to a sharply increased wear rate.

6.3.3 Abrasive wear of engineering materials

We have seen how abrasive wear may occur by mechanisms dominated either by plastic deformation (Section 6.3.1) or by brittle fracture (Section 6.3.2). In the first case the hardness of the counterface is an important factor in determining its wear resistance, whereas in the second the fracture toughness is more important, although hardness still plays a role. Even in materials with low fracture toughness, fracture will not always occur; it is favoured by severe contact conditions, for example by large, hard, angular abrasive particles under high normal load. Under conditions where fracture is not important, a fair correlation is found between the resistance of a material to abrasive wear and its indentation hardness. Figure 6.9 showed this behaviour for pure metals and some steels, and Fig. 6.18 illustrates the performance of a much wider range of materials. It must be stressed that the trends indicated in Fig. 6.18 are found only when the abrasive particles are 'hard' compared with the material being abraded; as we saw in Section 6.2.1, in practice this means that their indentation hardness must be at least about 1.2 times that of the surface.

It is evident from Fig. 6.18 that while there are correlations between wear resistance and hardness within each generic group of materials, the same correlation is not found in different groups. For example, a grey cast iron, a

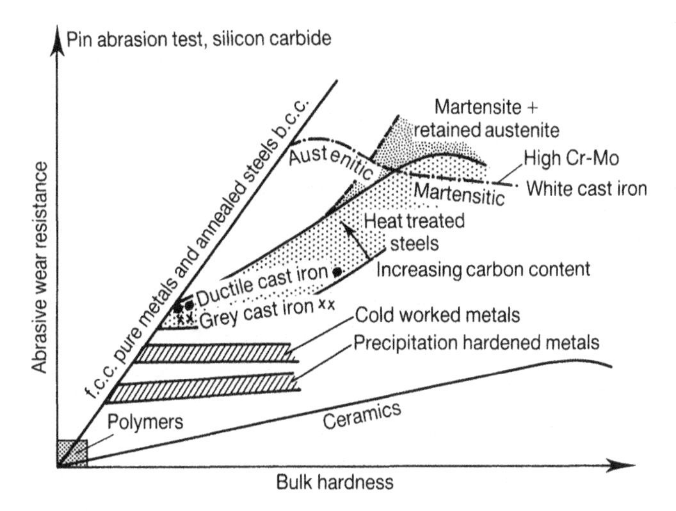

Fig. 6.18 Two-body abrasion resistance (1/volume wear rate) of various materials plotted against bulk hardness (from Zum Gahr K-H, *Microstructure and Wear of Materials*, Elsevier, 1987)

white cast iron and a ceramic may have the same bulk hardness but exhibit very substantial differences in wear resistance. We shall briefly examine the reasons for these differences.

We saw in Section 6.3.1 that if the wear mechanism is dominated by plastic deformation, an important factor in determining the wear rate (via the factor K in equation 6.5) is the fraction of the plastic groove volume removed as wear debris. This fraction varies between different classes of material. Low values of the ratio E/H (the ratio between elastic (Young's) modulus and hardness; see Section 6.3.1) favour deformation by cutting rather than ploughing, which leads to correspondingly large fractions of the groove volume removed as wear debris. We can therefore understand why ceramic materials, for which E/H is lower than for metals, show higher rates of wear at the same hardness. Polymers also have low values of E/H, and thus also have lower wear resistance than metals of the same hardness. As we noted in Section 5.11.2, there is generally a poorer correlation between abrasive wear rate and hardness for polymers, as shown in Fig. 6.19, than with the product of ultimate tensile strength and elongation (as illustrated in Fig. 5.35).

Pure metals of the same crystal structure generally show close proportionality between E and H, and so have effectively constant values of E/H. Thus under given abrasion conditions the value of K varies little for these pure metals, and the wear resistance ($1/Q$) is directly proportional to H as predicted by equation 6.5.

In heat-treated steels, where the elastic modulus E varies insignificantly with composition and microstructure, an increase in hardness leads to a decrease in E/H, and hence to a higher value of K. The benefits gained from hardening steels are therefore not so high as would be expected if K remained constant. A more detailed representation of the behaviour of steels is shown

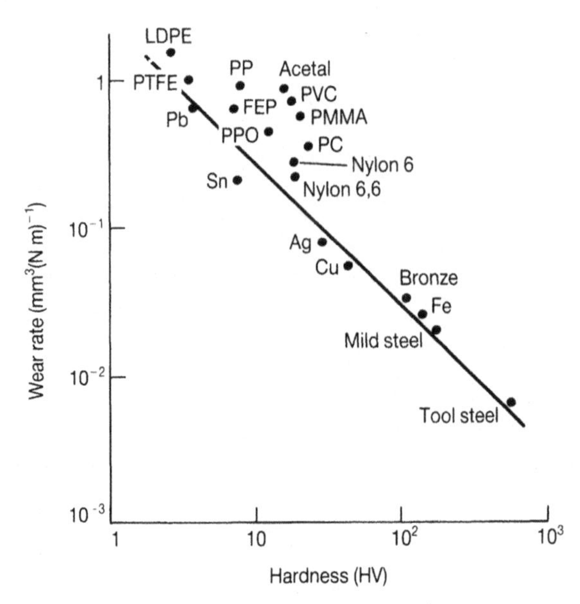

Fig. 6.19 Correlation between two-body abrasive wear rate (with silicon carbide abrasive) and bulk hardness for several polymers and metals (from Evans D C and Lancaster J K, in Scott D (Ed.) *Treatise on Materials Science and Technology*, Academic Press, **13**, 85–139, 1979)

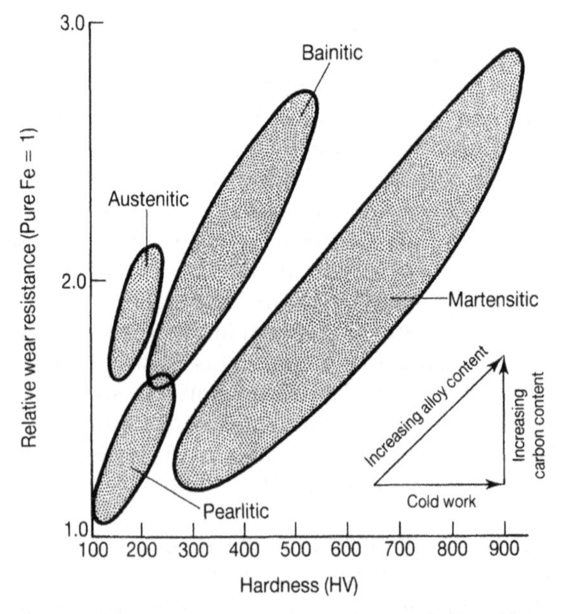

Fig. 6.20 Effect of structure, heat treatment and alloy content on the relative two-body abrasive wear resistance of steels against 90 μm alumina particles (from Moore M A, in Rigney D A (Ed.), *Fundamentals of Friction and Wear of Materials*, ASM, 1981, pp. 73–118)

in Fig. 6.20. As was explained in Section 6.3.1, hardening by cold work has no effect on abrasion resistance, because the wear process itself introduces even more severe surface strains. In contrast, an increase in carbon content, even if the hardness remains the same, leads to higher abrasion resistance. Different trends in the dependence of wear resistance on hardness are seen for steels of different microstructure. Austenitic steels show greater wear resistance at the same hardness than pearlitic or bainitic steels, whereas martensitic steels exhibit lower resistance. These observations can be understood in terms of the ductility and strain-hardening rates of the various microstructures: the lower ductility of martensite compared with austenite leads to removal of a greater fraction of the groove volume as wear debris, and hence to a higher value of K. Pearlitic and bainitic steels show intermediate behaviour. Retained austenite, for example in a martensitic or bainitic structure, is beneficial for resistance to abrasion by hard particles since it strain hardens during abrasion and exhibits high ductility.

Some materials in Fig. 6.18, notably the white cast irons and the ceramics with high hardness, show a decrease in wear resistance with increasing bulk hardness. We can understand this by recalling that under severe contact conditions in brittle materials wear by fracture may occur (see Section 6.3.2). In these circumstances maximum wear resistance will be achieved in a material with intermediate values of hardness and toughness since, in general, materials with high hardness have low fracture toughness, and vice versa. Figure 6.21, which although schematic is supported by experimental data for a wide range of materials, illustrates this point. The figure shows the relationship between fracture toughness and both wear resistance ($1/Q$) and hardness, under severe abrasion conditions. Materials with low fracture toughness, such as ceramics, tend to be hard, and their resistance to abrasive wear by brittle fracture increases with their toughness. Materials of high toughness, such as metals tend, in contrast, to be softer and suffer abrasive

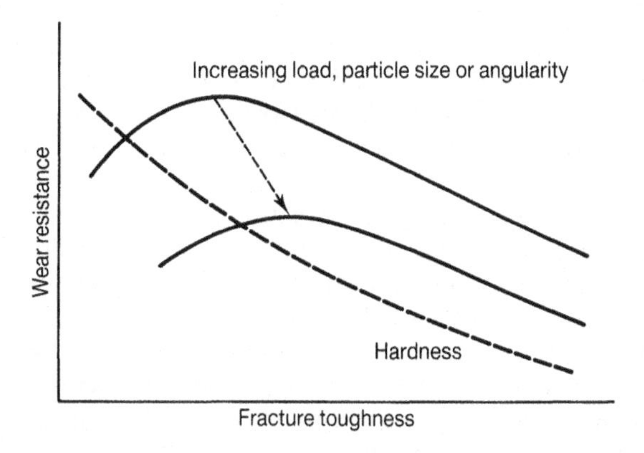

Fig. 6.21 Relation between fracture toughness and resistance to 'hard' abrasive wear for metallic and ceramic materials (from Zum Gahr K-H, *Microstructure and Wear of Materials*, Elsevier, 1987)

wear by plastic deformation rather than brittle fracture. For these materials, wear resistance increases with increasing hardness, and so falls with increasing toughness. At intermediate values of hardness and toughness, in materials such as hard tool steels and white cast irons, both hardness and toughness play important roles. The simplified picture represented by Fig. 6.21 also shows how increasing the severity of the contact conditions, for example by increasing the size of the abrasive particles, their angularity or the normal load, leads to an increase in the extent of the fracture-dominated regime, and to a consequent shift in the peak in wear resistance towards higher values of fracture toughness.

Many materials exposed to abrasive wear contain hard phases within a softer matrix. Such microstructures may be developed during casting or forging, or by heat treatment, or the materials may be composites synthesized from two or more distinct components. For example, hard transition metal carbides (e.g. VC, NbC, M_3C, M_6C, M_7C_3 etc) form important constituents of hard steels, white cast irons (e.g. Ni–Cr, high Cr or high Cr–Mo irons), cobalt based alloys (e.g. 'Stellites') and cemented carbides (cermets, e.g. WC/Co). All these materials are often used for their wear resistance. Composite materials, with polymer, ceramic or metal matrices and harder reinforcing constituents, are also sometimes used in applications where they are exposed to abrasive wear.

The response of all these materials depends on the size of the hard phase regions compared with the scale of the deformation caused by individual abrasive particles. The scale of this deformation can be described either by the width or the depth of the indentation formed by each particle (i.e. a or x in Fig. 6.7). If this dimension is substantially greater than the size of the hard particles (or fibres, in a fibre-reinforced composite) and their separation, then the material will behave very much like a homogeneous solid. This situation is illustrated schematically in Fig. 6.22(a). A finely dispersed hard second phase causes an increase in the flow stress of the matrix (e.g. by dispersion or precipitation hardening) which leads, in general, to increased wear resistance. Thus, there is a direct correlation between carbide volume fraction in quenched and tempered steels and their resistance to abrasive wear. Furthermore, for a given carbide volume fraction the wear resistance is improved by a finer carbide distribution, leading to a shorter interparticle spacing. Coherent or semi-coherent hard phases (formed, for example, by precipitation hardening) have a greater effect on abrasion resistance than incoherent particles.

This general rule that a high volume fraction of hard second phase particles is desirable for abrasion resistance is not, however, universally true. The matrix must also possess adequate toughness. In some materials where the matrix itself is brittle and the particle–matrix interface is weak, hard particles can act as internal stress-concentrators; cracks initiate at the interface and then propagate through the matrix. A high volume fraction of the second phase in such a material then leads to enhanced wear by fracture mechanisms. Behaviour of this type is seen in martensitic tool steels, where the abrasion resistance falls with increasing carbide content.

If the hard phase particles are comparable in size with the scale of the

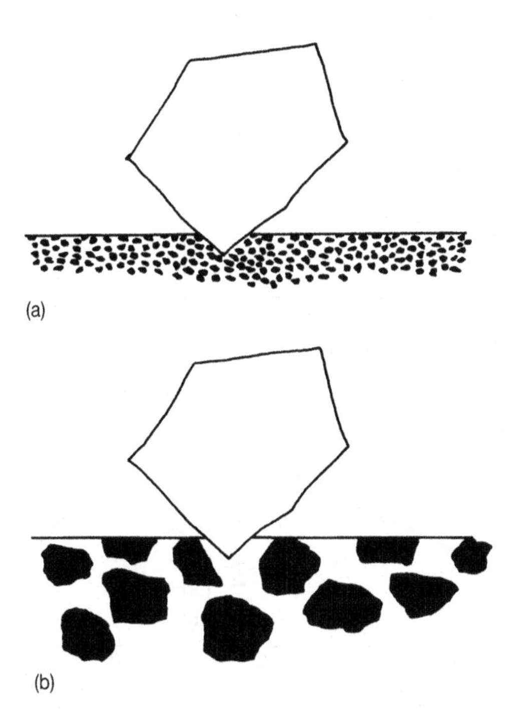

Fig. 6.22 Illustration of the importance of the relative sizes of the particle contact zone and the hard phase regions in the abrasive wear of a two-phase (or composite) material. In (a) the material responds in a homogeneous manner, whereas in (b) the response is heterogeneous (from Hutchings I M, in *Proc. 2nd European Conf. on Advanced Materials and Processes* **2**, Institute of Materials, pp. 56–64, 1992)

abrasion damage, or larger, then the material will respond heterogeneously (see Fig. 6.22(b)). The action of an abrasive particle on a region of the hard phase can lead to plastic flow or to fracture, depending on the load carried by each abrasive particle and on its geometry, as well as on the sizes and the mechanical properties of the abrasive and hard phase particles and of the matrix. The resultant wear rate will also depend on the strength of the interface between the hard phase and the matrix.

An example of the complexities possible in these circumstances is provided by the behaviour of white cast irons. Figure 6.23 shows how the abrasive wear resistance of a range of Cr–Mo white cast irons varies with carbide content. The series of alloys contained varying volume fractions of $(Fe_3Cr_4)C_3$ carbides in an austenitic matrix. The bulk indentation hardness of the alloys increased almost linearly with carbide volume fraction (Fig. 6.23(a)). Pin-on-disc abrasion tests (see Section 6.3.5) with garnet abrasive particles showed fair correlation between wear resistance and bulk hardness (Fig. 6.23(b)): an increase in hardness led to lower wear rate as predicted by equation 6.5. With silicon carbide abrasive, however, (Fig. 6.23(c)) an inverse correlation was found: the wear rate *increased* with carbide content and with bulk hardness. Under these two-body abrasion conditions the hard silicon carbide particles

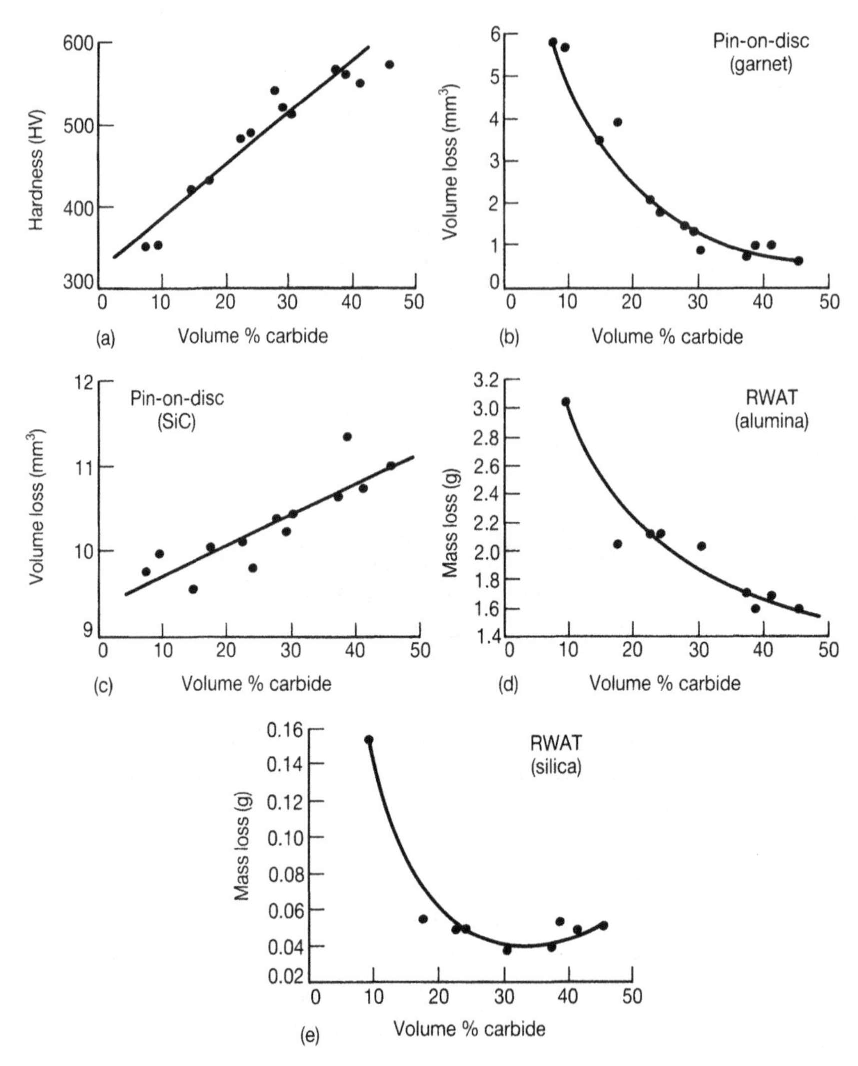

Fig. 6.23 Abrasive wear rates of a range of high Cr–Mo white cast irons, containing different volume fractions of carbides: (a) bulk microhardness of the alloy; (b) pin-on-disc (two-body) tests with 100 μm garnet particles; (c) pin-on-disc (two-body) tests with 90 μm silicon carbide particles; (d) rubber wheel (effectively three-body) tests with 250 μm alumina particles; (e) rubber wheel (effectively three-body) tests with 250 μm quartz particles (data from Zum Gahr K-H and Eldis G T, *Wear*, **64**, 175–194, 1980, and Fulcher J K, Kosel T H and Fiore N, in *Wear of Materials 1981*, Rhee S K, Ruff A W and Ludema K C (Eds), ASME, 1981, pp. 432–438)

penetrated the brittle carbide phase, causing it to crack, whereas the softer garnet particles did not cause fracture. Such behaviour is not simply a function of the hardness of the abrasive material; it also depends on the conditions of the test. In a three-body rubber-wheel abrasion test (see Section 6.3.5) the wear rates of exactly the same white irons due to alumina abrasive fell steadily with increasing carbide content (Fig. 6.23(d)), but with softer quartz abrasive the wear rate reached a minimum at about 30% carbide volume fraction and then rose again. This rise in wear rate was again associated with fracture of the carbides. In this case it was probably caused by preferential wear of the softer matrix by the quartz, which exposed the massive carbide particles and rendered them more vulnerable to fracture.

Cemented carbides (cermets) are often used to resist abrasive wear and form another class of composite materials containing both hard and soft phases: in this case a high volume fraction (typically 60 to 95%) of very hard carbide particles in a softer and more ductile metallic binder. The most commonly used carbide is tungsten carbide, with a hardness of 1900 to 2100 HV, in a cobalt matrix. The bulk (average) hardness of such cermets ranges from 800 to 2000 HV. High hardness is associated with a low binder content and small carbide grains. With such high bulk hardnesses, the abrasion of cemented carbides can often be considered as 'soft' abrasion (see Section 6.2.1). Wear by quartz, for example, with a hardness of about 800 HV, can fall into this category. Appreciably harder abrasive particles, such as silicon carbide or diamond, lead to 'hard' abrasion. Both the mechanisms of wear and the resulting wear rates depend strongly on the ratio of hardness between the abrasive particles and the cermet, and transitions between hard and soft abrasion with different abrasive particles are commonly encountered. Under hard abrasion conditions, abrasive particles cause plastic deformation in both the carbide particles and the binder phase. The contact area for individual abrasive particles is usually much larger than the carbide particle size (typically 0.3 to 3 μm), and wear occurs predominantly by plastic ploughing and cutting, with some local associated fracture in the more brittle composites. A correlation is found between the abrasion resistance and the mean free path in the binder phase between the carbide grains, as shown in Fig. 6.24(a). A short mean free path, due to a high volume fraction of fine carbide particles, gives the highest abrasion resistance. It also leads to high bulk hardness. Figure 6.24(a) also illustrates the effect of the strength of the metallic binder phase. An iron–nickel alloy, with higher strength than cobalt, gives greater wear resistance for the same mean free path. Increasing the strength of the binder further by heat treatment (which forms martensite) causes a still further increase in wear resistance.

In the 'soft' abrasion regime, when the abrasive particles are softer than the cermet, the wear rate is lower and the mechanisms of wear are different. Figure 6.24(b) shows that the wear rate of cemented tungsten carbides by quartz is about one tenth of that due to silicon carbide in the hard abrasion regime. The soft abrasives cannot produce plastic grooves in the carbide particles. Instead, the cyclic normal and tangential forces applied to the carbide grains cause them to move slightly; these repeated small displacements cause gradual extrusion of the metallic binder from between the grains. The extruded binder can then be removed by the abrasive particles by plastic deformation. Removal of the binder leads either to detachment of the carbide particles or to their cracking, perhaps by a fatigue mechanism. As in the case of hard abrasion, there is a strong inverse correlation between wear resistance and the mean free path in the binder, presumably because this influences the ease of extrusion of the binder material.

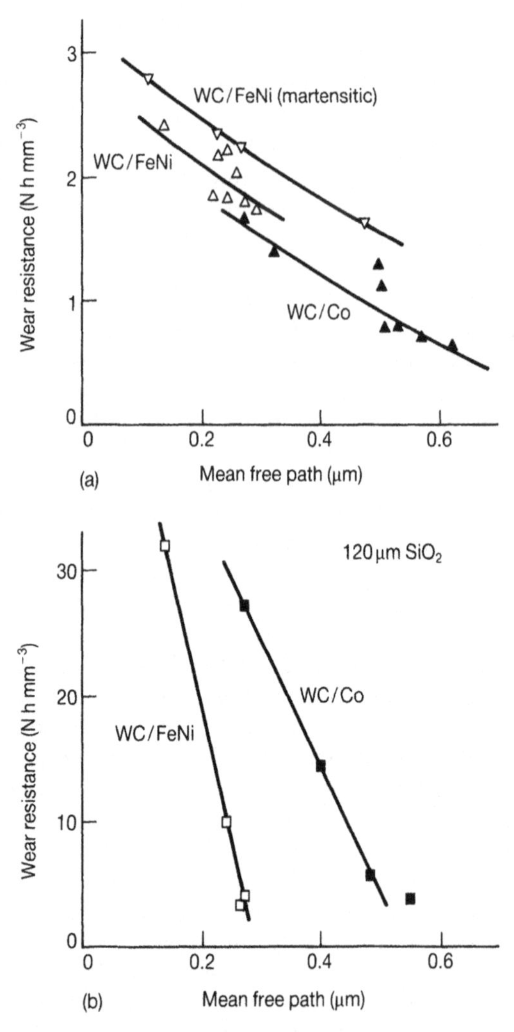

Fig. 6.24 Resistance of tungsten carbide cermets with cobalt and iron–nickel binders to two-body abrasive wear: (a) by silicon carbide particles ('hard' abrasion); (b) by quartz particles ('soft' abrasion) (from Larsen-Basse J, in Viswanadham R K, Rowcliffe D J and Gurland J (Eds), *Science of Hard Materials*, Plenum Press, 1983, pp. 707–813)

The complex behaviour of these composite materials illustrates clearly that 'wear resistance' can never be regarded as an intrinsic material property. The wear rate, and the mechanisms of wear, depend not only on the composition and microstructure of the material, but also to a very important extent on the conditions to which it is exposed.

6.3.4 Effects of lubrication and environment

Lubrication and the nature of the environment (e.g. atmospheric oxygen content or humidity) can both act as powerful influences on wear by abrasion.

The effect of lubrication on abrasive wear contrasts strongly with that on sliding wear. As we saw in Chapter 5, lubrication in sliding wear tends to reduce wear by lowering the tangential stresses on the surface and by decreasing the incidence and severity of asperity contact. Between relatively smooth surfaces, protective lubricant films are readily formed. Abrasive particles, however, will often be larger than the hydrodynamic or elastohydrodynamic film thickness, and the film cannot then prevent contact between the particle and the counterface. In abrasive wear, therefore, lubrication will not cause the large reduction in wear rate seen in the absence of hard particles, and often actually results in an increase. For example, the abrasive wear rates of metals under two-body conditions may be increased by factors of up to three when lubricating oil is applied. The main reason for this effect has already been outlined in Section 6.3.1: the lubricant reduces the friction between an abrasive particle and the metal surface, thereby widening the range of attack angles for which the cutting mode of deformation will occur, and thus leading to increased efficiency of material removal. Lubrication may also have the subsidiary effect of inhibiting adhesion of wear debris to the abrasive particles, so slowing the degradation of abrasivity under multiple-pass conditions by capping or clogging (see Section 6.3.5). For these reasons, enhancement of wear rate by lubrication would be expected whenever plastic abrasion mechanisms are dominant.

The effect of lubrication in brittle materials is less consistent. Although a reduction in the tangential force by lubrication may reduce the tendency to cracking, lubricants may also, through local chemical effects at the crack tip, lead to enhanced crack growth rates. Environmental factors can also provoke a transition under otherwise constant abrasion conditions from fracture-dominated wear to plastic processes. An illustration of the effects of lubricants and other liquids on abrasive wear in a ceramic is provided by a series of tests on alumina (Fig. 6.25). The wear rate was found to vary by a factor of up to ten between different liquid environments; the wear rates in water-soluble oil and in water were lower than in air, whereas those in tertiary amyl alcohol were significantly higher.

For plastics materials and elastomers, the influence of lubrication depends on the dominant wear mechanism. A hard thermoplastic, suffering abrasion by plastic processes, will respond in a similar way to a ductile metal, and lubrication may well increase the wear rate. If the material is an elastomer, and the wear mechanism is one of progressive crack growth driven by surface tractions (see Section 5.11.2), then lubricants will decrease the rate of wear by reducing the frictional forces acting. For this reason, the abrasive wear rates of rubber vehicle tyres in wet conditions are much lower (typically one tenth the value or less) than when dry.

Atmospheric composition can also influence abrasive wear rates, with oxygen content and humidity being potent factors. The effects of the gaseous environment are often complex; even in ductile metals the effects can vary in magnitude and even in direction. In general, oxygen provides a lubricating action in the abrasion of metals. Rapid oxidation of the freshly exposed metal surface at the tip of the abrasive particle produces a thin oxide film with low shear strength and reduces adhesion between the particle and the metal. A

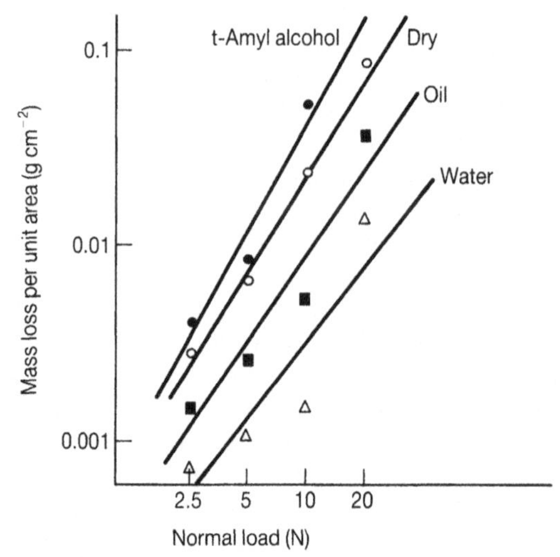

Fig. 6.25 Comparison of two-body abrasive wear rates due to silicon carbide particles of sintered glass-bonded alumina in different environments (from Hines J E, Bradt R C and Biggers J V, in Glaeser W A, Ludema K C and Rhee S K (Eds), *Wear of Materials 1977*, ASME, 1977, pp. 462–467)

similar effect is observed in conventional metal cutting, where aggressive small molecules such as oxygen, chlorine or carbon tetrachloride can react rapidly with clean metal both on the workpiece and on the cutting tool itself and reduce the frictional force on the tool. The abrasive wear of metals in the absence of oxygen is usually less severe than in its presence; not only is cutting less efficient, but in two-body wear, increased adhesion between metallic debris and the abrasive particles also leads to a progressive reduction in wear rate, due to capping and clogging (as discussed in the next section).

There is good evidence for the significant dependence of abrasion rates in air on the level of atmospheric humidity. An extreme example is seen in the abrasion of magnetic recording heads by tape coated with iron oxide (γ-Fe_2O_3) particles, where an increase in wear rate by a factor of up to ten has been noted with an increase in the relative humidity of the surrounding air from 1% to 50%. Much smaller effects occur in laboratory two-body abrasion tests, which are probably related to the influence of humidity on the mechanical properties of the abrasive particles, as well as to a lubricating action similar to that of oxygen. As we saw in Section 5.10.4, environmental factors play an important role in the sliding wear of ceramic materials, and similar effects would be expected to occur just as readily when the ceramic constitutes the abrasive particles as when it forms the abraded surface.

6.3.5 Testing methods for abrasive wear

The most commonly used laboratory tests for abrasive wear employ either a pin-shaped specimen sliding against fixed abrasive, or a rotating wheel sliding

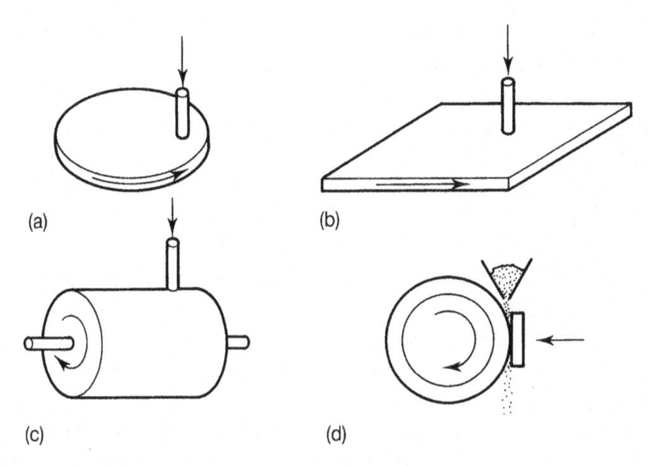

Fig. 6.26 Schematic illustration of four common methods used to measure abrasive wear rates of materials: (a) pin on abrasive disc; (b) pin on abrasive plate; (c) pin on abrasive drum; (d) rubber wheel abrasion test

against a plane specimen with loose abrasive particles being continuously fed between the two.

Figure 6.26(a) to (c) shows three common variants of the method in which a specimen pin slides against fixed abrasive particles, giving two-body wear. Commercial bonded-abrasive paper or cloth is usually used for the counterface, carrying evenly distributed grit particles of a narrow size distribution, bonded to the substrate by a strong resin. The wear rate due to such fixed abrasive particles decreases with repeated passes of the specimen over the same track. Several mechanisms are responsible for this progressive reduction in abrasivity: fracture of particles, leading to a decrease in the number of cutting points; removal of whole particles from the binder resin ('shelling'); rounding of the contacting areas of particles by chemical, mechanical or thermal mechanisms ('attrition'); adhesion of wear debris to the tips of particles ('capping'); and accumulation of wear debris in the spaces between particles so that it carries part of the applied load ('clogging'). The rates and relative importance of these mechanisms vary with specimen material, and also with load, sliding speed, atmosphere and other factors. In order to avoid the problems introduced by degradation of the abrasive, it is often ensured that the specimen always slides against fresh abrasive. In the pin-on-disc geometry (Fig. 6.26(a)), this can be achieved by moving the pin radially on the disc during the test, so that it describes a spiral track. Alternative geometries involve rectilinear sliding over a rectangular sheet of abrasive paper (Fig. 6.26(b)), or moving the pin axially along a rotating cylinder covered with abrasive particles (Fig. 6.26(c)). Apparatus of the latter type is specified in a German national standard procedure for abrasion testing of rubber (DIN 53516), widely used internationally. In each of these cases, a constant load is applied to the pin, often by a dead weight. The wear rate is usually measured by weighing the pin before and after the test, although in some designs an electrical transducer is used to monitor the position of the pin continuously, and hence measure its instantaneous length. In this way a continuous record of wear can be obtained during the test.

Figure 6.26(d) illustrates schematically the second common type of abrasive wear test. The specimen is in the form of a plate or block, pressed under constant load against the rim of a rotating wheel. In its usual form, which has been adopted as a US standard (ASTM G65: *Standard Practice for Conducting Dry-Sand/Rubber Wheel Abrasion Tests*), the wheel consists of a rubber rim of defined hardness (i.e. elastic modulus), moulded on to the surface of a steel disc. In the standard specification the thickness of the rubber rim is 12.7 mm, and the wheel has a width of 12.7 mm and an overall diameter of 228.6 mm; it rotates at 200 r.p.m. The force pressing the specimen against the wheel is also specified. Silica (quartz) particles of a narrow size distribution and from a specified source are fed at a constant rate into the contact region. Wear is measured by weighing the specimen; the standard procedure prescribes the duration of the test, as well as other test variables such as the load, sliding speed, hardness of the rubber and particle feed rate.

The two types of test shown in Fig. 6.26 subject the specimens to rather different conditions. The rubber wheel test involves loose abrasive particles, which indent the compliant wheel during abrasion. Although the particles probably do not roll during contact, they can certainly rotate significantly. In doing so they will tend to move away from 'cutting' orientations in order to minimize the frictional energy dissipation, so that conditions in this test are more similar to three-body abrasion than to two. Measured wear rates ($K \approx 5 \times 10^{-4}$ for metals) certainly compare more closely with values from three-body abrasion tests.

The tests which involve pin specimens sliding on fixed abrasives, in contrast, produce true two-body conditions. In many cases, probably because they are readily available, alumina or silicon carbide abrasive papers are used. As we have seen (in Section 6.2.1), these abrasives are much harder than naturally occurring abrasive particles, and care must be taken in applying the results of wear tests with such hard particles to service conditions in which the abrasive particles are usually much softer. The hardness, shape and size of the abrading particles will all affect the wear rate, for the reasons outlined in the previous sections: they will control the relative importance of the various mechanisms of wear, and the apparent wear resistance of materials may change markedly when the results from different types of wear test, employing different abrasive particles and different conditions, are compared.

A further effect, which contributes to a difference between the results of different types of abrasion test, concerns the compliance of the support provided to the abrasive particles. In the rubber wheel test (Fig. 6.27(a)), particles in contact with harder phases in a two-phase material or composite can penetrate more deeply into the rubber and thus carry less of the total load than if the support is more rigid, as is commonly the case in two-body abrasion tests. There will therefore be a greater tendency for hard brittle phases to fracture with a rigid abrasive support (Fig. 6.27(b)) than with a more compliant backing.

Quite apart from effects due to the mechanical design of the test method,

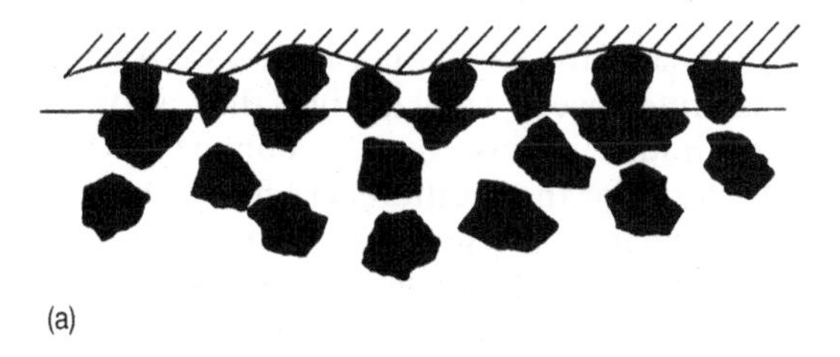

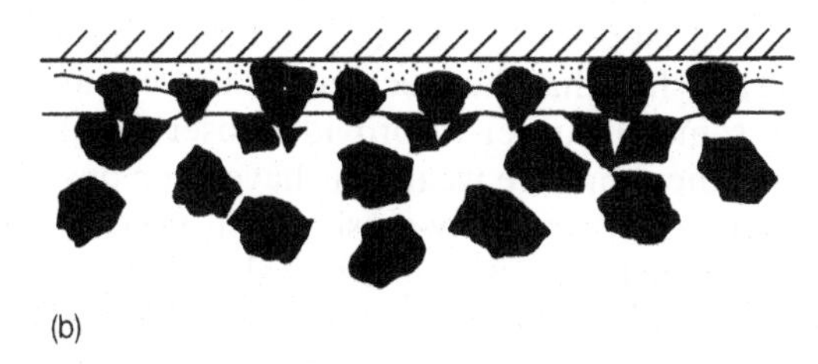

Fig. 6.27 Illustration of the effect of the compliance of the abrasive support in abrasive wear. If the support is compliant (a) the particles in contact with harder regions of the counterface can deflect the support and thus carry less of the total load than if the support is rigid, as at (b) (from Hutchings I M, in *Proc. 2nd European Conf. on Advanced Materials and Processes* **2**, Institute of Materials, pp. 56–64, 1992)

there are strong effects of the abrasive material alone, as shown earlier in Fig. 6.23 and further illustrated in Fig. 6.28. This shows the wear rates for two martensitic ferrous alloys, abraded in a pin-on-plane test (Fig. 6.26(b)). The two materials were a plain carbon steel (0.9% carbon) and a 27% chromium

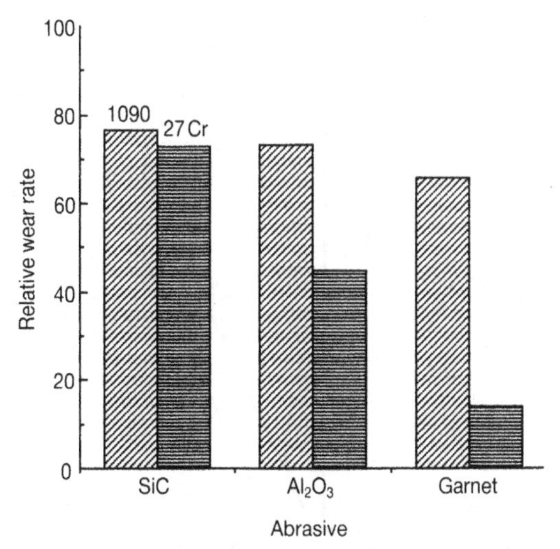

Fig. 6.28 Relative wear rates of two martensitic ferrous alloys (AISI 1090 steel with 0.9% carbon, and 27% Cr 2% Mo white cast iron) with similar values of bulk hardness (735 and 765 HV respectively), measured in pin-on-plane two-body abrasion tests with silicon carbide, alumina and garnet abrasives (from Gundlach R B and Parks J L, in Glaeser W A, Ludema K C and Rhee S K (Eds), *Wear of Materials 1977*, ASME, 1977, pp. 211–216)

white cast iron, with nearly the same bulk hardness. Against silicon carbide abrasive their wear rates were similar, but they showed very different performance when abraded against garnet; alumina gave intermediate behaviour. The main reason for these differences lies in the response of the two materials, and in particular of the hard chromium carbide phase in the cast iron, to abrasives of different hardness. Extensive fracture of the brittle carbide occurs with silicon carbide abrasive (which is harder than the carbide—see Table 6.1), whereas garnet, being softer, does not cause fracture.

As in the case of sliding wear discussed in Chapter 5, care must be taken in applying the results of laboratory tests to practical applications. At the very least, it must be established that the mechanisms of wear under the two sets of conditions are the same. In order to provide closer simulation of particular applications, many other abrasive wear tests have been devised. One example is provided by a laboratory-scale jaw-crusher test, used to simulate the wear by gouging abrasion which occurs in larger-scale mineral processing plant in the mining industry, and adopted as a US standard (ASTM G81). In the jaw-crusher, two flat plates (the jaws) are moved relative to each other in a complex motion. Figure 6.29 shows the movement of the plates in one design of jaw crusher: the ellipses show the motion of the moving plate. The wear rate of the specimen plate is determined by weighing it after a fixed amount of standard rock (typically up to 50 mm in size) has been crushed. The dimensions of the apparatus and the operating conditions must of course be standardized. In order to compensate for the inevitable slight variability in

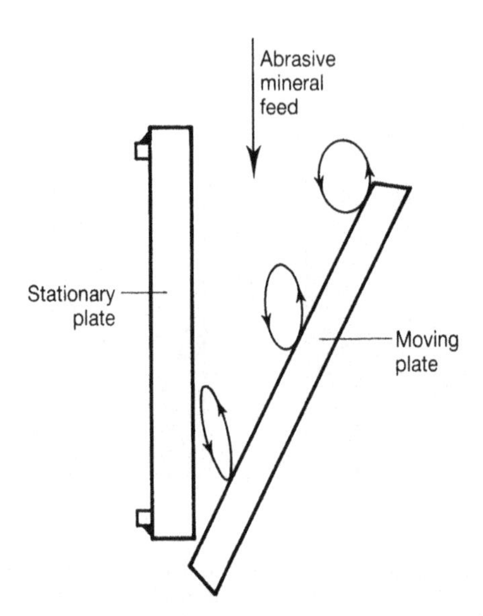

Fig. 6.29 Side view of the moving and stationary plates in one design of jaw crusher wear test apparatus. The ellipses show the motion of the moving plate at the points where they touch the surface (from Borik F and Sponseller D L, *J. Mat.*, **6**, 576–589, 1971)

test conditions, the other plate is made from a standard alloy, and its wear rate is also measured. In this way, the wear rates of the test specimen and the standard material can be determined simultaneously and then expressed as a ratio.

The results of all these types of abrasion test are influenced by many experimental variables, and for test results to be reproducible between different laboratories careful attention must be paid to measuring and standardizing all conditions. In the ASTM rubber wheel abrasion test, for example, which is closely defined, a series of wear rate measurements under the standard conditions on the same material with the same apparatus will typically show a distribution of wear rates with a standard deviation of some 5% of the mean value. Between tests carried out under nominally identical conditions in different laboratories, the standard deviation may be 10% or more. At least part of this variability can be ascribed to the abrasive particles. Particle shape in particular has a strong influence on wear rate (see Section 6.2.2), and some variation in abrasivity will inevitably be found between different batches of abrasive even from the same source. In many cases the use of a standard reference material to calibrate the test conditions, as outlined above for the jaw-crusher test, may be valuable.

6.4 EROSION BY SOLID PARTICLE IMPACT

The wear process known as *solid particle erosion* occurs when discrete solid particles strike a surface. It differs from three-body abrasion, which also involves loose particles, primarily in the origin of the forces between the particles and the wearing surface. In abrasion the particles are pressed against the surface and move along it, usually because they are trapped between two sliding surfaces (Fig. 6.1(b)). In erosion (Fig. 6.1(c)), several forces of different origins may act on a particle in contact with a solid surface. These are shown in Fig. 6.30. Neighbouring particles may exert contact forces, and a flowing fluid, if present, will cause drag. Under some conditions, gravity may be important. However, the dominant force on an erosive particle, which is

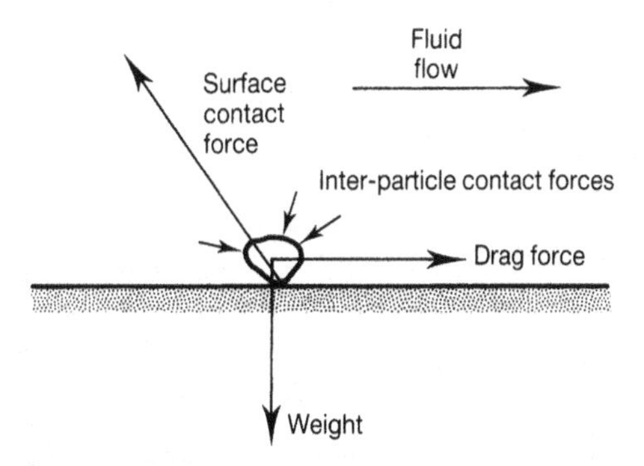

Fig. 6.30 Diagram showing the forces which can act on a particle in contact with a solid surface (from Hutchings I M, *Chem. Eng. Sci.* **42**, 869–878, 1987)

mainly responsible for decelerating it from its initial impact velocity, is usually the contact force exerted by the surface. In abrasive wear, we saw that the amount of material removed depends on the normal load pressing the particles against the surface and on the distance slid (Sections 6.3.1 and 6.3.2). In erosion, the extent of wear depends instead on the number and mass of individual particles striking the surface, and on their impact velocity.

As in the case of abrasion, mechanisms of erosive wear can involve both plastic deformation and brittle fracture. We shall examine these in the following sections. Erosion of metals usually involves plastic flow, whereas more brittle materials may wear predominantly either by flow or by fracture depending on the impact conditions, as discussed below.

6.4.1 Erosive wear by plastic deformation

Before looking in detail at the mechanisms of erosion involving plastic deformation, it is helpful to examine the behaviour of a single hard particle striking a softer surface at normal incidence (Fig.6.31).

We shall make the simplest possible assumptions: that the particle does not deform and that the problem can be analysed quasi-statically (i.e. by ignoring dynamic effects such as wave propagation and strain-rate sensitivity). The only force assumed to be acting is the contact force exerted by the surface. We shall further assume that the deformation of the surface is perfectly plastic, with a constant indentation pressure (hardness) H. At time t after initial contact, the particle of mass m will have indented the surface to a depth x; the cross-sectional area of the indentation at the surface will be $A(x)$, where $A(x)$ is determined by the shape of the particle. The upward force decelerating the particle will be that due to the plastic flow pressure acting over the area $A(x)$, and the equation of motion of the particle can therefore be written as

$$m \frac{d^2x}{dt^2} = -HA(x) \tag{6.15}$$

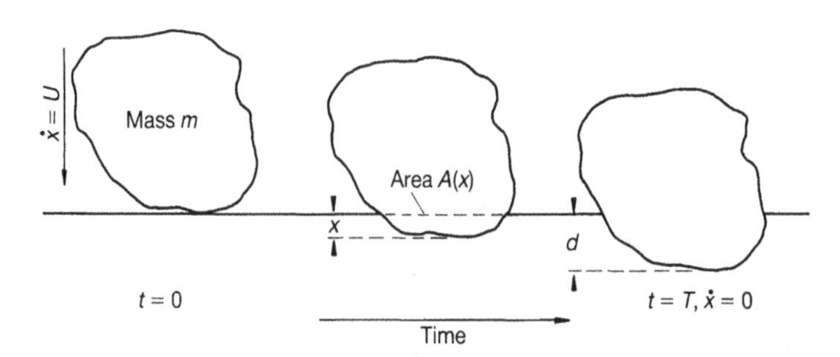

Fig. 6.31 The process of penetration of a rigid particle into the plane surface of a plastically deforming material, at normal incidence. Initial contact occurs at time $t = 0$, and the particle comes to rest at time $t = T$

For simple particle shapes this equation can readily be solved analytically, but for our purposes we wish to know only the final volume of the indentation, when the particle comes to rest at a depth d (Fig. 6.31). At this point the work done by the retarding force will be equal to the initial kinetic energy of the particle, which is assumed to have an initial velocity U:

$$\int_0^d HA(x)\mathrm{d}x = \frac{1}{2}mU^2 \tag{6.16}$$

The final volume of the indentation V is given by

$$V = \int_0^d A(x)\mathrm{d}x \tag{6.17}$$

and therefore, since H is assumed to be constant

$$V = \frac{mU^2}{2H} \tag{6.18}$$

The material displaced from the indentation can suffer several possible fates: it may be accommodated by elastic deformation of material away from the indentation, it may form a rim of plastically deformed material around the indentation, or it may be removed in some way as wear debris. We shall discuss below possible mechanisms by which material can be removed, but for the present shall assume only that some fraction K of the material displaced from the indentation is removed as wear debris. We can therefore write

$$\text{mass of material removed} = K\rho\frac{mU^2}{2H} \tag{6.19}$$

where ρ is the density of the material being eroded and K is a dimensionless factor.

Summation of equation 6.19 over many impacts suggests that the total mass of material removed from the surface should be proportional to the total mass of the erosive particles which have hit it. Figure 6.32 shows how the mass lost from a surface varies with the total mass of erosive particles which have struck it. For some materials, particles may become embedded in the surface and cause an initial mass gain, as shown by curve (b). After this incubation period, which is observed mainly with soft target materials and tends to be more pronounced at high angles of incidence, the erosion proceeds linearly with the mass of grit particles striking the surface. For most ductile target materials and most types of erodent particles, however, any incubation period is negligible and the mass lost from the surface is closely proportional to the total mass of erodent particles which have struck the surface: line (a) in Fig. 6.32 is followed. The linear relationship observed in steady-state erosion allows a simple definition of erosion, E, to be used:

$$E = \frac{\text{mass of material removed}}{\text{mass of erosive particles striking the surface}} \tag{6.20}$$

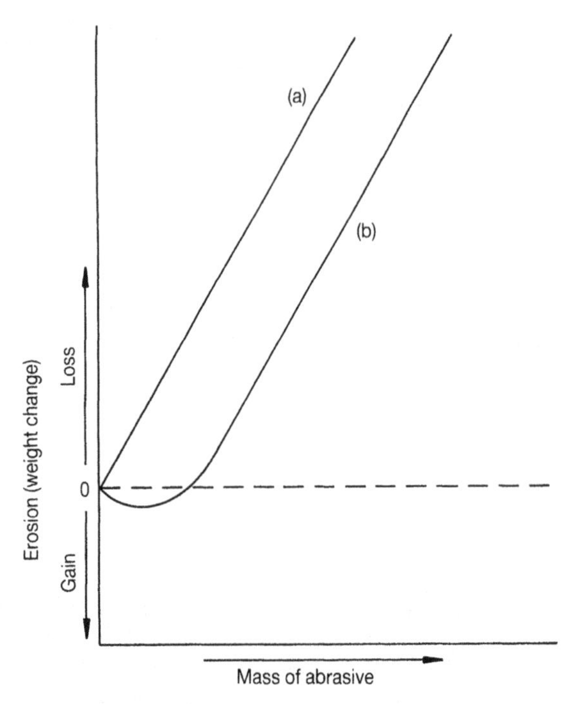

Fig. 6.32 Typical dependence of mass lost from the surface on the total mass of erodent particles which have struck it. Line (a) corresponds to linear erosion behaviour, with no incubation. Under some conditions, however, incubation behaviour (as in curve b) is observed

E is dimensionless, and equation 6.19 becomes

$$E = \frac{K\rho U^2}{2H} \tag{6.21}$$

A comparison of equation 6.21 with equation 6.5 for abrasive wear is instructive. Both predict wear rates which are inversely proportional to H, the hardness of the surface. The role of the applied normal load in abrasive wear (W in equation 6.5) is played in erosion by the quantity $\rho U^2/2$. In both cases, the severity of wear is determined by a dimensionless *wear coefficient* K, which as we saw in Section 6.3.1 is analogous to the Archard coefficient for sliding wear. K can be thought of as a measure of the efficiency of the material removal process; if all the material displaced by the erosive particle were removed, K would be unity. For the erosion of metals, K lies typically in the range 5×10^{-3} to 10^{-1}, very similar to the values observed in two-body abrasion.

Equation 6.21 provides only a crude estimate of the factors controlling erosive wear; it ignores, for example, any variation of erosion rate with impact angle. To improve our understanding we must examine in more detail the interaction between a hard particle and the surface of a ductile material.

The geometry of the deformation due to the impact of a hard particle depends on the impact velocity, on the shape and orientation of the particle and on the impact angle. Impact angles in erosion are usually defined relative to the plane of the surface, as shown in Fig. 6.33. For normal impact, $\theta = 90°$, while at glancing incidence θ tends to zero. The erosion of ductile materials (e.g. most metals) depends strongly on impact angle, as illustrated in Fig. 6.33 (curve a), typically showing a maximum at 20° to 30° and falling to one half to one third of the peak wear rate at normal incidence.

Studies of the impact of single particles on to metals at 30° impact angle show three basic types of impact damage, illustrated in Fig. 6.34. Rounded particles deform the surface by *ploughing*, displacing material to the side and in front of the particle (Fig. 6.34(a)). Further impacts on neighbouring areas lead to the detachment of heavily-strained material from the rim of the crater or from the terminal lip. This type of deformation embraces both the ploughing and wedge-forming modes of abrasion shown in Fig. 6.11(b) and (c). The deformation caused by an angular particle depends on the orientation of the particle as it strikes the surface, and on whether the particle rolls forwards or backwards during contact. In the mode which has been termed *type I cutting* (Fig. 6.34(b)), the particle rolls forwards, indenting the surface and raising material into a prominent lip, which is vulnerable to removal by subsequent nearby impacts. If the particle rolls backwards (Fig. 6.34(c)), a true machining action can occur, in which the sharp corner of the abrasive grain cuts a chip from the surface. This is *type II cutting* and occurs over only a narrow range of particle geometries and impact orientations.

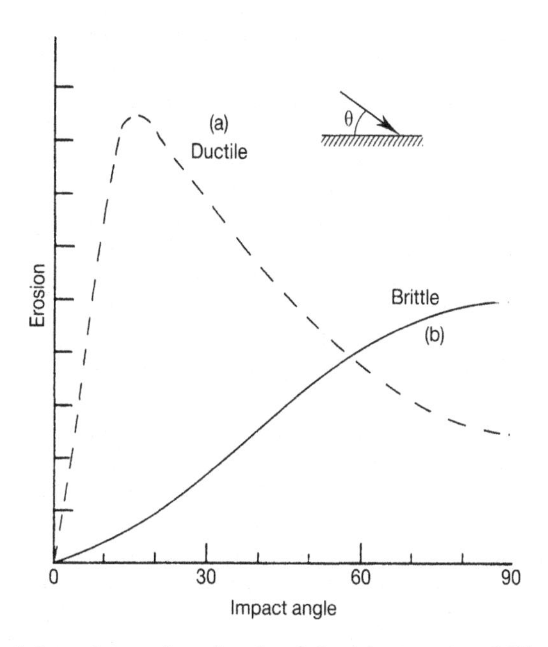

Fig. 6.33 Typical dependence of erosion (as defined by equation 6.20) on impact angle θ (defined as the angle between the impact direction and the surface). Ductile metals commonly show peak erosion at a shallow impact angle (curve a), while brittle materials often show maximum wear for normal incidence (curve b)

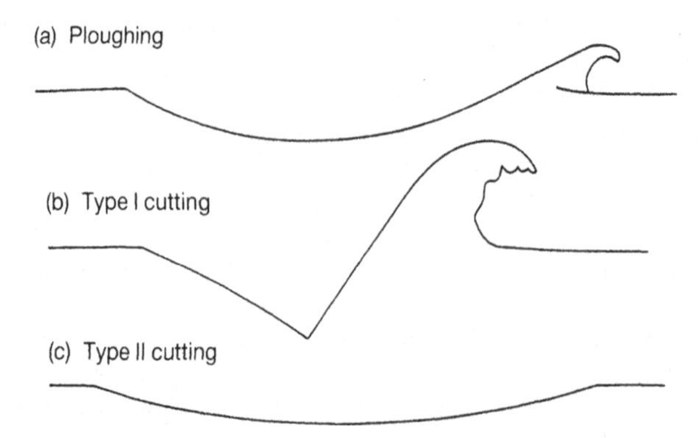

Fig. 6.34 Sections through impact sites formed by hard particles on a ductile metal, showing typical shapes. The impact direction was from left to right. (a) Ploughing deformation by a sphere; (b) type I cutting by an angular particle, rotating forwards during impact; (c) type II cutting by an angular particle, rotating backwards during impact (from Hutchings I M, in Adler W F (Ed.), *Erosion: Prevention and Useful Applications*, ASTM Sp. Tech. Pub. 664, 1979, pp. 59–76)

While clear distinctions can be drawn between these three types of deformation when single particles strike a plane surface, classification is not so simple when randomly oriented particles of irregular shape impinge on a previously eroded and thereby roughened surface. Nevertheless, features similar to those caused by single particle impacts can be distinguished on surfaces eroded by the impact of many particles. Figure 6.35(a) shows, for example, a steel surface eroded by silicon carbide particles. Erosive wear is associated with the detachment by plastic rupture of metal displaced from the impact sites into raised crater rims and lips. Although each impact displaces material from the indentation formed, it will often not become detached as wear debris until it has experienced several cycles of plastic deformation and become severely work-hardened.

The shape of the abrasive particles influences the pattern of plastic deformation around each indentation (as in abrasive wear—Section 6.3.1) and the proportion of the material displaced from each indentation which forms a rim or lip. More rounded particles lead to less localized deformation, and more impacts are required to remove each fragment of debris. An increased impact angle has a similar effect. In the extreme case of spherical particles at normal incidence, material is removed only after neighbouring impacts have imposed many cycles of plastic deformation, and the surface looks very different from one eroded by angular particles (Fig. 6.35(b)).

Despite the evidence that most of the wear debris formed in erosion has been deformed by several impacts, many theoretical models advanced for erosion are basically models for the impact of a single particle striking a plane surface, with empirical extrapolation of the results to the practical case of multiple impact. One such approach extends to oblique impact the simple model for normal impact described above, by solving the equations of motion for a rigid angular particle striking a perfectly plastic material at a shallow

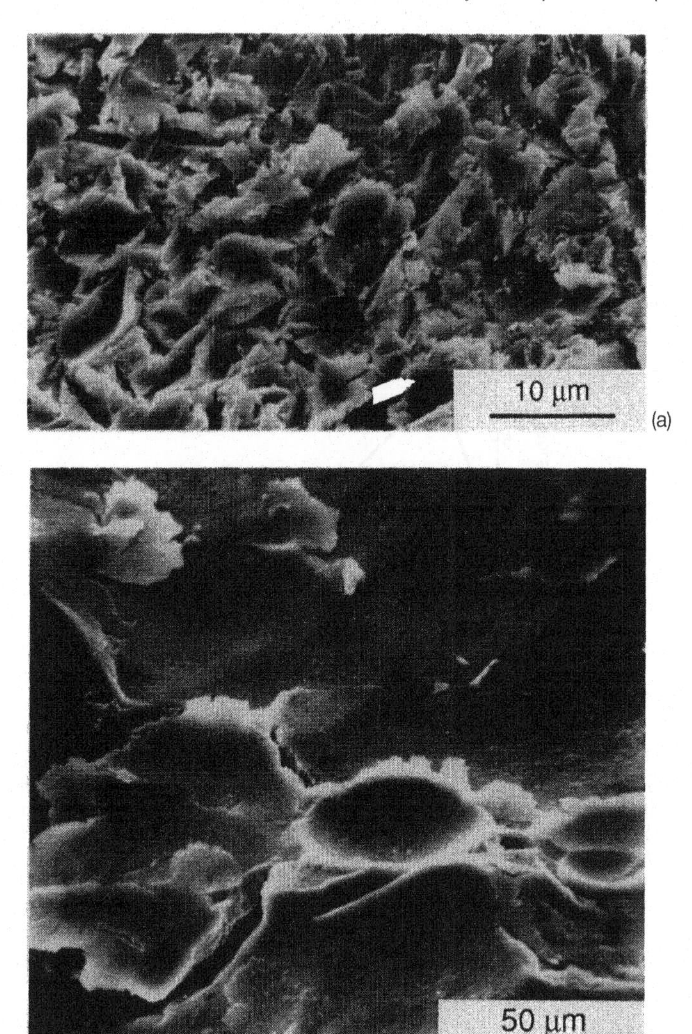

Fig. 6.35 Metal surfaces after erosion by hard particles: (a) mild steel eroded at 30° impact angle and 55 m s^{-1} by angular silicon carbide particles; (b) aluminium eroded at 90° by spherical glass beads at 60 m s^{-1} (from Cousens A K and Hutchings I M, in *Proc. 6th Int. Conf. on Erosion by Liquid and Solid Impact*, Cavendish Laboratory, Cambridge, UK, 1983, paper 41; and *Wear* **88**, 335–348, 1983)

angle, and assuming that material is removed only by a cutting action. Figure 6.36 shows an idealized two-dimensional model of a rigid particle cutting into the plane surface. The volume of material removed is taken to be that swept out by the motion of the particle tip (as distinct from the total volume displaced by the indentation), and the model therefore predicts zero erosion for normal incidence. Various simplifying assumptions are possible: one

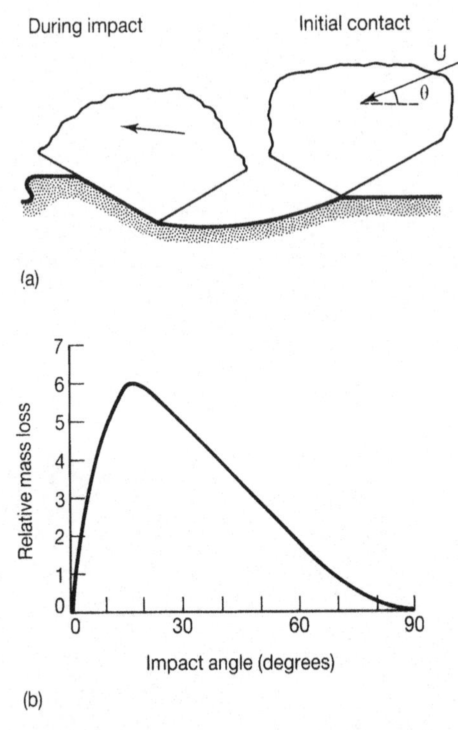

Fig. 6.36 Contact geometry assumed in a theoretical model for the erosion of ductile materials by a cutting mechanism: (a) shows the particle at initial contact and also during impact; (b) shows the angular dependence $f(\theta)$ predicted by the simplest form of the theory (from Finnie I and McFadden D H, *Wear* **48**, 181–190, 1978, and Finnie I, *Proc. 3rd US National Congress of Applied Mechanics*, 1958, 527–532)

method of analysis, in which the forces on the particle are assumed to act at its extreme tip, leads to the expression

$$E = \frac{K\rho U^2}{H} f(\theta) \tag{6.22}$$

The value of K depends on the geometry of the particle and on the fraction of particles actually cutting in an idealized manner (or, alternatively, on the fraction of the volume swept out by the particle tip which is actually removed as wear debris). The function $f(\theta)$ predicted by the theory is shown in Fig. 6.36, and is similar to the experimentally observed curve (a) in Fig. 6.33, although it falls to zero for normal incidence.

A similar but more realistic model, in which the point of action of the forces on the particle is allowed to move during the impact, leads to a more complex expression for E which can be approximated by

$$E = \frac{K_1\rho U^n}{H} f_1(\theta) \tag{6.23}$$

where the velocity exponent n lies typically between 2.0 and 2.5, and is itself a function of the impact angle θ.

The models which lead to equations 6.22 and 6.23 assume that material is removed by individual particles in a cutting action (similar to type II cutting in Fig. 6.34(c)), and are valid only for shallow angles of incidence. At high impact angles debris becomes detached only after repeated deformation, and models which take account of this fact are more applicable. The extreme case of erosion at normal incidence by spherical particles, in which cutting can play no role, may be modelled in two ways: by assuming that surface material becomes detached when the accumulated plastic strain reaches a critical value, or by treating the problem as one of low-cycle fatigue caused by the cyclic plastic deformation associated with successive particle impacts. Both approaches, with suitable assumptions about the mechanics of impact, lead to similar conclusions despite the differences in their initial assumptions. The erosion rate should follow the equation

$$E = \frac{K_2 \rho \sigma^{1/2} U^3}{\varepsilon_c^2 H^{3/2}} \tag{6.24}$$

where σ is the density of the spherical erosive particles, and ε_c is the critical plastic strain at which detachment of wear debris occurs. Equation 6.24 differs from equations 6.22 and 6.23 notably in the higher exponent of the velocity U, and in the fact that two distinct properties of the surface material determine its erosion resistance: not only its hardness H, but also a failure strain ε_c, which can be thought of as a measure of the material's ductility under erosion conditions.

In practice, the erosion of metals does show strong sensitivity to particle impact velocity. The dependence is often expressed in the form

$$E \propto U^n \tag{6.25}$$

and values of n between about 2.3 and 3.0 are commonly reported. The velocity exponent n is nearly always greater than the value of 2.0 predicted by the simpler models, and often lies around 2.4 for ductile metals at impact angles close to that of maximum erosion. There is some suggestion that higher values of n are associated with steeper angles of impact, as shown in Fig. 6.37 for the erosion of copper. This increase may be associated with an increase in the number of particle impacts needed to remove each fragment of wear debris, and to a consequent change in the mechanism from one dominated by single impact events to one better described as fatigue or accumulation of plastic strain.

All the theoretical models outlined above predict that erosion rates due to mechanisms involving plastic deformation should be inversely proportional to the hardness of the material, raised to a power of either 1 (equations 6.21, 6.22 or 6.23) or 3/2 (equation 6.24). Equation 6.24 also predicts a dependence on the ductility of the material. Since these equations also predict that the mass removed per unit mass of particles, E, should be proportional to the

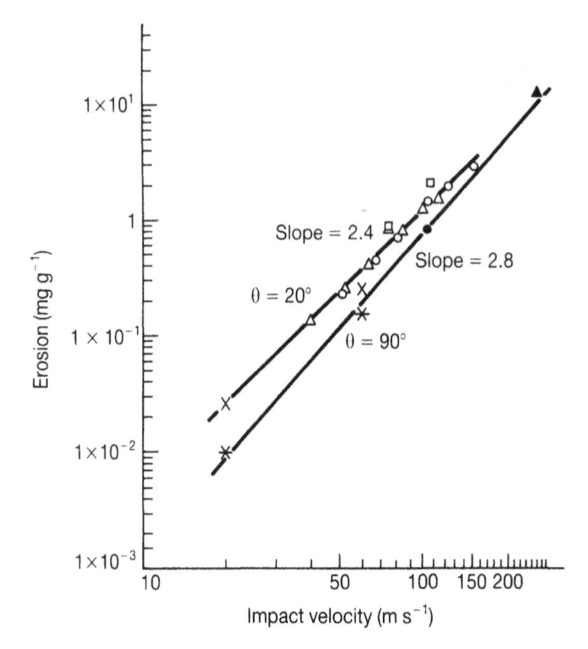

Fig. 6.37 Measurements of erosion rate of copper, from a range of different investigations, plotted against impact velocity (log scales) (from Ives L K and Ruff A W, in Adler W F (Ed.), *Erosion: Prevention and Useful Applications*, ASTM Sp. Tech. Pub. 664, 1979, pp. 5–35)

density of the material ρ, it is helpful when investigating the dependence on hardness alone to plot the quantity E/ρ against hardness. E/ρ represents the volume of material removed by unit mass of particles, and is a particularly useful measure of erosive wear in the context of design (see Chapter 7). Figure 6.38 shows the volume erosion, defined in this way, for several pure metals, plotted against their hardness in the annealed state as well as against the hardness of the eroded surfaces. The surface material becomes heavily work-hardened by particle impacts, and this is reflected in the difference between the two sets of hardness values. Although a fair correlation is found between volume erosion and (hardness of the annealed metal)$^{-1}$, better correlation is seen with (hardness of the eroded surface)$^{-0.6}$. It is the hardness of the eroded surface, of course, which would be expected to determine the steady-state erosion rate according to equations 6.21 to 6.24.

Similar results are shown in Fig. 6.39, in which ρ/E, the reciprocal of the volume wear rate and a measure of erosion resistance, is plotted against indentation hardness. Once again, the linear behaviour predicted by equations 6.21 to 6.23 is found only for some annealed metals. The dependence for work-hardened metals is not linear, and strikingly, the hardening of steels produces no corresponding increase in wear resistance. The results in this figure can be compared with those for abrasive wear shown in Fig. 6.9. Pure metals generally show quite good correlation between hardness and resistance to wear by both abrasion and erosion, although there are exceptions

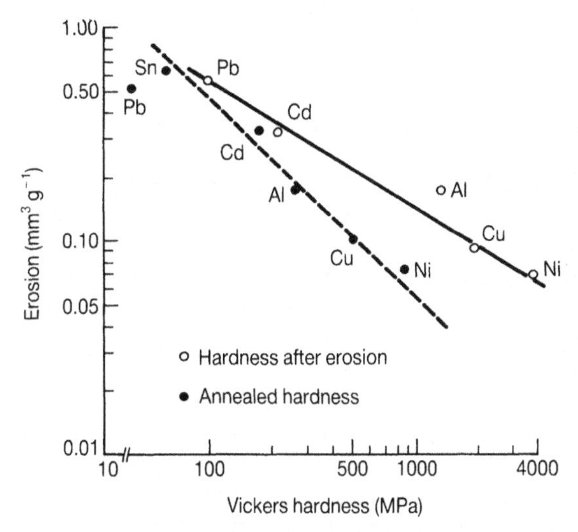

Fig. 6.38 Volume erosion (volume removed per unit mass of erodent particles) for a range of metals, plotted against two different measures of hardness: the hardness of the annealed metal, and the hardness of the surface material after erosion. The erosion tests were performed with 60 mesh silicon carbide particles at 75 m s^{-1}, at an impact angle of 20° (from Sheldon G L, *Trans. ASME: J. Eng. Mat. and Tech.* April 1977, 133–137)

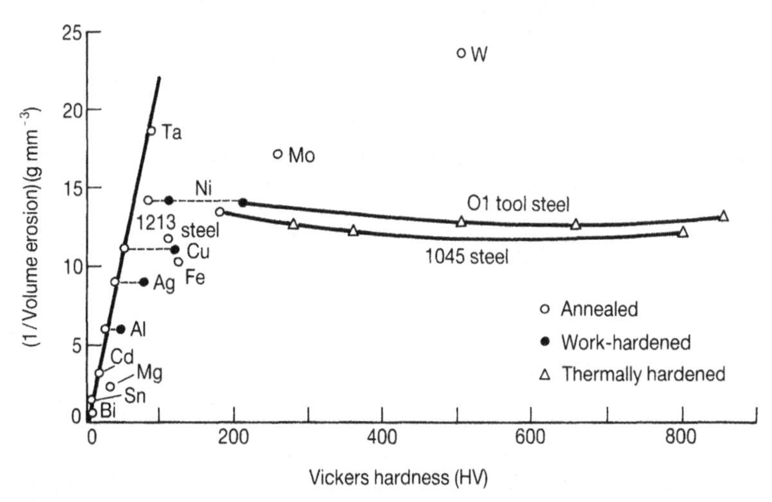

Fig. 6.39 Dependence of erosion resistance (1/ volume erosion in mm^3 g^{-1}) on Vickers hardness for several pure metals and steels, for various states of work hardening and heat treatment (data from Finnie I, Wolak J and Kabil Y, *J. Mat.* **2**, 682–700, 1967)

(molybdenum and tungsten, for example, in Fig. 6.39). Alloys, including steels, show a weaker dependence of abrasion resistance on hardness; against erosive wear they show almost no increase, and in some cases even a decrease, in resistance with increasing hardness.

There are several reasons why bulk hardness is a poor predictor of the erosion resistance of metals, some of which also apply to wear by abrasion and have already been discussed in Section 6.3.1. Better correlation is found between erosion resistance and the hardness of a surface after work-hardening by erosion, as might be expected, but the dependence is weaker than that suggested by theory. The degree to which plastic flow is localized around each particle impact site, which will influence the susceptibility of displaced material to removal, is probably important. This in turn is related to the work-hardening rate of the metal, and also to the ratio between elastic modulus and hardness, as discussed earlier for abrasion. In the case of alloys, the effect of an increase in hardness may well be offset by a decreased strain-hardening rate, which leads to removal of a larger proportion of the metal displaced by each impact.

It is important to remember that the impact of an erosive particle occurs not only on a surface which is heavily strain-hardened, but also over a very short time scale, leading to extremely high strain rates in the deformed material. Some idea of the mean strain rates can be derived from the simple model for normal impact discussed at the beginning of this section. For the normal impact of a sphere on to a rigid-plastic surface, it can be shown that

$$\dot{\varepsilon} \approx \frac{2^{3/2}\,U^{1/2}}{5\pi r}\left(\frac{3H}{2\sigma}\right)^{1/4} \tag{6.26}$$

where $\dot{\varepsilon}$ is the mean plastic strain rate, σ is the density of the sphere and r is its radius. Figure 6.40 shows strain rates derived from equation 6.26 for the impact of silica spheres on to mild steel. The deformation caused by erosive particles will typically occur at strain rates of about $10^6\,s^{-1}$. At such high strain rates, conventional metallurgical strengthening mechanisms such as precipitation, solid solution and grain boundary hardening have much less influence than at lower rates. It is likely that the negligible differences in erosion rate between members of the same alloy family, despite widely different quasi-static hardness, are at least partly ascribable to the very high strain rates involved in erosion.

6.4.2 Erosive wear by brittle fracture

When the impact of an erosive particle causes brittle fracture, material is removed from the surface by the formation and intersection of cracks. As we saw in the case of abrasive wear in Section 6.3.2, even if the mechanism responsible for the detachment of the wear debris is brittle fracture, there will often also be some plastic flow in the material around the point of contact of an angular particle. Although crack patterns caused by impact differ in detail from those produced by quasi-static indentation, the general morphology shown in Fig. 6.14 is nevertheless observed when the material is homogeneous and brittle and the particle is sufficiently hard and angular. More rounded or softer particles tend to cause purely elastic deformation and conical Hertzian fractures of the form shown in Fig. 5.28.

The extent of cracking due to particle impact is most severe when the impact direction is normal to the surface, and erosion under these conditions

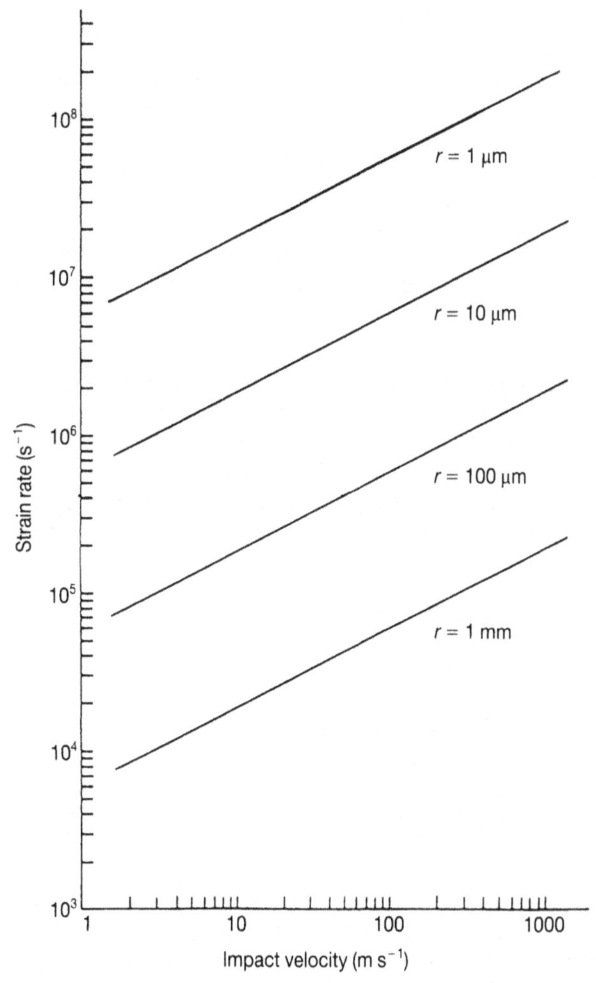

Fig. 6.40 Mean plastic strain rates predicted by equation 6.26 for the impact of silica spheres of different radius on to a mild steel surface (from Hutchings I M, in Levy A V (Ed.) *Proceedings of Corrosion/Erosion of Coal Conversion Materials Conference*, NACE, 1979, 393–428)

is then most rapid. In Fig. 6.33, curve (b) shows the dependence of erosion on impact angle for a typical case where wear occurs by brittle fracture. Erosion falls monotonically with angle away from 90°, in contrast to the behaviour when wear occurs by plastic deformation (curve a). Models for erosion by brittle fracture have therefore been developed mainly for impact at normal incidence. Most of these models assume crack patterns of one of the two types discussed above, and lead to expressions of the form:

$$E \propto r^m U^n \tag{6.27}$$

where r is the radius of the particle.

Models for material removal by the formation and intersection of Hertzian cone cracks suggest that the values of the exponents m and n should be related to the Weibull constants for the material, which describe the statistical distribution of fracture stress among samples or areas of the surface. Such models lead to values of m typically around 1, and a velocity exponent n in the range from 2.6 to 3.0.

For the angular particles which are more commonly encountered in practice, models for erosion by elastic–plastic fracture of the types illustrated in Fig. 6.14 are more applicable. Material is assumed to be removed by the intersection of lateral cracks with each other and with the surface. These models estimate the contact force exerted on the particle during impact, either by assuming that the pressure resisting penetration is the quasi-static hardness of the surface (as in the model leading to equation 6.21), or by using a dynamic wave-propagation model to determine the impact pressure. The contact force is then used in a semi-empirical analysis to predict the extent and depth of the lateral cracks formed, and hence the volume of material removed. The two different methods of estimating the contact force lead to slightly different predictions of erosion rate. The first (quasi-static) method predicts a volume erosion (volume removed per unit mass of erodent particles, E/ρ) given by

$$\frac{E}{\rho} \propto r^{0.7}\, U^{2.4} \frac{\sigma^{0.2}\, H^{0.1}}{K_c^{1.3}} \tag{6.28}$$

The dynamic model leads to the result

$$\frac{E}{\rho} \propto r^{0.7}\, U^{3.2} \frac{\sigma^{0.6}}{K_c^{1.3}\, H^{0.25}} \tag{6.29}$$

There is little practical difference between equations 6.28 and 6.29. Both predict a rather weak dependence of erosion on particle density. The most important material property determining erosion resistance is seen to be fracture toughness (K_c), with hardness (H) being much less significant. In contrast to the models for erosion by plastic deformation, these models predict a dependence of erosion on particle size, r.

In experimental studies of the erosion of brittle materials the mass or volume loss from the surface depends linearly on the total mass of erodent particles, as shown in Fig. 6.32 and discussed in the previous section. Some incubation behaviour may be observed before the steady-state regime. Velocity exponents in the range from 2 to 4 are commonly found, encompassing the range predicted by the theories discussed above. Substantial dependence of erosion on particle size is also seen, with exponents for r commonly lying between 0.7 and 1. Considerable variation of erosion between different materials is found, and attempts have been made to find a correlation with values of the fracture toughness K_c and hardness H measured in conventional mechanical tests. The results do not conform well with the predictions of any one theory. Figure 6.41, for example, shows experimental data for the erosion of a range of brittle materials by angular silicon carbide particles, plotted against the predictions of equations 6.28 and 6.29. Although in both plots there is a fair correlation between the measured erosion and the relevant combination of K_c and H, the slopes of the lines are greater than the value of 1.0 predicted theoretically. For these data the dynamic theory (equation 6.29)

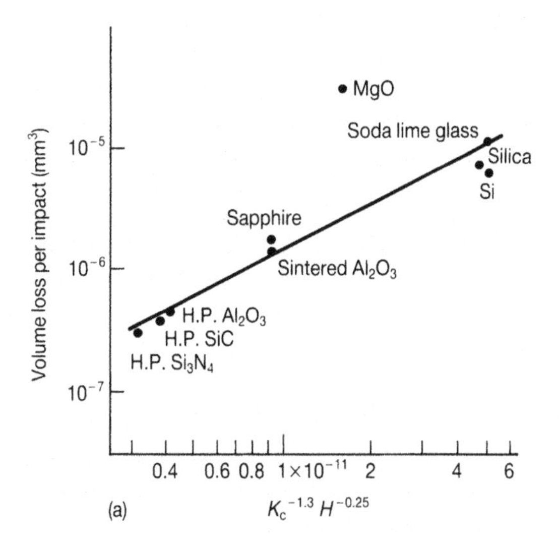

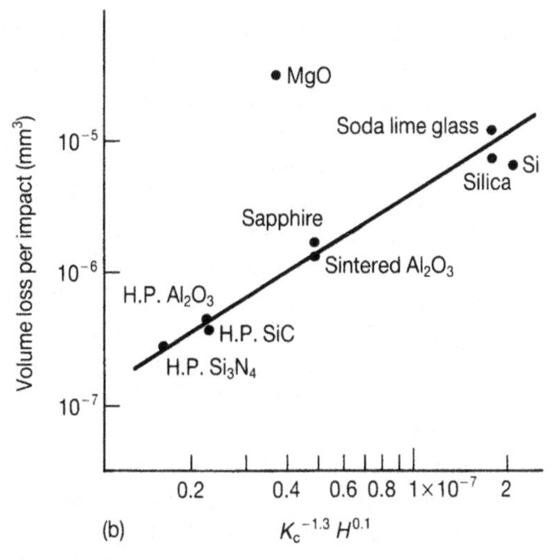

Fig. 6.41 Comparison with theoretical predictions of the measured values of erosion rate (expressed in terms of volume loss per impact) for a range of ceramic materials eroded at normal incidence by 150 μm silicon carbide particles at 63 m s^{-1}: (a) predictions from dynamic theory (equation 6.29); (b) predictions from quasi-static theory (equation 6.28) (from Wiederhorn S M and Hockey B J, *J. Mat. Sci.* **18**, 766–780, 1983)

gives slightly better agreement than the quasi-static theory (equation 6.28), although the difference is probably not significant.

The models so far discussed assume that the material being eroded is homogeneous and isotropic, but some important materials which erode by brittle mechanisms are heterogeneous on the scale of the damage caused by the erosive particles. Examples are cemented carbides (cermets), other composite materials with ceramic or some polymeric matrices, two-phase ceramics such as reaction-bonded silicon carbide, and ceramic materials containing significant numbers of microcracks, pores or inclusions, or weak grain boundaries. So far, no satisfactory predictive models for the erosion of these materials have been developed, and even the mechanisms of material removal are in many cases unclear. The models based on lateral cracking fail to provide accurate predictions of erosion rate. In Fig. 6.41 for example, the inconsistency between the erosion rate of polycrystalline MgO and that of the other materials is associated with the nature of the impact damage in the MgO samples, which cracked readily along the grain boundaries and did not form lateral cracks.

6.4.3 Erosion of engineering materials

We have seen that the dependence of erosion, defined by equation 6.20, on impact angle varies with the mechanism by which wear occurs. If plastic deformation dominates, then maximum wear occurs at a shallow angle (Fig. 6.33 curve (a)), whereas erosion by brittle fracture is most rapid for normal incidence (Fig. 6.33 curve (b)). These two types of behaviour are so characteristic that they are often described as *ductile* and *brittle* erosion behaviour. It is important to note, however, that the angular dependence of erosion is not a characteristic of the material alone, but depends also on the conditions of erosion. For example, although most metals eroded by hard angular particles do show typical 'ductile' behaviour (Fig. 6.33 (a)), erosion by spherical particles, even of a ductile metal such as mild steel, can lead to apparently 'brittle' angular dependence, although wear still occurs by purely plastic processes. Alloys of high hardness and low ductility may also show their maximum erosion rate at normal incidence; an example, for a low alloy bearing steel after different heat treatments, is shown in Fig. 6.42. At comparatively low hardness and high ductility, the steel shows characteristic 'ductile' behaviour, while with high hardness it shows apparently 'brittle' behaviour, although microscopic examination shows no sign of brittle fracture. The terms 'brittle' and 'ductile' in the context of erosion should therefore be used with caution.

Materials conventionally thought of as brittle, such as ceramics, glasses and some polymers, usually suffer peak erosion at normal incidence. However, on erosion by very small particles, nominally brittle materials can show truly ductile behaviour, with material being removed only by plastic deformation and maximum erosion occurring at a shallow angle. This transition from brittle to ductile mechanisms is illustrated in Fig. 6.43 for the erosion of soda-lime glass by silicon carbide particles. The reason for this behaviour is

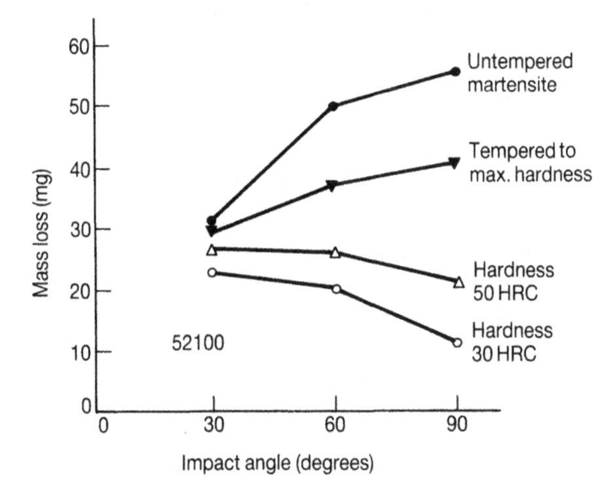

Fig. 6.42 Erosion rate of AISI 52100 steel samples (1% C, 1.5% Cr) with different microstructures, as a function of impact angle. Silica particles were used, at 153 m s^{-1} (from Gulden M E, in *Proc. 5th Int. Conf. on Erosion by Liquid and Solid Impact*, Cavendish Laboratory, Cambridge, UK, 1979, paper 31)

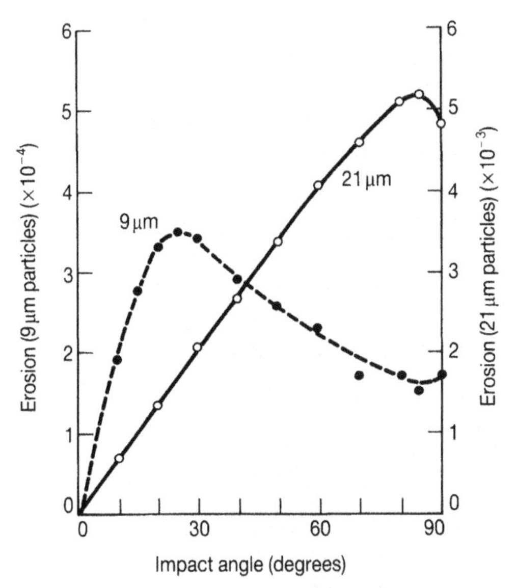

Fig. 6.43 Dependence of erosion rate on impact angle for soda lime glass eroded by 9 μm and 21 μm silicon carbide particles at 136 m s^{-1} (from Sheldon G L, *Trans. ASME B: J. Eng. for Industry* **88**, 387–392, 1966)

exactly the same as that discussed in Section 6.3.2 in the context of abrasive wear. Fracture occurs only when the indentation size produced by each particle exceeds a certain threshold. The indentation size is determined by the particle size and the impact conditions; by reducing the particle size or the impact velocity, it is possible to ensure that the impact events remain plastic, and the material then erodes in the same way as a ductile metal. As we saw in Section 6.3.2, the quantity H/K_c provides a measure of the brittleness of a material, and this ratio has dimensions (length)$^{-1/2}$. The dimensionless group K_c^2/rH^2, where r is the radius of the erodent particle, provides a useful guide to the nature of the dominant erosion mechanism. For low values of K_c^2/rH^2 the material response will be dominated by fracture, while for high values extensive plastic flow will occur.

Metallic alloys show remarkably little difference in erosion rate between members of the same alloy system, despite substantial differences in hardness. Data for two steels were shown in Fig. 6.39; such observations have been confirmed in numerous other studies. Figure 6.44, for example, shows erosion rates under the same conditions for ranges of ferritic, martensitic and austenitic steels, aluminium alloys, nickel alloys and copper alloys. The lack of variation within each alloy system is striking, as is the small range of variation between the different alloy systems. The general conclusion that alloy selection can be of little benefit in reducing erosive wear is true provided the hardness of the alloy is less than that of the erodent particles themselves. If an alloy harder then the erodent can be found, erosion rates may be reduced by the effect discussed in Section 6.2.1 and illustrated in Fig. 6.2.

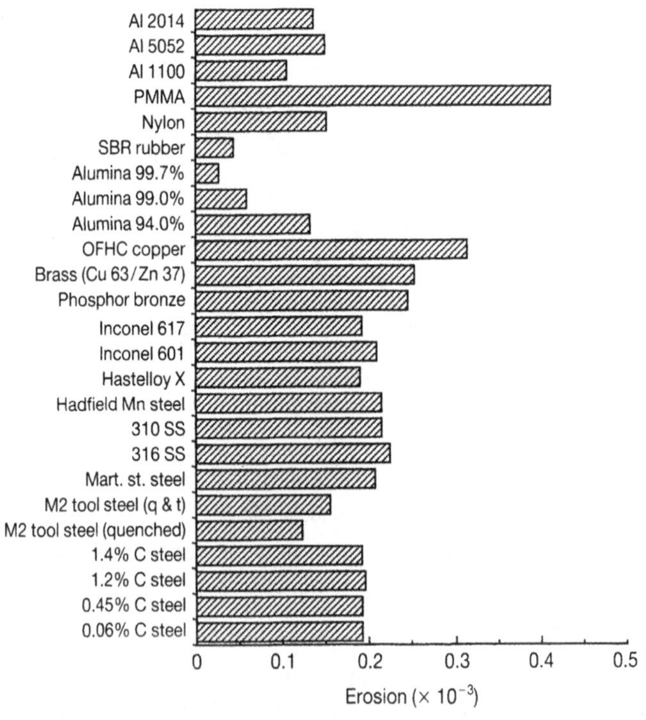

Fig. 6.44 Erosion rates (mass loss per unit mass of erodent) for a range of materials exposed to olivine sand particles (Mohs hardness 6.5–7), 350–500 μm in size, at 45° impact angle and 66 m s^{-1} (data from Söderberg S, Hogmark S, Engman U and Swahn H, *Tribology International*, December 1981, 333–343)

This effect probably explains the relatively low erosion rate of the quenched tool steel shown in Fig. 6.44.

Ceramic materials show considerably greater variation in erosion resistance than metals. We have seen that erosion rates predicted theoretically from measured values of hardness and fracture toughness may be grossly in error if the mechanism of erosion is different from that postulated in the model. Most ceramics are harder than quartz, but nevertheless show significant erosion rates when eroded by this material. Data for the three different polycrystalline sintered aluminas in Fig. 6.44 typify the range of behaviour for these materials. The worst sample showed a rather lower mass loss than the steels, but the corresponding volume loss (because of its lower density) was actually higher. The best alumina ceramic gave a volume loss less than 20% of that of the carbon steels. In these tests the erosion resistance of glass-bonded sintered aluminas correlated with alumina content, but this rule does not hold generally; in other work a material with 88% alumina content has been found to give higher erosion resistance than one with 97.5%. More important, probably, are the grain size, the nature and strength of the grain boundary phase and the proportion and distribution of porosity.

Binder phase content has been found to be important in the erosion resistance of other ceramics and cermets. For example, reaction-bonded silicon carbide containing some 10% free silicon is eroded by alumina particles some eight or nine times as fast as hot-pressed silicon carbide of >99% theoretical density. Similar behaviour is seen in silicon nitride: reaction-bonded silicon nitride containing residual porosity may erode at ten times the rate of pressureless sintered and hot-pressed specimens. At present, although broad trends can be discerned it is not possible to predict the erosion resistance of ceramic materials with any precision. Materials of nominally the same composition and fabrication method produced by different manufacturers via the same route may show completely different behaviour in an erosion test.

Cemented carbides (cermets) have been widely studied. As for the case of abrasive wear (Section 6.3.3), the dominant mechanism of erosion depends on the scale of individual particle contacts. If the erosive particles are small enough, wear can occur by preferential erosion of the metallic binder phase, leading to undercutting and eventual removal of whole carbide grains. If the binder regions are too small to allow this type of localized attack, then overall erosion of the carbide and binder occurs, usually by ductile mechanisms if the erodent particles are small. The combination of these mechanisms leads to a strong and monotonic dependence of erosion rate on binder content, as illustrated in Fig. 6.45 for the erosion of WC–Co cermets by a fine silica slurry (with a particle size down to 1 μm). With larger erodent particles, fracture of the brittle carbide grains can become important, and can lead to a peak in the erosion rate at an intermediate value of binder content. Figure 6.46 illustrates this behaviour for WC–Co cermets eroded by 100 μm silicon carbide particles. At shallow angles of incidence, a general trend of increasing erosion rate with binder content is seen, although not so strong as that shown in Fig. 6.45. For normal incidence, however, a sharp peak in erosion rate is seen at a cobalt binder content of ~10 wt.%. Under these conditions the dominant

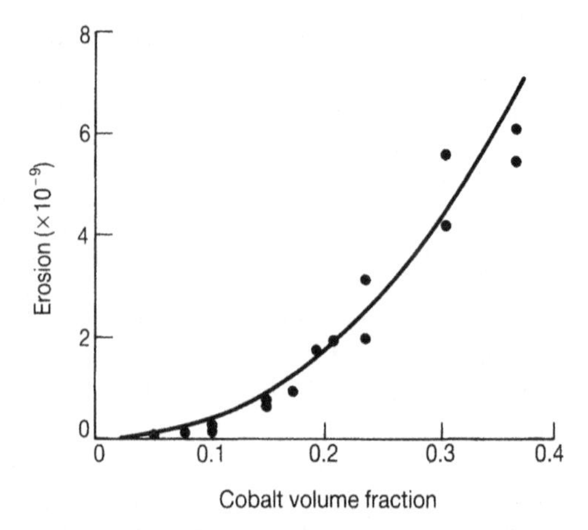

Fig. 6.45 Dependence of erosion rate on volume fraction of cobalt binder phase for a series of tungsten carbide/cobalt cermets eroded by a slurry of 10 μm silica particles in oil, at 133 m s^{-1} and a nominal impingement angle of 90° (from Wright I G and Shetty D K, in Field J E and Dear J P (Eds.), *Proc. 7th Int. Conf. on Erosion by Liquid and Solid Impact*, Cavendish Laboratory, Cambridge, UK, 1987, paper 43)

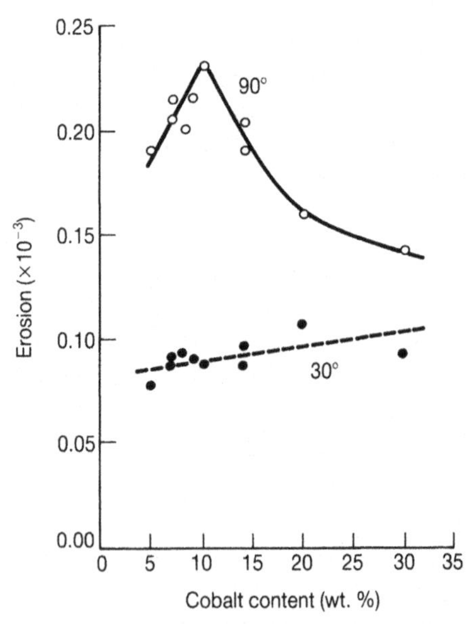

Fig. 6.46 Dependence of erosion rate on volume fraction of cobalt binder phase for a series of tungsten carbide/cobalt cermets eroded by airborne 100 μm silicon carbide particles at 40 m s^{-1} and at impact angles of 30° and 90° (from Pennefather R C, Hutchings R and Ball A, in Field J E and Dear J P (Eds.), *Proc. 7th Int. Conf. on Erosion by Liquid and Solid Impact*, Cavendish Laboratory, Cambridge, UK, 1987, paper 60)

mechanism of erosion is fragmentation of the carbide grains, which is most prevalent at this level of binder content.

Polymers, including plastics materials and elastomers, show very wide variation in erosion resistance. Four polymers are included in Fig. 6.44. Their erosion rates, in terms of mass loss under the conditions of the test, range from some three times that of the steels (in the case of acetal) to only one quarter (styrene-butadiene rubber). The mechanisms and angular dependence of erosion vary correspondingly widely. Ductile thermoplastics, such as nylon, acetal, polycarbonate, polypropylene and polyvinylchloride erode by mechanisms involving plastic deformation, as for metals, and experience maximum wear at shallow impact angles. More brittle thermoplastics and thermosets such as polystyrene, PMMA and some epoxy resins erode by brittle fracture, and show maximum erosion rates at high angles. As for other materials eroding by brittle fracture, a transition may occur, with sufficiently small erosive particles, to wear by plastic mechanisms.

Some elastomers show excellent resistance to erosion at normal incidence and moderate velocities, but they show greater sensitivity to both impact angle and velocity than either conventionally ductile or brittle materials. Figure 6.47 illustrates the dependence of erosion on impact velocity for a natural rubber compound, which shows a very steep rise in erosion above about 50 m s^{-1}. For this material the erosion at 90° impact angle is around one quarter that at 30°. The mechanisms of erosion in rubbers are similar to those of abrasive wear (see Section 5.11.2), with tearing, crack propagation and fatigue all playing a role. Attempts to correlate erosion rate with tensile strength are generally unsuccessful, nor does the Ratner–Lancaster correla-

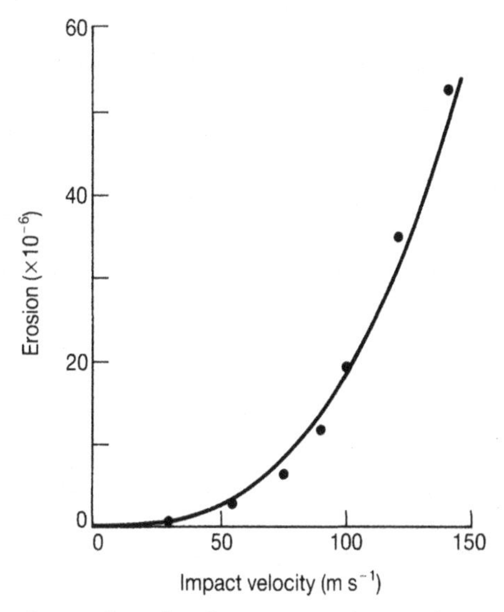

Fig. 6.47 The dependence of erosion (mass removed per unit mass of erodent) on impact velocity for natural rubber, eroded by 120 μm silica sand particles at 30° impact angle (from Arnold J C and Hutchings I M, *J. Natural Rubber Research*, 1992)

tion (Section 5.11.2) appear to be useful. High rebound resilience, however, (defined as the relative height of rebound of a hard sphere striking the rubber surface) has been found to be associated with good erosion resistance in a range of unfilled rubber compounds, and in some systems a low elastic modulus also leads to a low erosion rate.

6.4.4 Testing methods for erosive wear

Laboratory-scale erosion testing is performed for several reasons: to provide data on absolute and relative wear rates under specific conditions, to examine the validity of theoretical models, and to study the mechanisms of wear. The first of these objectives gives information of direct value to the design engineer, while the others are of more value in improving our understanding of erosive wear. For the results of a test to be useful, the impact conditions (particle velocity, flux and impact angle) must be closely defined, and the particles and material being tested should be well characterized.

Methods commonly used for laboratory erosion testing can be divided into those in which the particles are accelerated in a gas or liquid stream, and those where circular motion is used to achieve the impact velocity. Figure 6.48 shows schematic diagrams of four types of testing method. In the method shown in (a), particles are accelerated in a fluid stream along a nozzle which may be either parallel-sided or of more complex shape. They then strike the target material which is held some way from the end of the nozzle at a fixed angle. This test, often called the *jet impingement* or *gas-blast* method, can be used with gas-borne particles (usually with air as the carrier gas) or with a liquid slurry, often with oil or water as the fluid medium. The few national

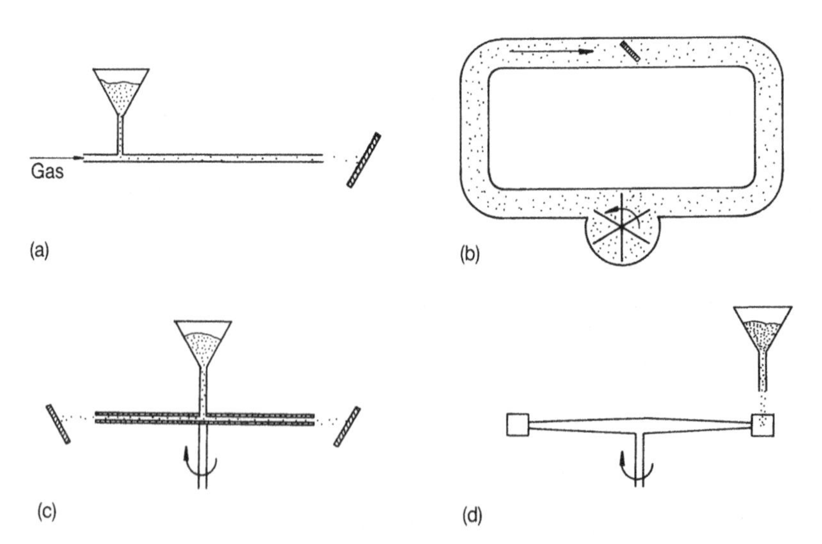

Fig. 6.48 Schematic illustration of four different methods of exposing specimens to erosive wear in laboratory tests: (a) jet impingement method; (b) recirculating loop; (c) centrifugal accelerator; (d) whirling arm rig

standards for erosion testing by airborne particles (e.g. ASTM G76, DIN 50332) all employ gas-blast methods. Nozzles of widely different dimensions can be used for different applications: practical tests have covered the range from about 1 mm to 50 mm diameter. A small nozzle results in a correspondingly small area of wear on the specimen, but requires only low feed rates of fluid and particles. A larger nozzle is more suitable for testing materials with heterogeneous microstructures of large scale, such as some coarse-grained ceramic materials and composites, but demands a correspondingly greater supply of fluid and erodent. Impact velocities in the range from 10 to 100 m s^{-1} are readily achievable, although with long acceleration distances in the nozzle higher speeds are possible for gas-borne particles. In most tests of this type, particles are used only once, obviating any problems of particle degradation by impact. With very large nozzles, however, the quantity of erodent required may make recirculation of the particles essential, and care must then be taken to ensure that the erosion rate does not fall during the test due to fragmentation or attrition of the erodent particles.

Problems of particle degradation also occur in the *pumped loop* or *recirculating loop* test shown in Fig. 6.48(b). Here, a two-phase flow of particles and fluid (gas or liquid) is driven around a loop of pipework. The method is valuable for establishing the wear rates of pipework components such as bends or valves in pneumatic or hydraulic conveying systems. It can also be used more directly to examine the behaviour of materials by completely immersing specimens in the flow. Although the schematic diagram suggests that the flow passes through the pump, in some practical designs wear in the pump is avoided by separating the particles from the fluid and then recombining them after the fluid has been pumped.

The method shown in Fig. 6.48(c), sometimes called a *centrifugal accelerator*, uses circular motion to generate a continuous stream of particles; it is usually used in air or vacuum. The erodent particles are fed into the centre of the rotor, and move outwards along radial tubes or channels, leaving the rotor at a speed governed by the peripheral speed of the rotor. Their motion is not tangential to the rotor because by the time they reach the rim they have acquired some radial velocity, although in some designs steps are taken to reduce the radial velocity to negligible proportions. Stationary specimens are arranged around the rim of the rotor, and the method can be used to compare the erosion behaviour of up to twenty or so different specimens simultaneously.

In the apparatus shown in Fig. 6.48(d), two specimens at the ends of a balanced rotor move at high speed through a slowly falling stream of particles, striking them at the peripheral speed of the rotor and at an angle determined by the orientation of the specimens. This *whirling arm* test is usually carried out in vacuum to eliminate aerodynamic effects on the particles and reduce the power needed to turn the rotor. A similar principle is used in the *slurry-pot* test in which a rotor carrying specimens is immersed in a tank containing a slurry of liquid and particles; in this type of test, however, the impact angle and velocity are difficult to define, and it is better suited to comparison of materials than to absolute measurement of erosion rates.

As we have seen in the previous sections, the particle impact velocity is the most important variable which influences erosion rate. In erosion testing it is particularly important that the velocity be held constant, and preferably known accurately. In slurry erosion tests, it is sometimes reasonable to assume that the particle velocity is the same as that of the liquid, which can often be measured by simple methods. In the jet impingement test, for example, the volume of slurry leaving the nozzle of known dimensions in a fixed time provides a direct measure of the exit velocity of the slurry jet. In most test methods for airborne particles, however, independent measurement of the particle velocity is necessary. Multiple flash photography, laser Doppler velocimetry (LDV) and the rotating disc method are all potentially useful. For reasons of economy the third of these methods, illustrated in Fig. 6.49, is often used. The stream of particles passes through a slot in one disc of the pair, which rotate together at high speed on a common shaft. The particles strike the second disc and make a mark. A second mark is made by allowing the particle stream to pass through the slot with the discs stationary. The displacement of the first mark relative to the second is then simply related to the speed of rotation of the discs and the time of flight of the particles over the distance separating the discs. The random error in velocities measured in this way is typically $\pm 10\%$, although there can be a systematic error due to the aerodynamic influence of the rotating discs themselves on the particle stream. This error may be $\sim 10\%$ or even greater; it will be more severe with small particles and with particles of low density. Figure 6.50 shows the systematic error in the double-disc measurements for 50 μm particles of silicon carbide and alumina, determined by more accurate measurements of particle velocity by LDV. The rotating disc method consistently underestimates the true velocities of these small particles by some 10 to 15%. In a jet-impingement apparatus the distance separating the specimen from the end of the nozzle can also change the impact velocity, since the gas flow close to the specimen surface becomes deflected and will change both the speed and direction of motion of the particles. Here again, LDV provides a much better estimate of the true impact velocity than the double-disc method.

Uncertainty in impact velocity is the source of the greatest error in erosion tests, which even when repeated under nominally identical conditions with

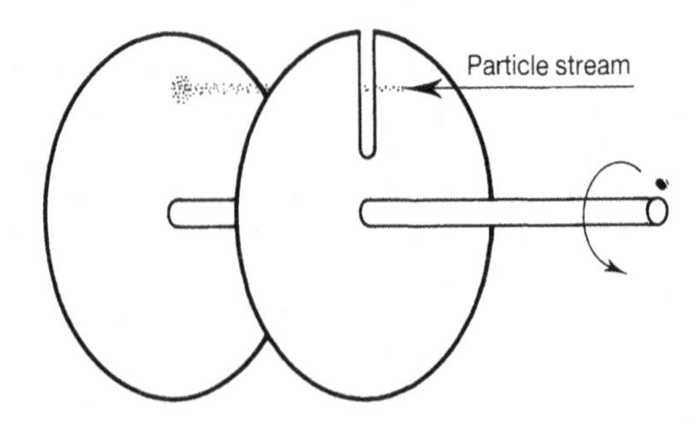

Fig. 6.49 Schematic illustration of the double-disc method of measuring particle velocity in gas-borne particle erosion experiments (after Ruff A W and Ives L K, *Wear* **35**, 195–199, 1975)

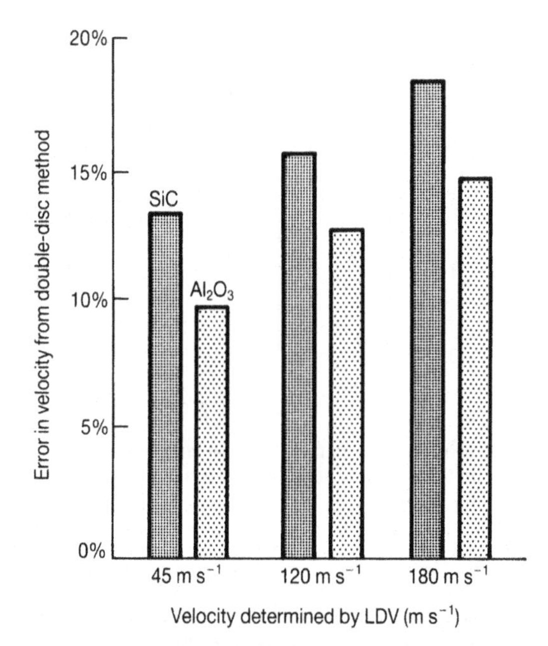

Fig. 6.50 Comparison of particle velocities measured by the double-disc method and by laser Doppler velocimetry, for 50 μm silicon carbide and alumina particles accelerated in a gas-blast apparatus. The percentage error in the double-disc measurements is plotted against the mean velocity derived by LDV (from Ponnaganti V, Stock D E and Sheldon G L, in Bajura R A (Ed.), *Polyphase Flow and Transport Technology*, ASME, 1980, pp. 217–222)

the same materials in different laboratories can give significantly different results. Figure 6.51 shows the results from a series of well-controlled tests on a low-carbon steel carried out in this way by five different research groups. The mean difference in measured erosion between the five laboratories is 24%, most of which can probably be accounted for by uncertainties in velocity measurement. A variation of only some 8 to 10% in velocity would account for a variation of this order in erosion.

A variable which is important and sometimes ignored in erosion testing is the flux of particles striking the surface, either measured as the number of particles, or their total mass, impinging on unit area per unit time. Since erosion tests are usually conducted for a fixed time or with a fixed mass of erodent particles, the flux will determine the number of impacts to which an individual point on the surface is exposed, and also the time interval between successive impacts. In practical cases of erosive wear the particle flux may be low; a laboratory test may nevertheless be carried out at a higher flux in order to obtain measurable wear in a reasonable time. This type of test is called an *accelerated test*. If the results of an accelerated test are to be applied to predict erosion under practical conditions, it must be established that the effect of particle flux on erosion is not significant. For most materials at moderate fluxes this is true. However, at very high fluxes erosion may be affected by interactions between particles striking the surface and those rebounding from

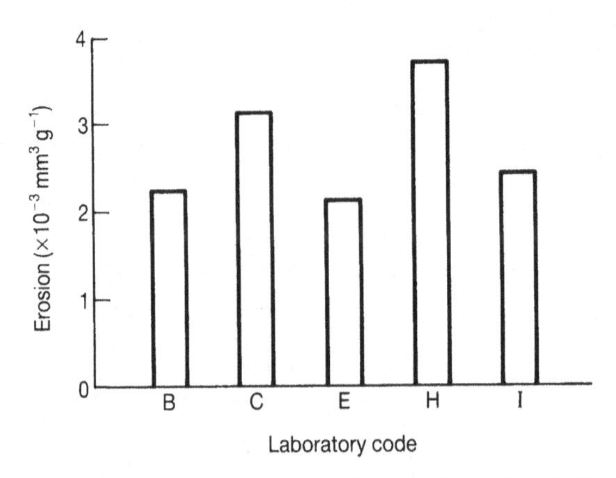

Fig. 6.51 Results of erosion tests carried out under nominally identical conditions in five different laboratories, on 0.2% carbon steel with 50 μm alumina particles at 30 m s^{-1} (from Ruff A W, *Wear* **108**, 323–335, 1986)

it, or possibly by thermal effects as the surface heats up under a high rate of kinetic energy dissipation. Such effects may become important at fluxes above about 1 kg m^{-2} s^{-1} on steels, and at rather lower fluxes in materials with lower thermal diffusivity. High fluxes should therefore be avoided in laboratory erosion tests.

A further effect of particle flux on erosion measurements may occur when chemical reactions take place on the eroded surface. These reactions may, but need not, be conventional corrosion processes. For example, the erosion of natural rubber has been found to depend strongly on particle flux at low fluxes, probably due to local oxidative attack of the material at the particle impact sites; this dependence can be eliminated by the incorporation of an anti-oxidant into the rubber formulation.

Incubation behaviour (see Section 6.4.1) can lead to misleading results in erosion testing if it is not recognized and allowed for. It is dangerous, for example, to test a material for a single time period, or with a single quantity of erodent, and consider the resultant mass or volume loss to represent a trustworthy measure of erosion. Test procedures should be used which will identify any incubation period, and provide an accurate measurement of the rate of wear in the steady-state regime after incubation.

The erodent particles used in a laboratory test should be selected carefully if the results are to be used in a practical application. Not only should the particle size be the same as that in the practical case, but the particle material and angularity should also be the same. As we saw for abrasive wear (Section 6.3.5), the material of the particles can influence the wear mechanisms as well as the rate of wear; in an extreme case even the ranking of materials in terms of relative erosion rate can be changed by the use of different erodent particles.

Further reading

Briscoe B J and Adams M J (Eds), *Tribology in Particulate Technology*, Adam Hilger, 1987
Friedrich K (Ed.), *Friction and Wear of Polymer Composites*, Composite Materials Series, Vol. 1, Elsevier, 1986
Larsen-Basse J, The role of microstructure and mechanical properties in abrasion, *Scripta Met. and Mat.*, **24**, 821–826, 1990
Miller J E and Schmidt F (Eds), *Slurry Erosion: Uses, Applications and Test Methods*, Sp Tech. Pub. 946, ASTM, 1987
Misra A and Finnie I, A review of the abrasive wear of metals, *Trans. ASME: J. Eng. Mat. and Tech.* **104**, 94–101, 1982
Preece C M (Ed.), *Erosion, Treatise on Materials Science and Technology*, **16**, Academic Press, 1979
Rabinowicz E, *Friction and Wear of Materials*, John Wiley, 1965
Rigney D A (Ed.), *Fundamentals of Friction and Wear of Materials*, ASM, 1981
Ritter J E (Ed.), *Erosion of Ceramic Materials*, Trans Tech Publications, 1992
Zum Gahr K-H, *Microstructure and Wear of Materials*, Tribology Series No. 10, Elsevier, 1987

7

Wear and design

7.1 INTRODUCTION

Wear, as a factor in the performance of an engineering system, is often neglected in design. Yet, as we saw in Chapter 1, wear leads in many cases to important maintenance or replacement costs and to associated loss of production or availability, which also carry economic penalties. In addition, wear often causes reduced efficiency in the operation of a machine before it has progressed far enough to justify replacement or overhaul. Wear should, therefore, always be considered early in the design process, and not just as an afterthought once service trials have shown it to be a problem.

The designer has two main concerns in the context of wear: to establish whether significant wear will occur in service, and, if so, to take steps to reduce the rate of wear to acceptable levels within the economic and other constraints imposed on the design. The meanings of the terms 'significant wear' and 'acceptable levels' clearly depend on the details of the system being designed, as also do the steps which can be taken to reduce wear. In order to propose methods of reducing wear, the designer needs to establish the mechanism by which it will occur, and also to understand the factors which control wear by that mechanism. One of the aims of the earlier chapters of this book has been to provide a basic understanding of such wear mechanisms.

The preceding two chapters have described several types of wear which commonly occur in mechanical systems. Where solid surfaces slide over each other, wear can occur by several different mechanisms, depending on the state of lubrication, the nature of the sliding surfaces, the chemical environment and the operating conditions, such as normal load and sliding speed. As we have seen, the relative displacement of the surfaces need not be intentional or at all large: even when they are nominally fixed, a slight relative oscillatory motion may occur leading to fretting and associated wear. Wear under all these conditions has been discussed in Chapter 5.

When the contact predominantly involves rolling rather than sliding, as in rolling element bearings, surface damage and wear occur by rolling contact fatigue, which is discussed in Chapter 9 together with materials for all types of bearing application. If hard particles are present between sliding surfaces the wear mechanisms are somewhat different and were discussed in Chapter 6, together with the similar processes which occur when hard particles impinge on to a surface, causing wear by erosion.

In the present chapter we shall review the methods available to estimate wear rates in design, and then examine the measures which can be taken to reduce wear in a range of situations.

7.2 ESTIMATION OF WEAR RATES

Three methods are available to the designer to predict rates of wear. The first, not a useful method in initial design but nevertheless valuable as a means of predicting ultimate lifetimes of operating systems, is to measure the wear rate in an actual system in service. For example, measurements of erosive wear in a slurry transport pipeline under service conditions can be used to estimate the time before the pipe wall will be reduced to a certain critical thickness (perhaps determined by its ability to withstand internal pressure, in conjunction with a suitable safety factor) and hence to predict its useful life.

There are pitfalls in such an approach. As we saw in Chapters 5 and 6, wear rates in some systems are not constant but change with time: they may either decrease after a running-in period (as in many cases of lubricated sliding wear) or they may increase after an initial incubation period (as in solid particle erosion under certain conditions). To predict lifetimes by extrapolation, the designer must be confident that the measured wear rate will remain effectively constant, and that the operating variables of the system will not change, affecting the wear rate, over its lifetime. Further, some knowledge of the mechanism of final failure of the system must exist, and the wear rate measured must be related to that mechanism. In the case of the slurry pipeline, for example, wall perforation by wear usually occurs on the outside curve of a bend, and it is the wear rate at that point which will determine the lifetime of the system, not the average wear rate at other locations.

The second method of estimating wear rates in service is to obtain data from tests on components of the system, under conditions which simulate those expected in service. This method is widely used to predict wear in bearings, which can be exposed in laboratory test rigs to loads, speeds, temperatures and states of lubrication which closely simulate service conditions. If an accelerated test is used to produce wear more rapidly than it occurs in service, for example by increasing the sliding speed of a bearing, then care must be taken (as remarked in Section 5.2) to ensure that the wear mechanism is not changed and that the other relevant operating conditions remain the same. As with wear measurements on complete systems in service, data from rig tests on components must be extrapolated only with caution.

Theoretical or empirical equations relating wear rate to the operating variables such as load and speed provide the third means of predicting wear rates. Simple equations for sliding wear (the Archard equation), abrasion and erosion were given in Chapters 5 and 6. These can provide only a crude estimate of wear rate, but may well be of value in initial design calculations provided that values of the operating variables are known. In the case of

Table 7.1 Typical values of dimensionless wear coefficient K for different types of wear

	K
Sliding wear: metals and ceramics	
mild	10^{-6}–10^{-4}
severe	10^{-4}–10^{-2}
Abrasive wear: metals	
three-body	5×10^{-4}–5×10^{-3}
two-body	5×10^{-3}–5×10^{-2}
Erosive wear: metals	5×10^{-3}–10^{-1}

equations 5.7 for sliding wear, 6.5 for abrasion and 6.21 for erosion, the absolute value of the wear rate depends on the value of the dimensionless wear coefficient K. Typical values of K have been discussed in Chapters 5 and 6, and are summarized here in Table 7.1. Although the ranges listed here are wide, sufficient experimental data exist for rather narrower ranges to be quoted for some specific materials and conditions. Sources of such information are listed at the end of this chapter.

7.3 THE SYSTEMS APPROACH

Where relative motion takes place between surfaces, wear can never be completely eliminated, although in some circumstances it may be reduced to an insignificant level. If the expected (or, in the case of a redesign of a functioning system, actual) rate of wear is unacceptable, the designer must change the system in some way to reduce it. It is important to appreciate that the wear rate in a particular system is determined by the interaction of many factors. In the so-called *systems approach* to wear analysis, these influences are recognized and divided into two groups: the *structure* of the mechanical system, and the *operating variables* imposed on that structure.

The structure can be described by the materials making up the surfaces which are in relative motion, the nature of any interfacial material present (including any lubricants and abrasive particles), the environment (e.g. a surrounding gas or liquid), and by the geometrical relationships between these components. The operating variables are the conditions imposed on the system during use, such as operating speed, load and temperature. All these factors can influence the wear rate, and the designer can control most of them. We shall consider, in the remainder of this chapter, how wear in a system can be reduced by changing the operating variables, by lubrication, and by suitable selection of materials.

7.4 REDUCING WEAR BY CHANGING THE OPERATING VARIABLES

There may be only limited scope for alteration of the operating variables, since these are usually determined by the overall purpose of the system. The simple equations given in Chapters 5 and 6 indicate the likely effect of changing the operating variables on several mechanisms of wear, and can provide some guidance. For example, in sliding wear with or without the presence of abrasive particles, equations 5.7 and 6.5 suggest that the wear rate will be reduced by lowering the normal pressure (applied load per unit nominal contact area); this may be achieved either by reducing the load (which will often be impractical within the constraints of a particular design) or by increasing the area over which it is carried. As we saw in Chapter 5, there is also the possibility that a reduction in normal pressure, or in sliding velocity, may lead to a transition in the mechanism of sliding wear. This can lead to a beneficial reduction in wear rate, but may alternatively (as illustrated in Section 5.6) cause a catastrophic increase. Knowledge of the regime in which wear is occurring is clearly very helpful in this context, and if the system is so well characterized that a wear mechanism map can be constructed, this will provide an invaluable tool. Future development of such maps is likely to be of great benefit in the design process.

Fretting wear results from relative movement of two surfaces in contact (see Section 5.9). Two classes of fretting problem can be identified: *displacement-controlled* and *stress-controlled*. In the former, which occurs most commonly between parts which are required to move, such as bearing surfaces, the relative displacement cannot be eliminated and the designer must try to reduce the surface traction produced by the displacement. Lowering the normal force between the surfaces or the coefficient of friction between them will have this effect, as will design measures which reduce any concentration of contact stress caused by geometrical factors. In the second case, described as stress-controlled, the object of measures to reduce wear must be to reduce or eliminate the relative movement. Here an *increase* in the clamping force or coefficient of friction will lead to reduced displacement, and thus to a lower rate of fretting wear. Judicious use of suitable adhesive bonding or sealing materials can be helpful in these respects, as well as in preventing the access of the atmospheric oxygen necessary for the fretting wear mechanism to operate. The optimum design solution for a particular fretting problem therefore depends on the origins of the movement. In all cases, major reductions in fretting wear can be achieved by attention to the source of the fretting motion, whether it be caused by differential thermal expansion, by out-of-balance forces in rotating components, or by other sources of vibration. Unlike many other types of wear, fretting can often be completely eliminated by the use of suitable design measures.

Surface damage by contact fatigue, common in gear teeth, cam followers and rolling bearings (see Section 9.2.2), occurs as a result of cyclic surface loading. Reduction of the number of loading cycles to which the system is exposed is usually impractical, but steps can sometimes be taken to reduce the maximum contact stress through changes in geometry or reduction in load, with beneficial results.

If wear is due to hard particles, as in abrasive or erosive wear, then the phenomena discussed in Chapter 6 suggest ways in which the wear rate may be reduced. In some systems, particles can be removed, and this will usually lead to reduced wear. Since larger particles tend to cause proportionately more wear than smaller ones, methods of separation which remove the larger particles selectively (such as filtration or inertial separation) can be especially useful. If the particles are present as contaminants in a fluid lubricant, then reducing the maximum size of the particles present to below the minimum lubricant film thickness will result in greatly reduced wear.

In some systems the presence of abrasive particles is inevitable, and there is no scope for reducing their concentration or size. In wear by solid particle erosion (see Section 6.4), the most powerful factor controlling the wear rate is the particle impact velocity, and any measures to reduce this (if possible within the design limitations) will be beneficial. Wear in pipelines carrying gas-or liquid-borne particles (e.g. slurries) tends to be concentrated at the outside radius of bends (the 'extrados') or at constrictions (e.g. valves), since it is at these points, where the flow changes direction, that particles strike the surface rather than sliding along it. As we have seen, the angle of impact which causes the most rapid wear depends on the nature of the surface material; the impact angle of particles at a right-angled bend lies typically in the range 20° to 45°, depending on the shape of the bend, the nature of the particles and of the carrier fluid. There is evidence that, due to these geometrical factors, the wear rate is particularly severe in slurry pipe bends with a ratio of bend radius to pipe diameter of between 2 and 3.5, and wear rates can be reduced by using bends of different, preferably larger, radii. In the pneumatic transport of powders the use of a blind T-piece rather than a conventional bend is often successful; particles pack into the redundant arm of the bend and protect the material from damage.

7.5 EFFECT OF LUBRICATION ON SLIDING WEAR

Lubrication provides a powerful method of reducing wear in many sliding systems. Full-film hydrodynamic lubrication leads to the lowest rates of sliding wear, as we saw in Chapter 4, with a corresponding value of K typically less than 10^{-13}. Such wear rates are truly insignificant. But hydrodynamic conditions cannot be maintained for ever, and under the conditions of boundary lubrication which will pertain during starting or stopping the value of K may rise as high as 10^{-6}, depending on the properties of the lubricant used. If conditions are so severe that the boundary film is penetrated and sliding occurs between essentially unlubricated surfaces, then the wear coefficient may become 10^{-4} to 10^{-2}, values which would be unacceptable in nearly all engineering applications. Values of K applicable to these and other sliding conditions are listed in Table 7.2.

Clearly, hydrodynamic (or squeeze-film) lubrication is the most desirable state for a sliding system, and if feasible the designer should try to ensure that the system remains in this regime under all operating conditions. The most important factor determining the regime of lubrication is the minimum lubricant film thickness compared with the roughness of the surfaces, as discussed in Section 4.5. The value of λ, defined by equation 4.16, can be increased either by reducing the roughness of the sliding surfaces or by increasing the thickness of the lubricant film. The surface roughness is

Table 7.2 Typical values of wear coefficient K for lubricated sliding wear (from Peterson M B, in Peterson M B and Winer W O (Eds.), *Wear Control Handbook*, ASME, 1980, pp. 413-473).

Type of lubrication	K
Hydrodynamic and squeeze film	$<10^{-13}$
Elastohydrodynamic	10^{-13}–10^{-9}
Boundary	10^{-10}–10^{-6}
Solid lubricants	$\sim10^{-6}$
Unlubricated (severe wear)	10^{-4}–10^{-2}

determined by the final surface finishing process used in manufacturing (see Section 2.4), as modified by wear during the running-in process. The lubricant film thickness can be raised, for a given load and sliding speed, by increasing the bearing area or the viscosity of the lubricant (although at the expense of greater frictional power dissipation). The design of fluid film bearings is a complex subject and outside the scope of this book; sources of further information are listed at the end of Chapter 4, while materials for all types of bearings are discussed in Chapter 9.

7.6 SELECTION OF MATERIALS AND SURFACE ENGINEERING

7.6.1 Introduction

The choice of materials from which the components of a system are made is frequently circumscribed by factors which have little or nothing to do with tribology. Cost, for example, is a vital factor in many applications. Overall weight may be important, and so may corrosion resistance. Mechanical properties such as strength, stiffness and toughness are usually of primary concern in most mechanical engineering applications. Electrical or magnetic properties are important in other cases. Although these requirements may limit the range of usable materials, they usually provide some scope for choice. Furthermore, since most of the properties listed above (except perhaps corrosion resistance) are determined by the bulk of the material, there is ample scope for modifying those surface properties which are of major concern to the tribologist. The modification or coating of a surface in order to achieve a combination of properties in both the surface and the underlying bulk which could not otherwise be achieved is known as *surface engineering*; methods for modifying the wear and frictional behaviour of surfaces form the subject of Chapter 8. The numerous processes which are available should be considered as an integral part of the process of overall design and material selection. Figure 7.1 outlines the steps involved in designing a tribological system, although it must be appreciated that for the most effective design some of the steps will be iterative.

Metals form by far the most common choice of materials for mechanical components. The compositions and microstructures of metallic alloys are largely standardized, often internationally, and their properties can be more or less accurately predicted. Non-metallic materials are less well standard-

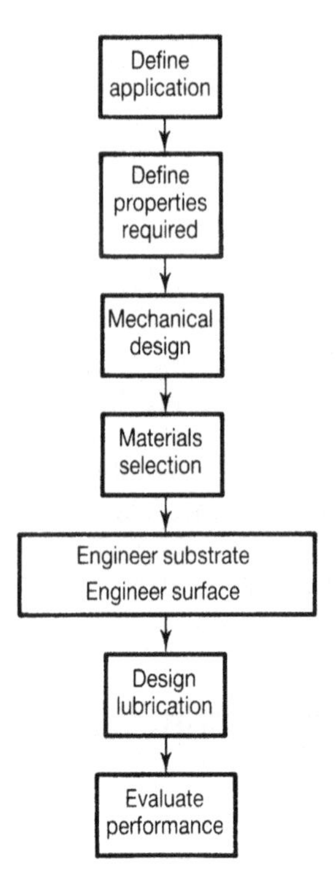

Fig. 7.1 Sequence of steps in designing a tribological component involving surface engineering (adapted from Bell T, *Metals and Materials* 7, 1991, 478-485)

ized, and the properties of these materials, even with nominally identical compositions, tend to vary more widely than those of metals. The extent of tribological information available for all these materials varies in a similar way. For metals, more data are generally available, and these data show less variability, than those for other materials. Unlike conventional mechanical and physical properties, however, the response of materials in tribological applications often cannot be expressed by simple numbers. As we saw in Chapters 5 and 6, there is no single quantity representing the 'wear resistance' of a material which is applicable over a range of mechanisms of wear, or even in many cases for a single mechanism over a wide range of operating conditions. The selection of materials and surface engineering methods for tribological performance is therefore inevitably less quantitative, and more dependent on generalized rules and orders of merit than, for example, selection for fracture toughness, specific stiffness or yield stress.

The relative performance of different materials varies in different tribological applications, depending largely on the particular mechanism of wear which dominates. In the following discussion we shall concentrate on the selection of materials for resistance to wear by sliding (lubricated and unlubricated, and including fretting), and to abrasion and erosion. The properties required in the important tribological context of bearings (both plain and rolling element) are rather different, and this topic is treated separately in Chapter 9.

7.6.2 Sliding wear

The relative ranking of materials for resistance to sliding wear depends on the precise conditions under which sliding occurs. As we saw in Chapter 5, sliding wear in all types of material can occur by several different mechanisms, and transitions between different dominant mechanisms can be induced by small changes in normal load, sliding speed, temperature or environment. General statements can, however, be made about the sliding wear behaviour of materials. Most of the experimental data available relate either to relatively low wear rates, often associated with some lubrication and with the mild wear regime, or alternatively to severe wear and surface damage under heavy load and low sliding speed (i.e. *galling*: see Section 5.1).

The Archard wear equation (see Section 5.3), which relates the wear rate Q (volume removed per unit sliding distance) to the normal load W

$$Q = \frac{KW}{H} \tag{7.1}$$

provides only a crude indication of how the properties of a material affect its resistance to sliding wear. The only material property which explicitly enters the equation is the hardness, H, of the softer of the two surfaces. The other properties of the material which also influence its wear rate are hidden in the dimensionless wear coefficient K. The value of K also depends strongly on the state of lubrication of the system. The effects of these two factors are separated in Fig. 7.2, which shows how the value of K depends on the

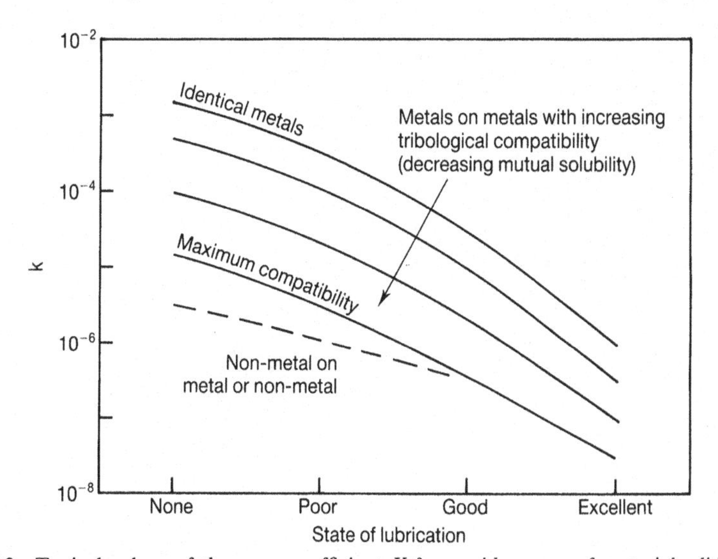

Fig. 7.2 Typical values of the wear coefficient K for a wide range of materials sliding under varying conditions of lubrication (data from Rabinowicz E, in Peterson M B and Winer W O (Eds), *Wear Control Handbook*, ASME, 1980, pp. 475–506)

material of the sliding surfaces and on the quality of lubrication. As discussed in Section 7.5 above, the potential benefits of lubrication are large.

In general, higher values of K are found for metals sliding against metals than for non-metals sliding against either metals or non-metals. Most metals sliding against counterfaces of the same metal show high values of K. If the metals are dissimilar, the value of K is lower and depends on the degree of *tribological compatibility* of the two metals. The term 'compatibility' in this context denotes a reluctance of the opposing surfaces to form a strong interfacial bond, which would lead to a high wear rate. The tribological compatibilities of metals do not correlate perfectly with other properties, although the extent of mutual solid solubility, deduced from the equilibrium phase diagram for the materials of the sliding couple, is often suggested as a guide.

Figure 7.3 represents in chart form the mutual solubilities of pairs of pure metals. Those combinations marked as completely insoluble show negligible solid solubility in each other, and also even form two distinct co-existing phases in the liquid state; they usually form tribologically compatible pairs. Identical pairs of metals are, of course, completely mutually soluble and show poor compatibility. Other pairs show varying extents of solid solubility, as indicated on the chart. In general, sliding pairs with high mutual solubility

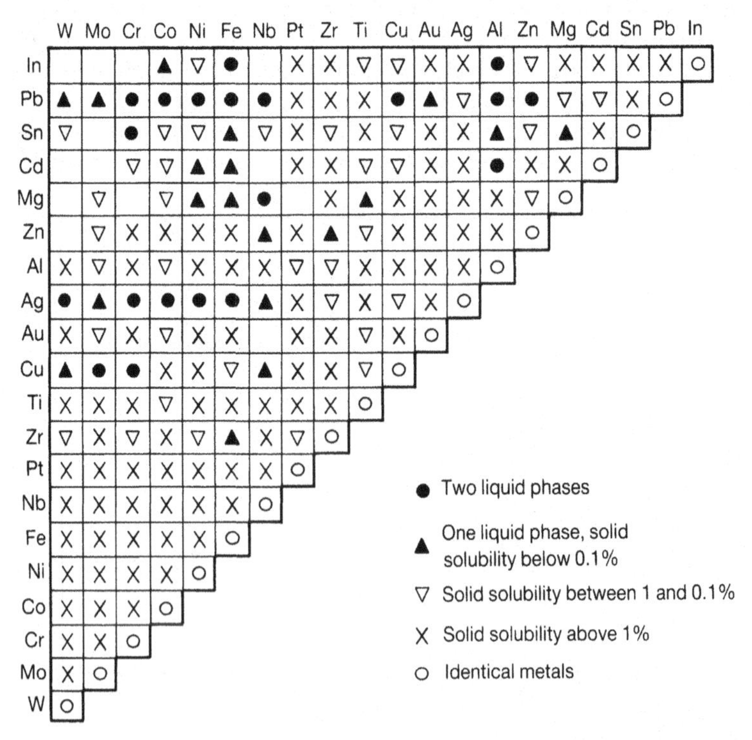

Fig. 7.3 Chart showing the relative mutual solubilities of pairs of pure metals, deduced from their binary phase diagrams (from Rabinowicz E, in Peterson M B and Winer W O (Eds.), *Wear Control Handbook*, ASME, 1980, pp. 475-506)

show low tribological compatibility and hence relatively high values of K; low mutual solubility, leading to good compatibility, is needed for low wear rates.

Mutual solubility is not the only factor influencing tribological compatibility. It is also associated with the properties of the surface films (usually oxide) on the sliding materials. The absence of appreciable oxide films on the *noble metals* such as gold, silver, platinum and rhodium tends to be associated with low wear rates for these metals. Oxidative mechanisms probably play an insignificant role. Values of K are then typically about one third of the values predicted from Fig. 7.2 for other metals.

Some metals with hexagonal close-packed structures also show anomalous behaviour, associated both with their limited ductility compared with the cubic metals, and also with chemical factors. Titanium, zirconium and hafnium, for example, show the behaviour expected from Fig. 7.2 when sliding unlubricated against themselves, but exhibit only slightly reduced wear rates when lubricated by most hydrocarbon lubricants. Other h.c.p. metals generally show lower wear rates than those predicted by Fig. 7.2.

As we saw in Section 5.6, the hardness of steels and other metals which form an oxide film during sliding is important in determining the stability of that film and hence the dominant mechanism of wear. If the metal is hard enough to provide sufficient mechanical support for the oxide, mild wear may occur at a comparatively low rate by an oxidative mechanism (see Section 5.7.2). In this way, hardness can have a strong influence on the sliding wear of some metals. However, although increasing the hardness of a particular alloy usually leads to a reduction in wear rate, hardness is not a good predictor of the relative wear resistance of different alloys. Other factors, especially the presence of microstructural features such as carbides in steels or graphite in cast irons, are often much more important.

The resistance of metals to severe adhesive wear and associated surface damage (galling) under high normal loads does not always correlate with their wear resistance under less severe conditions. Several factors have been found to control the susceptibilities of metals to galling: the effectiveness of surface films in preventing adhesion, the strength of the adhesion once these films break down, and the extent of junction growth (see Section 3.4.3) which occurs on further sliding. Mutual solid solubility, as an indicator of the strength of the adhesive force, plays some role; metals which adhere strongly are more liable to galling damage. Hexagonal metals with a restricted number of operative slip systems show lower galling tendencies than cubic metals, presumably because of their lower ductility. Studies of the effect of stacking-fault energy in alloys have suggested that metals with a low stacking-fault energy are less inclined to gall than those with a high energy. Other investigators have found high work-hardening rates to be beneficial in several cobalt-, nickel- and iron-based alloys, although the work-hardening rate is not an infallible indicator of resistance to galling; austenitic stainless steels, for example, are notoriously susceptible to galling yet work harden rapidly by the formation of martensite. Hardness alone is a poor indicator of galling resistance; in steels, for example, a high concentration of carbides or nitrides is desirable, and confers greater wear resistance than similar hardness levels achieved with a lower volume fraction of these hard and brittle precipitates.

Hard coatings or diffusion layers which are also of very limited ductility generally confer good resistance to this type of wear. Rough surfaces, preferably of a random nature (e.g. as produced by abrasive grit blasting) generally increase galling resistance because junction growth is limited; metals with polished surfaces are, in contrast, much more liable to suffer galling.

As we have seen in Section 5.10 and Fig. 7.2, ceramics undergoing mild sliding wear can exhibit wear coefficients as low as, or even lower than, those for dissimilar metals. This fact, coupled with their greater hardness, indicates that ceramics can offer significantly lower sliding wear rates than metals. However, there are drawbacks to the use of bulk ceramics in tribological applications. Their mechanical properties (especially fracture toughness) may not be adequate for the demands placed upon them, they may be difficult to fabricate in the shape required, and the possibility of small-scale surface fracture leading to severe wear (see Section 5.10.3) calls for cautious design. Nevertheless, bulk ceramic components can prove extremely durable for some tribological purposes: examples include alumina bushes and face seals in water-circulating pumps, silicon nitride valve components, and alumina femoral heads and cups for artificial hip joint implants.

Some of the disadvantages of bulk ceramics can be overcome by using the material in the form of a coating on a metallic substrate, and ceramic coatings produced by plasma spraying, physical vapour deposition (PVD) and chemical vapour deposition (CVD) form an important group of surface engineering methods. In all tribological uses of ceramics, lubrication is highly advantageous since by reducing the surface tractions the occurrence of local fracture leading to severe wear can be avoided. However, the possibility of chemical interaction between an unsuitable lubricant and the ceramic, causing enhanced wear, must always be considered (see Section 5.10.5).

Polymers are rarely thought of as wear-resistant structural materials, being usually used as bearing materials, often under conditions of marginal or dry lubrication (see Sections 9.4.2 and 9.4.3). Some polymers with sufficient strength may, however, be used as bulk components in tribological applications, notably nylons (polyamides), acetal, polyetheretherketone (PEEK) and polyethersulphone (PES), often reinforced with a suitable filler. Cages for rolling element bearings, for example, have been moulded from glass-filled nylon for some time (see Section 9.2.3), and this material, as well as acetal, is also commonly used for lightly loaded gears. Selector forks for racing car gearboxes fabricated from carbon-fibre reinforced PEEK combine good tribological performance with substantial weight savings compared with the conventional forged steel components.

The diversity of surface engineering methods available to the designer, and described in detail in Chapter 8, permits the tribological properties of a surface to be selected independently, at least to some extent, from the properties of the bulk of the material. Figure 7.4 illustrates the wide range of depth and hardness which can be achieved in the surface regions by these methods.

From the diagram it will be clear that different methods will be most suited to different types of application. The depth of the modified region or the

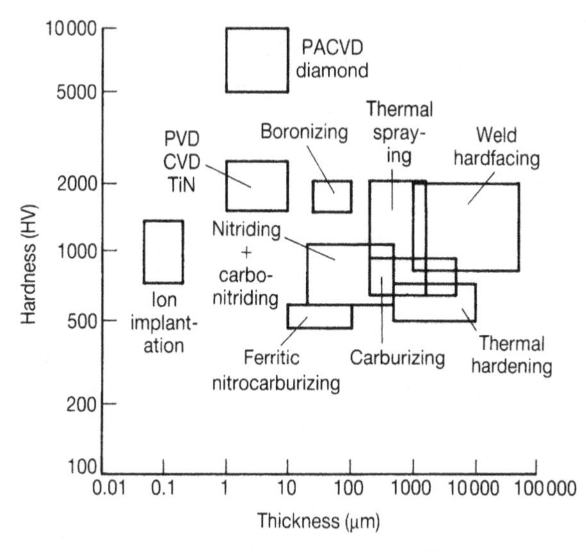

Fig. 7.4 Typical depths of treated layer or coating, and range of hardness achieved, for different methods of surface engineering. All the methods shown are applicable to steels, but the diagram is not exhaustive, and some of the coating methods can be applied to other substrates

thickness of the coating is of primary importance; methods which produce only thin layers with modified properties, such as the vapour phase coating methods (CVD and PVD) or ion implantation (all described in Section 8.3.3) will be useful only in applications where the maximum extent of wear will be comparably small, and where the stress imposed on the surface during use decays rapidly with depth into the surface. For such applications, e.g. small precision engineering components, milling cutters, twist drills and press dies, these methods can offer substantial benefits. PVD titanium nitride coatings, in particular, on high speed steel and cemented carbide tools are found to give greatly enhanced life; even better performance is claimed for similar coatings containing other elements (e.g. (Ti,Al)N or (Ti,Nb)N). Such technology has great potential for further development.

In other cases, where the surface stress imposed in service penetrates more deeply into the component, methods which produce a thicker modified layer or coating are required. In a heavily loaded gear wheel, for example, the surface material must have a high yield stress to ensure that it remains elastic when exposed to localized contact stresses and to reduce sliding wear. However, the core of the gear tooth and rest of the wheel require high fracture toughness and resistance to fatigue cracking, in order to withstand the cyclic loading, perhaps applied impulsively, to which the gear train is exposed. In a steel, these two requirements of highest yield stress or hardness (desirable in the surface) and the best fracture toughness (for the core material) cannot be achieved together. The problem of optimizing both at an economical cost can only be solved by modifying the surface material, perhaps by carburizing or nitriding, so that a suitably hard surface can be combined with core material of high toughness.

Further possibilities exist to combine different methods of surface treatment on a single component (by so-called *duplex surface engineering*). For example, nitriding, carburizing or thermal hardening may be carried out to produce a deep layer of modified material, followed by PVD or CVD coating or ion implantation to produce a thinner surface layer of yet different properties. Coatings can also be produced with graded compositions and properties, to improve the matching of mechanical properties and enhance adhesion at the interface with the substrate. It is clear that substantial development of these and other surface engineering processes will occur in the future.

Guidance on the selection of specific materials for particular applications would be out of place in a general text, and sources of such information are suggested at the end of this chapter. Computerized databases and expert systems are also becoming increasingly powerful and attractive as sources of information for the designer. However, some brief guidelines on the selection of materials for resistance to sliding wear can be given:

(1) There is no general correlation between wear rate and coefficient of friction, although lubricants, which may be externally supplied or present as constituents of one of the materials (e.g. graphite in cast irons or molybdenum disulphide in some filled nylon composites) will tend to reduce both wear and friction. Even poor lubrication is better than none in reducing wear.
(2) Surfaces of identical metals in sliding contact should be avoided. High tribological compatibility, which is usually associated with low mutual solubility, is desirable in dissimilar metals.
(3) High hardness is often beneficial and in metals can be achieved locally by suitable surface engineering methods (see item 6 below and Chapter 8). However, other important factors such as mutual solubility and work-hardening rate may mask the influence of hardness.
(4) In steels, high carbide or nitride contents are beneficial, even at the expense of slightly reduced hardness.
(5) A high work-hardening rate in a metal usually gives good resistance to severe wear and galling. The initial surface finish correlates with resistance to galling. Random rough surfaces (e.g. produced by abrasive blasting) provide greater resistance than smoother surfaces.
(6) Hard surface layers, for example those produced in steels by carburizing, nitriding or nitrocarburizing, or hard non-metallic coatings, as deposited for example by PVD or CVD processes (see Section 8.3.3) often give excellent resistance to sliding wear. High hardness and low ductility seem to be desirable mechanical properties in such coatings.

7.6.3 Fretting wear

As we saw in Section 7.4, fretting may be classified as either displacement-controlled or stress-controlled, and can often be effectively limited by correct design measures. The selection of materials can also play a part in reducing fretting wear.

Some metals, notably titanium and its alloys, are notoriously susceptible to fretting damage, and their use should be avoided if fretting is likely. If a reduction in friction is desirable, as in displacement-controlled fretting, then lubrication may be effective, either by means of an externally-supplied liquid lubricant (preferably a low viscosity oil which can penetrate between the surfaces) or a solid lubricant film (inorganic or polymeric) applied to one of the surfaces. A soft metallic coating, applied for example by electroplating or by physical vapour deposition (see Chapter 8) to a hard substrate, may also reduce the frictional force.

Methods of surface engineering leading to increased hardness can also reduce fretting wear, in both displacement-controlled and stress-controlled applications. Hardening the surface may also introduce compressive residual stress, with the further beneficial effect of reducing the tendency to fatigue crack growth from the fretting damage (fretting fatigue). However, some surface treatments cause an increase in friction, while others lead to a decrease. Care must therefore be taken to choose a hardening method which gives an appropriate level of friction for the particular application.

7.6.4 Wear by hard particles

We saw in Section 6.2.1 that the rate of wear of a material by abrasion or erosion falls sharply if its hardness can be made greater than that of the abrasive particles themselves. Desirable as this condition may be, it is not readily achieved with steel components since very few alloys are as hard as naturally occurring abrasive particles. Silica, for example, typically has a hardness of some 800 HV, a value attained only by martensitic steels, untempered, of high carbon content (see Fig. 8.4). Although the results discussed in Sections 6.3.3 and 6.4.3 can be used to select engineering alloys with optimum resistance to abrasive and erosive wear respectively, the designer seeking the lowest wear rates will be forced to use harder materials, either in bulk form or as coatings. Candidate bulk materials include white cast irons (usually containing chromium, with a high volume fraction of carbides in a martensitic matrix), cermets (e.g. tungsten carbide in a cobalt or nickel binder) and ceramics (e.g. alumina, silicon carbide, silicon nitride, boron carbide or sintered polycrystalline diamond). As discussed in Sections 6.3.3 and 6.4.3, these materials can offer excellent resistance to wear by hard particles and in many cases are available as prefabricated components (e.g. as tubes, tiles, slurry pump liners, grit blaster nozzles etc) which can readily be incorporated into new or existing designs.

For some applications, however, bulk wear resistant materials may not be suitable, perhaps for reasons of cost, overall weight, difficulty of fabrication or mechanical properties. Surface engineering methods can then be used to apply a coating of a wear-resistant material to a substrate with lower wear resistance but with the desired bulk properties. As in the case of sliding wear (see Section 7.6.2), the thickness of the modified surface layer is a most important consideration. Both the expected rate of wear and the depth to which significant stresses are induced by particle contact must be taken into account. Weld hardfacing processes (see Fig. 7.4 and Section 8.3.2) must be used to produce the thickest coatings: metallic matrices (typically iron- or cobalt-based) with a high content of chromium or tungsten carbides can be deposited by welding methods to provide resistance to high stress abrasion by large particles, in applications such as rock drills, excavator teeth and ore-crushing machinery. Rather thinner coatings of these materials can be produced by thermal spray processes, some of which can also deposit ceramic materials such as alumina, chromia, zirconia or titania. Boronizing of steels also leads to a surface layer with sufficient hardness to resist some types of abrasive wear, but the benefits of the other methods of surface modification of steels discussed in Section 8.2 are generally not great because of the limited increase in hardness which can be achieved. The very thin but hard coatings formed by CVD and PVD processes provide useful abrasive wear resistance only if the abrasive particles are sufficiently small. For example, PVD titanium nitride coatings can increase the life of moulds and extrusion dies used for polymer processing, by reducing the rate of abrasive wear due to fine filler particles in the polymer melt.

In certain cases, polymers can provide viable alternatives to very hard materials. For example, ultra-high molecular weight polyethylene (UHMWPE) can be used to line hoppers and chutes carrying powdered materials, and offers low sliding friction together with adequate abrasion resistance. Polyurethanes are also used to line powder-handling equipment, and in aqueous slurry pipelines provide dual protection against both corrosion and erosive wear. Other elastomers, especially natural rubber, can also show good resistance to erosive wear by airborne particles, especially for rounded particles at high impact angles and low velocities (see Section 6.4.3).

Further reading

ASM Handbook, 10th edition, Vol 3, Friction, Lubrication and Wear Technology, ASM International, 1992

Budinski K J, *Surface Engineering for Wear Resistance*, Prentice Hall, 1988

Czichos H, *Tribology: A Systems Approach to the Science and Technology of Friction, Lubrication and Wear*, Tribology Series No.1, Elsevier, 1978

Glaeser W A, *Materials for Tribology*, Elsevier, 1992

Lansdown A R and Price A L, *Materials to Resist Wear*, Pergamon, 1986

Neale M J (Ed.), *Tribology Handbook*, Butterworths, 1973

Peterson M B and Winer W O (Eds), *Wear Control Handbook*, ASME, 1980

UK Department of Trade and Industry, *Wear Resistant Surfaces in Engineering*, HMSO, 1986

Waterman N A and Ashby M F (Eds.) *Elsevier Materials Selector*, Vol 1, Elsevier, 1991

8

Surface engineering in tribology

8.1 INTRODUCTION

There are two common objectives in the use of surface engineering for tribological applications: to increase the wear resistance of the surface material, and to modify its frictional behaviour. In some cases, both are achieved together. Figure 8.1 summarizes the range of methods used in surface engineering. The microstructure of the surface material may be modified selectively without changing its composition, as in transformation hardening or melting followed by rapid solidification. Alternatively, both the composition and the microstructure may be changed together; this is often achieved by thermally-enhanced diffusion of a different chemical species into the surface. The changes in composition and microstructure which can be brought about by these methods are inevitably limited, and for many purposes coatings of completely different materials can be applied.

In this chapter we shall examine all the methods of surface engineering outlined in Fig. 8.1. Many are applicable only to metals, and some to steels alone, although others can also be applied to polymers and ceramics. A major distinction can be drawn between methods which involve modification of

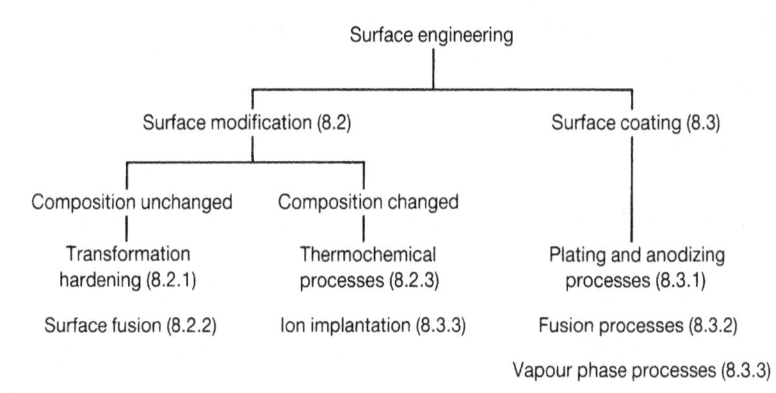

Fig. 8.1 A summary of the methods available for surface engineering, which also outlines the structure of Chapter 8

either the structure or the composition, or often both, of the substrate, and methods in which a coating of a completely different material is applied in some way. We shall deal with these two classes of method in turn.

8.2 SURFACE MODIFICATION

8.2.1 Transformation hardening

Undoubtedly the simplest method of producing a region of hard material at the surface of a softer carbon steel component is by *transformation hardening*, in which the surface material is rapidly and selectively transformed by heating to austenite, which is then quenched to form martensite and subsequently tempered. The method is restricted in principle to materials which show a suitable phase transformation and, in practice, is applied only to ferrous alloys.

The maximum hardness which can be achieved by this method is modest as we shall see below, but since it is rapid and relatively simple to implement, transformation hardening is widely used for such tribological components as gear teeth, camshafts and crankshafts, cutter blades and various bearing surfaces. The source of heat used in transformation hardening determines the depth of material affected and the properties of the hardened layer. Common methods use an oxy-acetylene or oxy-propane flame (*flame hardening*) or high frequency electrical induction heating (*induction hardening*), while new methods have been developed more recently employing laser or electron beams. In both flame and induction hardening, an external quenching medium is used to cool the workpiece after austenitizing: commonly, water jets or a water bath. The depth of the hard layer depends on the rate and method of heating. Typically, depths in the range 0.25 to 6 mm can be achieved by both methods.

Better control of depth is attained through induction heating, since the high frequency eddy currents responsible for heating are localized in the surface of the workpiece by the 'skin effect'. The depth to which the alternating current penetrates, δ, depends on its frequency f, and also on the temperature, which affects both the magnetic permeability μ and the resistivity ρ of the workpiece material. The skin depth δ is given by

$$\delta = \left(\frac{\rho}{\pi \mu f}\right)^{1/2} . \tag{8.1}$$

The permeability of iron falls sharply to unity at its Curie temperature of 770 °C when it ceases to be ferromagnetic; with δ in mm, the relationship becomes approximately

$$\begin{aligned} \delta &= \frac{20}{\sqrt{f}} \quad \text{at } 20\ ^\circ\text{C} \\ \delta &= \frac{500}{\sqrt{f}} \quad \text{at } 800\ ^\circ\text{C} \end{aligned} \tag{8.2}$$

where f is the frequency in Hz. Typically, frequencies between 3 kHz and 500 kHz are used to produce hardened layers 0.5 to 5 mm in depth with input power densities (power per unit surface area) of about 20 W mm^{-2}. Figure 8.2 indicates the approximate range of depths and process temperatures used in both flame and induction hardening of steels.

In *laser hardening* an infra-red beam from a high power (0.5–15 kW) carbon dioxide continuous laser is directed on to the steel surface. A coating (e.g. graphite, iron oxide, or various phosphate treatments) is usually applied to the surface to increase its infra-red absorption. The beam is focused to produce a spot 1–2 mm in diameter. By means of mirrors, and also by moving the workpiece, the spot can be made to traverse the surface in any desired pattern; large areas are treated by scanning the beam in a raster pattern. As the laser beam passes over a point on the surface, it is first rapidly heated (at a rate up to 10^6 K s^{-1}), then cooled by conduction of heat into the surrounding and underlying material. The cooling rate by conduction is so high (typically $>10^4$ K s^{-1}) that no external quenching is required. Figure 8.3 shows the range of power densities and heating times over which transformation hardening occurs. If the power density is increased (for example by focusing the beam more sharply) or the heating time is prolonged, surface melting will occur; processes in which this is used are discussed briefly in Section 8.2.2. The high power densities and short heating times used in laser transformation hardening result in high temperatures and shallow depths of hardening, as indicated in Fig. 8.2. The higher temperatures lead to more rapid dissolution of carbides than in flame or induction hardening.

Electron beam hardening is similar in many respects to laser hardening. The energy input is provided by an electron beam focused to a diameter of about

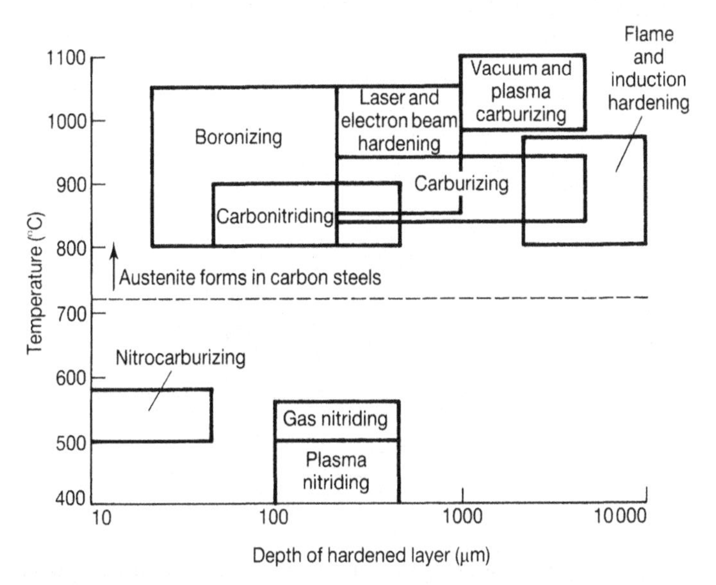

Fig. 8.2 Comparison of the process temperature and depth of hardened material produced by various methods of surface modification of steels

3 mm, giving a similar power density to that used in the laser process (1 to 10 kW cm^{-2}— see Fig. 8.3). The beam is moved over the surface by electromagnetic deflection, and the workpiece may also be moved in order to treat the desired area. The electron beam process differs from laser hardening in that no surface coating is needed to enhance absorption, and also in that it must be performed in a moderate vacuum (at a pressure of 1–10 Pa). Depths of hardening of up to about 2 mm can be achieved (see Fig. 8.2).

Figure 8.4 shows how the hardness of untempered martensite formed in plain carbon steels varies with carbon content. A further modest increase in hardness, typically up to 100 HV, may result from alloying additions in low alloy steels. A steel of high carbon content would, in principle, be desirable for maximum hardness after transformation hardening but, in practice, other factors limit the composition. Carbon contents above 0.5% give low toughness in the hardened layer, and lead to increased susceptibility to quench cracking. On the other hand, low carbon contents (<0.3%) require very rapid cooling rates (>400 K s^{-1}) in order for martensite to be formed, which cannot be achieved in flame and induction hardening. For these reasons, these two processes are usually applied only to steels within a narrow range of carbon content: typically 0.4 to 0.5%. Alloy steels may be used as well as plain carbon, as can cast irons with suitable matrix compositions.

The rapid cooling produced by self-quenching in laser and electron-beam hardening means that steels of lower carbon content can be satisfactorily hardened, although the hardness of the martensite formed will not be high (see Fig. 8.4). A further feature of the rapid cooling rate is retention of some austenite in the hardened layer, which becomes more prevalent in steels of higher carbon content; as we saw in Section 6.3.3, retained austenite can lead to enhanced resistance to abrasive wear.

All the transformation hardening processes are characterized by a short process time, and all are to some extent suited to localized treatment of small

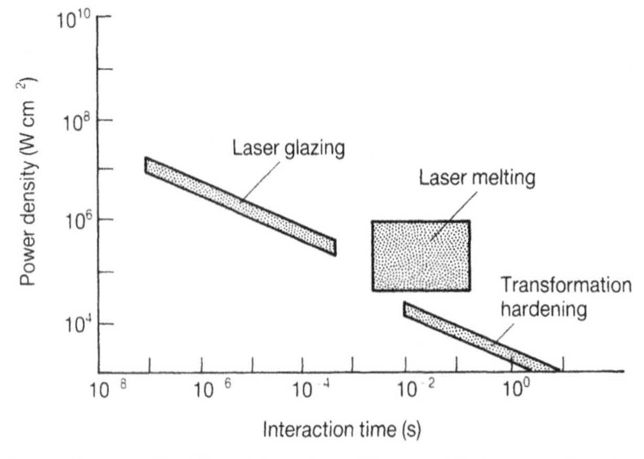

Fig. 8.3 Regimes of power density and exposure time used in laser surface treatment of steels (from Oakley P J, in Bucklow I A (Ed.), *Proc First Int. Conf. on Surface Engineering*, The Welding Institute, vol. 3, paper 35, 1986)

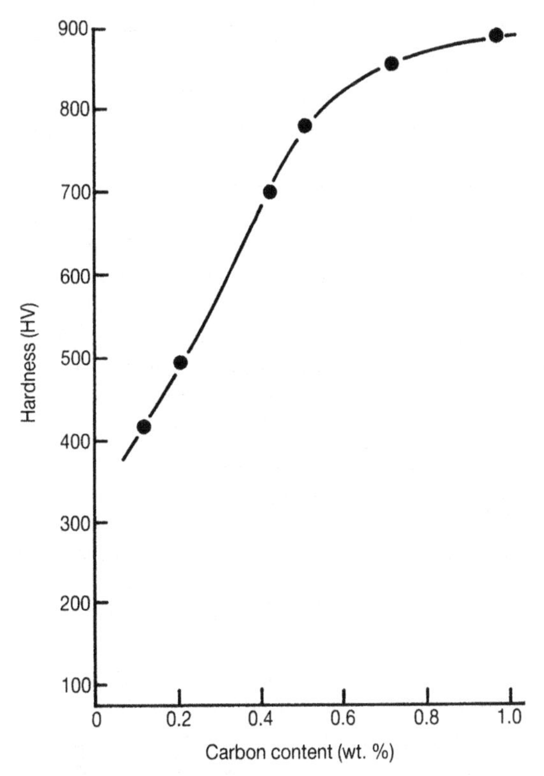

Fig. 8.4 Hardness of untempered martensite as a function of carbon content for plain carbon steels (data from Grange R A, Hribal C R and Porter L F, *Metallurgical Transactions* **8A**, 1775–1785, 1977)

areas of surface. All are limited in the range of materials which can be hardened, and in the maximum hardness which can be produced. Flame and induction hardening are well-established methods, while the laser and electron-beam processes are more recent and more specialized and involve greater initial investment. However, the latter two processes have some unique features: they can harden steels of low carbon content, bulk heating and consequent distortion of the component are absent, and the process can be applied selectively to precisely controlled areas.

8.2.2 Surface melting

A further method of modifying surface properties without change in composition employs the liquid–solid phase transformation, by locally melting material and allowing it to resolidify. If solidification is rapid, the process causes homogenization and refinement of the microstructure. It may also lead to supersaturation and the formation of non-equilibrium phases, and even to glass formation in suitable materials. In order to produce localized surface melting and a rapid cooling rate, a high input power density is needed. This is usually achieved by laser or electron-beam heating, as described above, although electric arc welding methods with non-consumable electrodes (e.g. the tungsten inert gas process—TIG) can also be used.

In laser and electron beam surface melting, higher power densities are employed than in transformation hardening (see Fig. 8.3). By adjustment of the power and process time the temperature profile can be controlled, which in turn determines the cooling rate once the power input ceases. Although simple melting results in no change in composition, the refinement of microstructure caused by rapid solidification leads generally to a significant increase in hardness in the treated surface compared with that of the bulk. In contrast to transformation hardening, the method can be applied to both ferrous and non-ferrous alloys and potentially even to non-metals. The effect of microstructural refinement in tribological applications may extend beyond a simple increase in hardness; for example, laser surface melting of cast aluminium–silicon alloys can reduce the size of the brittle silicon phase so far that the mechanism of wear is changed. Figure 8.5 shows microstructures of a hyper-eutectic aluminium–silicon alloy before and after surface melting; the primary silicon particles were reduced in size from >60 μm to <5 μm, and the bulk hardness was increased from 80 HV to 160 HV. More important, the mechanism of sliding wear of the treated surface was primarily by plastic deformation, whereas the untreated surface exhibited extensive fracture in the large silicon grains, associated with a much higher wear rate.

Low carbon steels are unsuitable for surface melting treatment, since soft δ-ferrite is formed and retained in the quenched surface layer. At carbon contents between 0.4% and 0.9%, however, the treated region consists of

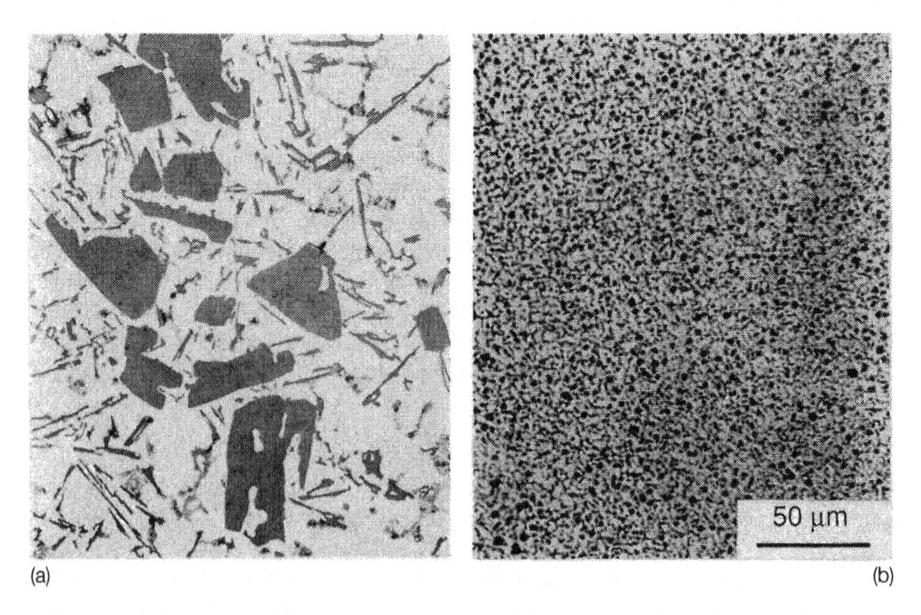

Fig. 8.5 Microstructures of an aluminium–17% silicon alloy (a) as cast and (b) after laser surface melting (from Coquerelle G, Bosch J P and Farges G, in Niku-Lari A (Ed.), *Advances in Surface Treatment*, vol. 5, Pergamon, 409–423, 1987)

martensite and retained austenite which can be hardened by further heat treatment, and the process yields significant benefits.

Cast irons respond well to surface melting, developing fine-grained structures similar to those formed in conventional chill-casting. The TIG melting process is widely applied to wearing surfaces on crankshafts, camshafts and cam followers, and allows a very hard surface to be combined with a tough core material. Laser and electron beam methods can also be used. A typical improvement in the dry sliding wear resistance of a pearlitic grey cast iron is illustrated in Fig. 8.6.

8.2.3 Thermochemical processes

The range of surface microstructures which can be achieved by transformation hardening and by surface melting (without alloying) is limited by the composition of the starting material, and the properties which can be achieved are therefore restricted. Methods in which the composition of the surface is locally altered, on the other hand, can produce microstructures and associated mechanical properties which are completely different from those of the substrate. Most processes for compositional modification take place in the solid state, although as we shall see below laser melting associated with surface alloying or particle injection can also be used for this purpose.

Two classes of solid state processes can be identified in which atomic transport by thermally-enhanced diffusion leads to surface modification. Both can be termed *thermochemical processes*. In the first class, diffusion of small atoms into the surface leads to the formation of an interstitial solid solution in the substrate material, and sometimes to the formation of compounds as very fine precipitates; the modified region, however, remains chemically similar to the bulk. *Carburizing* and *carbonitriding* are examples. In processes of the second type, chemical reaction occurs between the diffusing atoms and constituents of the substrate, causing the formation of a distinct layer of a new compound at the surface. Examples of this type of process are provided by *nitriding*, *nitrocarburizing*, *boronizing* and *chromizing*. In their end result these processes are similar to the coating processes discussed in Section 8.3,

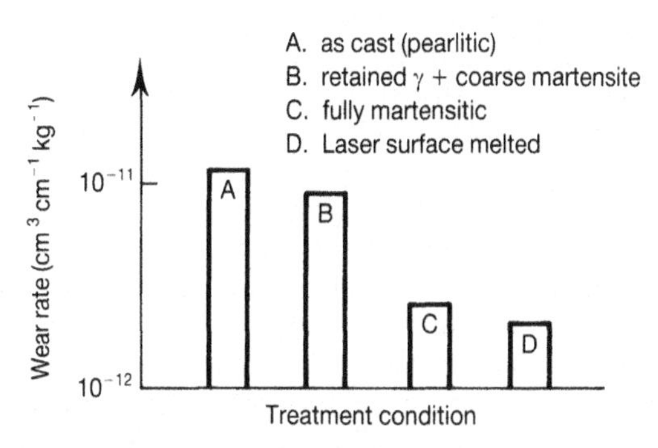

Fig. 8.6 Dry sliding wear rates of a grey cast iron after various heat treatments and laser surface melting, measured in a pin-on-disc test (from Mordike B L, in Niku-Lari A (Ed.), *Advances in Surface Treatment*, vol. 5, Pergamon, 381–408, 1987)

but a clear distinction can nevertheless be drawn in that the surface layer, while of different composition from the substrate, is formed by chemical reaction with the substrate material; for this reason the interface between the surface layer and the bulk material is often more diffuse, and of higher strength, than for an externally applied coating.

Carburizing and carbonitriding

The most widely used thermochemical process for surface hardening of steels is *carburizing*, sometimes called *case hardening*. Carburizing involves the diffusion of atomic carbon into steel from the surface, to produce a surface region or *case* of enhanced carbon concentration, typically up to several millimetres deep (see Fig. 8.2). The process is used with plain carbon or low alloy steels of low initial carbon content, typically 0.15 to 0.2% C, which provide a core which will remain tough even in the as-quenched condition. Carburizing is carried out in the austenite phase region, typically at temperatures of 900 °C or higher; at these temperatures the diffusion of carbon in austenite is rapid. The carbon concentration in the surface layer may be enhanced to 0.7–0.9% by carburizing, producing a maximum hardness of up to about 900 HV (see Fig. 8.4). In some methods the component is quenched immediately after the carburizing process, while in others it is cooled slowly and can then be machined to final dimensions in a soft condition before the final heat treatment to produce the hardened case of lightly tempered martensite.

In all methods of carburizing the martensitic transformation which provides the mechanism for hardening also causes distortion of the component due to the associated change in lattice volume. Although this distortion can be reduced by careful practice, it can never be entirely eliminated. Dimensional tolerances after carburizing are therefore lower than after those methods of surface treatment which are carried out in the ferrite phase region. The process is widely used for rotating shafts and bearing components, cam followers, gears and camshafts; in addition to the high surface hardness and consequent wear resistance which is of tribological interest, the martensitic transformation produces a compressive residual stress in the surface which gives substantially increased fatigue life.

The depth of hardening and the hardness achieved depend on the time, temperature and carbon activity at the surface during carburizing, and these in turn depend on the process used. *Gas carburizing* is a common method. Traditionally the components are heated to ~900 °C in an atmosphere of carbon monoxide, hydrogen and nitrogen; recent developments use a mixture of methanol and nitrogen. In *pack carburizing*, sometimes used to obtain very thick cases, the components are packed into sealed boxes with a granular medium containing charcoal and an 'energizer' (typically barium carbonate), and heated. Carburizing takes place in the same way as in the traditional gas process, through the action of carbon monoxide formed from the carbon and residual oxygen within the box. Since the diffusion of carbon in iron is thermally activated, at higher temperatures it becomes significantly more rapid. Despite the limitations imposed by furnace design and heating times, the temperature for gas carburizing can be raised to 1000 °C with substantial shortening of the process time.

Two more recent processes also operate at high temperature, about 1050 °C: *vacuum carburizing* and *plasma carburizing*. In vacuum carburizing, the components are heated in moderate vacuum, and methane or propane is then admitted to the furnace at a low pressure. Reaction of the gas at the surface of the hot steel provides the source of carbon. Following saturation of the surface with carbon, a further period under vacuum at high temperature is then allowed for it to diffuse inwards. In plasma carburizing, a glow discharge in methane at low pressure is used to deposit carbon on the surface of the hot substrate which is held at a negative potential. As with the vacuum process, a short carburizing period is then followed by a longer diffusion period. Both vacuum and plasma carburizing are energy-efficient processes and, because of the high process temperature, result in much deeper case hardening than lower temperature methods of the same duration. Figure 8.7 illustrates hardness profiles in a low carbon steel carburized for approximately the same time by the traditional gas method, by the vacuum method with two different gas pressures, and by the plasma method. It can be seen that a much longer carburizing time would be required for the lower temperature gas method to achieve the same depth of hardening as the other methods.

Carbonitriding is a very similar process to carburizing, involving the simultaneous diffusion of both carbon and nitrogen into austenite in a low

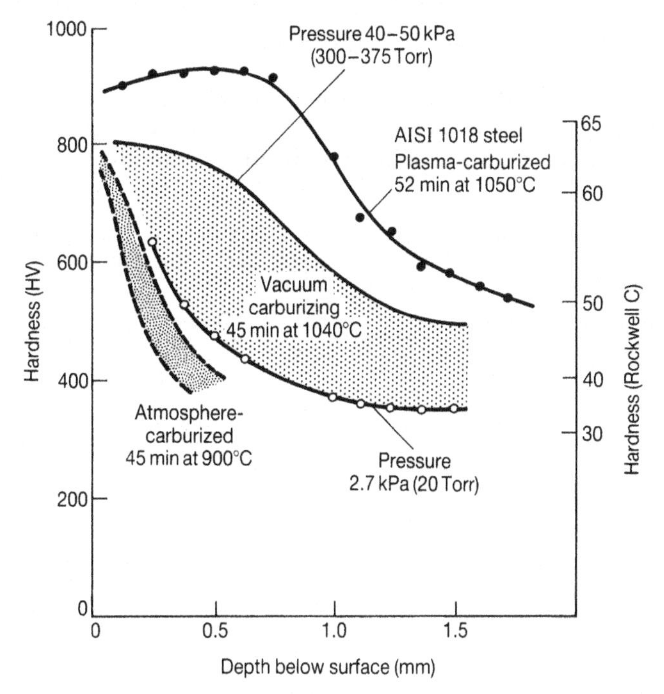

Fig. 8.7 Variation of hardness with depth in a plain 0.18% carbon steel carburized by different methods (from Grube W L and Gay J G, *Metallurgical Transactions* **9A**, 1421, 1978)

carbon steel (<0.25% C). The process is typically carried out at 800 to 900 °C to produce case depths from 0.05 to 0.75 mm (Fig. 8.2). The final nitrogen concentration in the hard layer is 0.5 to 0.8%, with accompanying carbon concentration similar to that achieved in carburizing. The effect of the nitrogen is to increase the hardenability of the surface layer; oil quenching rather than water quenching can therefore be used, reducing the risk of quench-cracking. The nitrogen also increases the resistance to tempering of the martensite formed. In general, steels hardened by carbonitriding exhibit greater resistance to sliding wear than those carburized to the same hardness.

Two methods are commonly used for carbonitriding: a gas phase method and a molten salt treatment. *Gas carbonitriding* is very similar to gas carburizing, and can be achieved simply by the addition of ammonia to the gas mixture used for carburizing. *Salt bath carbonitriding*, sometimes misleadingly called *liquid carburizing*, involves immersion of the components in a molten salt bath containing sodium cyanide (typically 45% NaCN, 40% Na_2CO_3, 15% NaCl) at 880 °C. Treatment times are up to 1 hour, and the parts may be quenched straight from the salt bath.

Nitriding, nitrocarburizing and metallizing

The second group of thermochemical processes consists of those in which a reaction occurs between a constituent of the substrate and an externally-supplied chemical species, either hardening the substrate through precipitation of the reaction product, or forming a hard layer of the reaction product at the surface. We shall examine four processes: *nitriding*, in which nitride compounds are formed as fine precipitates close to the surface; *nitrocarburizing*, which forms a layer of iron carbonitride; *boronizing*, in which iron boride layers are formed; and *chromizing* which forms a chromium carbide layer. The hardnesses resulting from these surface treatments are listed in Table 8.1, together with values for steels hardened by other methods, for comparison.

Table 8.1 Maximum hardnesses attainable in steels by various thermal and thermochemical methods of surface modification (data from Child H C, *Surface Hardening of Steel*, Engineering Design Guide No. 37, Oxford University Press, 1980)

Process (and compound formed)		Hardness (HV)
Thermal hardening (0.5% C steel)		700
Carburizing and carbonitriding		850–900
Nitriding:	Cr–Mo steel	650
	Cr–Mo–V steel	900
	Cr–Mo–Al steel	1100
Nitrocarburizing (forms $Fe_2(C,N)$)		500–650
Boronizing (forms FeB, Fe_2B)		1500
Chromizing (forms Fe_2Cr_3)		1500
Toyota diffusion (TD) process		3000–4000

Nitriding differs from the thermochemical hardening processes discussed in the previous section in that it is carried out at a lower temperature: 500 to 570 °C (see Fig. 8.2). In this range, carbon steels are ferritic. The process is applied to steels which contain the nitride-forming solute elements aluminium, chromium, molybdenum, titanium, tungsten or vanadium. The most effective hardening, giving values up to 1100 HV, is obtained in steels containing aluminium or titanium. In nitriding, atomic nitrogen is formed at the surface of the steel and diffuses inwards, reacting with the solute atoms to form very fine nitride precipitates typically 5 to 15 nm in size. The lattice strains associated with these precipitates are high enough to nucleate dislocations which, together with the precipitates themselves, have a strong hardening effect. The nitriding process is not restricted to ferritic steels, but can also be applied to austenitic stainless steels which contain nitride-formers. The hardening produced by nitriding is retained at temperatures up to 500 °C; in contrast, the hardness of the martensitic layers produced by carburizing starts to fall above 200 °C.

In addition to the high hardness which gives nitrided surfaces good tribological properties, nitriding also produces a compressive residual stress which enhances the fatigue strength. Because the process is operated in the ferritic phase region for carbon and low alloy steels, heat treatment of the core material must be carried out *before* nitriding. Steels of about 0.4% carbon content are therefore commonly used to give reasonable core strength after tempering at the nitriding temperature. The relatively low process temperature and the absence of phase transformations mean that there is very little distortion associated with nitriding. However, a disadvantage of the low temperature is the relatively sluggish diffusion of the nitrogen atoms into the surface, necessitating long treatment times and limiting the practical thickness of the hardened layer (see Fig. 8.2).

Two main methods are used for nitriding steels. In *gas nitriding* the parts are heated to 530 °C in a stream of ammonia gas (often mixed with other gases). The ammonia dissociates at the metal surface, forming nitrogen which diffuses inwards. Process times are long, measured in days rather than hours: four days may be needed to develop a hard layer 500 μm thick. In *plasma nitriding* or *ion nitriding* the steel component is placed in a chamber containing nitrogen and hydrogen at a pressure of 10 to 1000 Pa, and a plasma discharge is established at a potential of 500 to 1000 V with the workpiece as cathode. The electrical power dissipation heats the steel surface, which is bombarded with nitrogen ions. The process is energy-efficient and about three times as fast as gas nitriding at the same temperature. It can also be operated at lower substrate temperatures (down to 350 °C) and can therefore be applied to high strength steels, such as tool steels, which would be over-tempered at a higher process temperature.

In both gas nitriding and plasma nitriding a thin surface layer of iron nitrides (mixed ε-Fe_2N (h.c.p.) and γ-Fe_4N (c.c.p.)) may form, known as *white layer*. This usually has a deleterious effect since, although hard, it cracks easily and can promote fatigue failure in the substrate. Formation of white layer can be avoided by careful control of the gas composition (in gas nitriding) or the process conditions (in plasma nitriding); if formed it can be removed by grinding or by chemical dissolution.

In the process known as *nitrocarburizing* the formation of a compound layer very similar to the white layer is deliberately employed to provide a hard and wear-resistant surface on cheap steel substrates. In addition to the relatively thin hard layer, the method also produces significant hardening of the underlying steel through inward diffusion of nitrogen. The composition of the hard layer depends on the details of the process: it is usually described as ε-$Fe_2(C,N)$ although it may also contain some oxygen and sulphur. Like nitriding, nitrocarburizing is carried out on carbon steels in the ferritic condition; mild or very low alloy steels are commonly used, although the process can also be used with higher strength alloys. The hardness of the surface layer is lower than that produced by nitriding (typically 500 to 650 HV), but the process is much faster. Layer thicknesses up to about 20 μm are used, taking about 2 hours to form. Molten salt bath processes, often misleadingly referred to as *salt bath nitriding*, provided the earliest nitrocarburizing methods and were based on sodium cyanide and cyanate. The proprietary names of these processes (e.g. 'Tufftride') are sometimes used generically. The early bath compositions are now being replaced with thiocyanate baths of lower toxicity, and gas and plasma processes have also been developed, often using methane as the source of carbon, in addition to ammonia.

Boronizing, also known as *boriding*, involves the diffusion of boron into the surface of a metal to form a hard layer of metal boride. The process is predominantly applied to steels and is carried out above the austenite transformation temperature, between 800 and 1050 °C (see Fig. 8.2). In steels, boronizing forms compound layers up to 200 μm thick, often consisting of two distinct phases: an outer layer of FeB (orthorhombic), and an inner layer next to the substrate of Fe_2B (body-centred tetragonal). The iron boride layers are very hard (>1500 HV), and can provide excellent resistance to abrasive wear. The structure of the two-phase boride layers on a low-carbon steel is shown in Fig. 8.8.

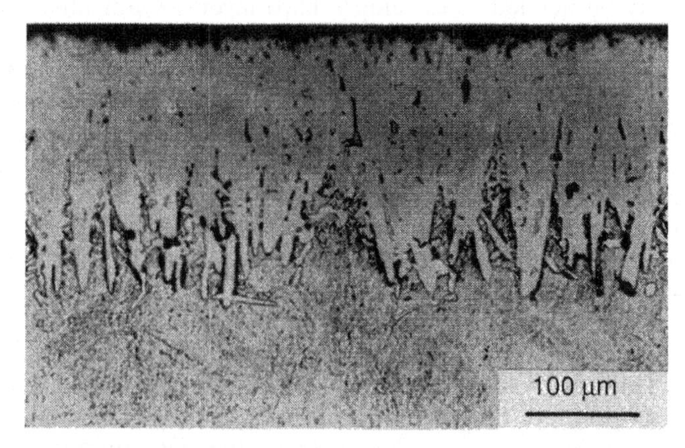

Fig. 8.8 Typical microstructure of the iron boride layer produced on a low-carbon, low-alloy steel by pack boronizing (courtesy of A J Ninham)

The two iron boride phases have different coefficients of thermal expansion and therefore develop differential residual stresses on cooling from the processing temperature. Cracking at the interface between the phases sometimes originates from this source. By control of the process conditions formation of FeB can be reduced or prevented, giving a preferable single-phase compound layer. The irregular, deeply indented interface between the Fe_2B layer and the steel substrate in Fig. 8.8 is characteristic of boride layers formed on low carbon steels, and provides a strong mechanical bond to the underlying steel. Alloying elements in the steel not only reduce the rate of boride layer growth (and hence the thickness for fixed treatment conditions), but also change the morphology of the interface. Chromium, nickel and carbon all have similar effects: increasing their concentration leads to a smoother interface. At a chromium content of 12% in a low carbon steel the interface is completely smooth.

The most common process is *pack boronizing*, carried out in a similar way to pack carburizing. The granular boronizing medium consists of a mixture of boron carbide (B_4C), an inert diluent such as silicon carbide or alumina, and an activator (e.g. KBF_4). The function of the activator is to transport boron in the gas phase to the steel surface; it is regenerated by reaction with the boron carbide. The boride layer formed on low alloy steels is typically 150 μm thick for a treatment time of 6 hours at 900 °C. Pack boronizing is an attractive process because it can be carried out with simple plant. Other methods involving either molten salt baths or gas phase processing (with BCl_3 as the boron source) are available, and plasma boronizing is under development. In common with other surface treatment processes carried out in the austenite region, boronizing may be followed by heat treatment in order to develop optimum properties in the core steel.

Boronizing is not confined to steels. The process can also be applied to tungsten carbide/cobalt cermets, in which cobalt borides (CoB, Co_2B and Co_3B) and tungsten carboborides ($W_2(C,B)_5$ and $W(C,B)_2$) are formed, and to titanium and its alloys which form titanium boride layers (TiB and Ti_2B) with hardnesses up to 2500 HV.

Boronizing is one of several processes sometimes collectively known as *metallizing* or *metalliding*, in which hard surface layers are formed by reaction between a constituent of the substrate and one or more metals or metalloids. *Chromizing* is another such process for the surface hardening of carbon and tool steels, in which chromium reacts with carbon from the steel to form a surface layer of chromium carbide. In this process the rate-controlling step is the diffusion of carbon towards the surface of the steel, and the process may have to be preceded by carburizing to ensure that the initial carbon concentration is sufficiently high. The surface layer formed by chromizing is characterized by very high hardness (>1500 HV) which is retained to high temperature (typically 700 °C). Pack, gas-phase and molten salt bath processes are all available, and treatment temperatures from 900 to 1000 °C produce chromium carbide layers up to 40 μm thick.

Other transition metal carbides, as well as chromium carbide, are formed in the salt bath process known as the *Toyota Diffusion (TD) process*. The method has been developed primarily to provide extremely hard wear resistant layers of chromium, niobium, titanium and vanadium carbides on tool and die steels. Layer thicknesses are typically 5 to 10 μm, formed by immersion for up to 10 hours in a salt bath containing molten borax and iron alloys of the appropriate carbide-forming element. The salt bath temperature, between 800 and 1050 °C, is usually chosen so that the steel can be quenched directly from the bath. Hardness values of 3200 to 3800 HV have been reported for vanadium and titanium carbides, and 2500 to 3000 HV for niobium carbide, formed by the TD process. These carbides can also be formed by other thermochemical processes; for example, a pack process conducted in a reactive gas (methane + argon) has been developed for chromium, titanium and vanadium carbides, and it is likely that other processes based on gas-phase and plasma treatments will become commercially important.

Although we have concentrated above on processes applied to steels, similar principles can be adopted to form hard surface layers on non-ferrous metals. Processes exist for surface hardening alloys of copper, aluminium and titanium, in which the electrolytic deposition of a metallic alloy on the surface is followed by heat treatment; interdiffusion during the latter phase leads to the formation of hard compounds, giving a hard surface layer typically 20 to 30 μm thick. For example, in the *Delsun* process for brasses and bronzes an alloy of antimony, cadmium and tin is electrodeposited, and then followed by a diffusion treatment at 400 °C. The surface layers formed by such processes are much softer than the layers formed on steels by the methods discussed above (e.g. 450 to 600 HV on brasses and bronzes, and 200 to 500 HV on aluminium alloys), but are nevertheless substantially harder than the substrate alloy and give significant tribological benefits.

8.3 SURFACE COATINGS

8.3.1 Plating and anodizing processes

One of the earliest processes for depositing a coating of one metal on the surface of another was *electroplating*, and the method is still widely used for both decorative and engineering purposes.

Electroplating or *electrodeposition* involves the reduction of metallic ions at the surface of the substrate, which is made the cathode in an electrolytic cell. The process is illustrated schematically in Fig. 8.9. The relevant cathodic reaction is:

$$M^{n+} + ne^{-} = M$$

Although many different metals and alloys can be electrodeposited from aqueous solution, and some ceramic materials can be electrodeposited from molten salt baths, the coatings of major tribological interest are chromium and nickel. These coatings are often termed *hard chrome* and *hard nickel*, to distinguish them from the much thinner coatings of different properties used for decorative purposes. Steel substrates are often plated with these materials. Because both are deposited from aqueous solution at low temperature

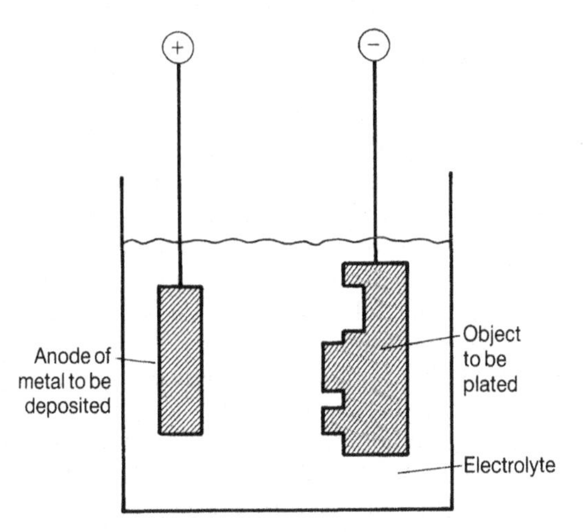

Fig. 8.9 Schematic illustration of an electrolytic cell used to form an electroplated coating

(below 70 °C), the properties of the steel are not influenced by thermal effects; however, in both cases steel substrates can suffer hydrogen embrittlement caused by absorption of atomic hydrogen generated during the plating process. Careful heat treatment is needed to avoid this.

Chromium electrodeposits for tribological purposes are hard (850 to 1250 HV as deposited) and show a low coefficient of friction in dry sliding against steel (typically 0.15 to 0.2). They are plated from a chromic acid bath, with thicknesses typically from 10 to 500 μm, much thicker than the submicrometre chrome plating used for decoration. The microstructure of electroplated hard chromium is extremely fine-grained (~10 nm), containing oxide inclusions and microcracks which give the coating some porosity.

Nickel is also electrodeposited from acidic solution, and has a hardness of up to 400 HV. Layers up to several millimetres thick can be plated, for example to salvage worn parts.

A process widely used to form nickel-based coatings for tribological applications is *autocatalytic* (or *electroless*) plating, in which the deposition of metallic nickel from a solution containing nickel ions and a reducing agent is catalysed by the substrate surface and no external current is applied. Reducing agents containing phosphorus and boron are used to produce autocatalytic coatings of nickel containing about 10% phosphorus or 5% boron respectively. Electroless nickel-phosphorus has a hardness of ~500 HV, and nickel-boron ~700 HV; coatings can be applied to a wide range of metallic and non-metallic substrates. As for electroplating, autocatalytic plating can cause hydrogen embrittlement in steels and suitable precautions must be taken. Autocatalytic plating offers a significant advantage over the electrolytic processes in producing a much more even thickness of deposit; in the electroplating of complex shapes it is, in contrast, often difficult to produce a coating of constant thickness.

Both electrolytic and autocatalytic processes can be modified to incorporate fine particles (typically 0.5 to 5 μm in size) into the growing film. Hard particles such as diamond, silicon carbide or alumina, as well as solid lubricants such as PTFE can be incorporated into the growing metallic matrix to provide coatings with enhanced wear resistance or lower friction. Several proprietary coating processes have been developed in this way.

An electrochemical process used to form hard coatings on aluminium is *anodizing*, in which a layer of partly hydrated alumina is grown on the substrate by anodic reaction in a sulphuric acid solution. Although other metals also form anodic films, aluminium and its alloys are the only ones for which practical use is made of the process.

The method known as *hard anodizing* produces coatings most suitable for tribological purposes, with thickness from 25 to 150 μm and hardness from 350 to 600 HV. Since the alumina layer is formed by oxidation of the substrate, anodizing should strictly be viewed as a surface modification process rather than a coating method; it is, however, convenient to discuss it here together with other electrochemical processes. Not all aluminium alloys are suitable for hard anodizing: alloys containing more than 10% silicon or 5% copper, for example, experience severe local attack in the anodizing bath. Low alloy contents usually lead to harder films. It should also be noted that the process causes some loss of fatigue strength (by up to 50%). Anodized coatings are porous, and use can be made of this porosity by impregnating the coating after it is formed with solid lubricants such as PTFE or MoS_2. Proprietary coatings made in this way can give very low friction (determined by the solid lubricant) and high hardness (determined by the alumina film).

8.3.2 Fusion processes

There are two classes of coating process in which the coating material is applied in the molten state, which we can conveniently consider together as *fusion processes*. These are *welding* and thermal spraying.

In the first, the coating material is melted in contact with the substrate by techniques very similar to those used to join materials by welding; the surface of the substrate is heated during the process to the melting temperature of the coating material. In the second, the coating material is melted some distance from the substrate, and projected towards it in fine molten droplets; the substrate remains relatively cool and the coating is formed by solidification of the droplets as they strike the surface. These two types of method differ in the temperature of the substrate during the coating operation, and also in the nature and thickness of the coating which can practicably be applied.

Welding methods, often known as *hardfacing* processes, are best suited to the application of thick coatings, from 1 to 50 mm or more, of metallic materials. All conventional fusion welding methods can be used to form coatings, with the coating material being fed into the fusion zone either as a filler rod, consumable electrode or previously applied paste. The methods commonly available, which are listed in Table 8.2, range from simple manual oxy-acetylene welding to fully automated submerged arc and electroslag methods. Laser methods, although not listed, are also being developed.

The thick coatings which can be applied by welding methods are most

Table 8.2 Comparison of hardfacing processes (from UK Department of Trade and Industry, *Wear Resistant Surfaces in Engineering*, HMSO, 1986)

Method	Form of filler	Approximate minimum deposit thickness (mm)	Dilution of deposit (%)	Usual mode of application	Typical deposition rate (kg h^{-1})
Oxy-acetylene, with welding rods	Bare wire, rod or tube	0.5	1 to 5	Manual	$\frac{1}{2}$ to 3
Oxy-acetylene with powders	Powder	0.08	1 to 5	Manual	$\frac{1}{2}$ to 7
Tungsten-inert gas (TIG)	Bare rod, wire or tube	1	5 to 10	Manual	$\frac{1}{2}$ to 2
Plasma transferred arc	Powder	0.25	5 to 30	Fully automatic	$\frac{1}{2}$ to 7
Shielded metal arc	Flux coated wire, rod or tube (manual), flux cored wire (semi-automatic)	2	10 to 30	Manual	1 to 3
				Semi-automatic	2 to 10
Open arc	Tubular wire, which may be flux-cored	2	15 to 25	Semi-automatic	2 to 10
Metal inert gas (MIG)	Bare wire or tube	2	10 to 25	Semi-automatic	2 to 10
Submerged arc	Bare wire, tube or strip	2	15 to 35	Fully automatic	2 to 70
Electroslag	Bare rod or tube	20		Fully automatic	50 to 350

suited to applications where the wear rate of the coating will inevitably be high, or where the applied loads will produce high stresses at a considerable depth beneath the surface. They are therefore widely used to provide resistance to high stress abrasive wear (see Section 6.1): for example, in mining, quarrying and agricultural applications. Welding methods can be used to apply hard facings to localized areas, and several techniques are portable and can therefore be used to apply or repair coatings under field conditions.

The restriction that hardfacing materials must be molten at the temperature of the welding process imposes limitations on the materials which can be used. Ceramics for example are, in general, too refractory. For this reason, welded coatings are usually metallic or contain a fusible metallic phase; these same materials can also be fabricated by casting into bulk components, and are often also used in that form. Hardfacings can be grouped together, in order of increasing hardness, as austenitic manganese steels, hard (martensitic and tool) steels, cast irons with high carbide content (e.g. chromium-containing austenitic and martensitic irons) and cemented carbides (usually tungsten carbide/cobalt cermets). Nickel- or cobalt-based alloys may also be used. The melting points of these materials limit the choice of substrate, since in the welding process its surface is heated to the same temperature as the coating material (typically 1400 °C). Considerations of miscibility in the liquid state, and the possible formation of intermetallic compounds by interdiffusion, also constrain the choice and, in practice, hardfacing methods are applied almost exclusively to steels.

In the *thermal spraying* processes, the second group we shall consider in this section, many of the restrictions of the welding methods are removed. Here, the temperature of the molten droplets of coating material is much higher than that of the substrate. Since the droplets solidify rapidly on striking the surface, to which the rate of heat transfer is much lower than in welding, substrate temperatures remain typically below 200 °C. Not only can coatings of refractory metals and ceramics be applied by these methods, the choice of possible substrate materials is also much wider than for hardfacing. Two sources of heat can be used in thermal spraying: combustion of a gaseous fuel (as in *flame spraying* or the *detonation-gun* process) or an electrical discharge (as in the *electric arc spray* process and in *plasma spraying*).

In *flame spraying* the coating material is fed as a wire or fine powder into a flame, usually of oxy-acetylene although propane or hydrogen is sometimes used as a fuel. Figure 8.10 illustrates the principle of operation of a wire-fed spraying torch; in this design an external supply of air is used to enhance the acceleration of the molten droplets once they are formed. The flame temperature is of the order of 3000 °C; the droplets are heated to over 2000 °C and strike the substrate at about 100 m s^{-1}. Metallic wires are readily sprayed; rods of ceramic materials and combustible tubes containing powders of ceramics or cermets can also be used. In other designs of flame-spraying gun the material is fed directly into the flame in powder form.

The *electric arc spray* process is closely similar in principle; an arc is struck between consumable wire electrodes giving a local temperature of >4000 °C, and a jet of gas (usually air) is used to project the molten droplets of the wire

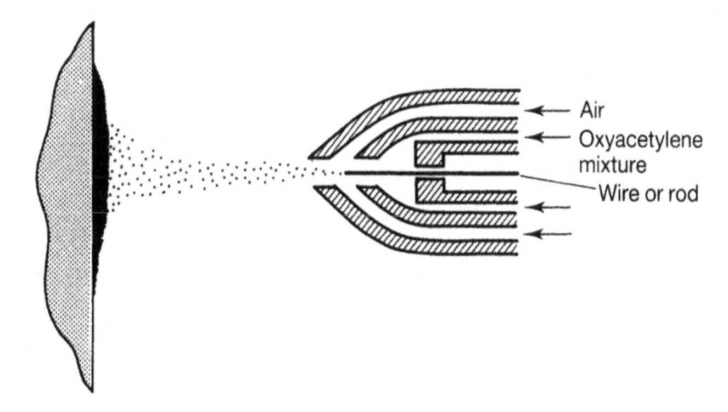

Fig. 8.10 Schematic diagram of a wire-fed flame spraying torch

material at the substrate. In all these thermal spraying processes carried out in air, significant entrapment of air occurs during deposition, which leads to porosity in the final coating. Inevitably some oxidation of the molten droplets also occurs in the spraying of metals. Porosity levels in the coating may be as high as 20%, reducing the mechanical strength, although for some tribological applications in lubricated sliding, controlled porosity can provide beneficial local reservoirs of lubricant. Porosity in metallic coatings applied by flame spraying can be reduced by heating the surface after spraying (e.g. with an oxy-acetylene flame or by r.f. induction) to fuse and consolidate the coating. Such *spray-fuse* processes can give effectively zero final porosity and good metallurgical bonding between the coating and substrate, but involve high substrate surface temperatures during the fusion process. In this respect they are similar to the welding methods described above.

Another method of coating in which thermal and kinetic energy is derived from combustion and which gives well-bonded coatings of low porosity is the *detonation-gun*, or *D-gun*, process. In this proprietary method, powdered coating material is fed into a tube in which pulsed detonations of an oxy-acetylene gas mixture take place. These rapidly repeated explosions (4 to 8 times per second) give a flame temperature of ~3000 °C and accelerate the molten droplets to ~800 m s^{-1}. At such high velocity the droplets form a dense well-consolidated coating on the substrate, with porosity levels as low as 0.5 to 1%.

Plasma-spraying methods are widely used; a typical design of plasma spray gun is shown schematically in Fig. 8.11. A plasma is formed in an inert gas (usually argon with a small proportion of hydrogen or helium) by a high-energy electric arc (typically 40 kW). The coating material is fed into the plasma as a fine powder, where it melts. The very high temperature of the plasma (15000 °C with argon) enables a wide range of coating materials to be sprayed. Rapid expansion of the hot gas accelerates the molten droplets to 250–500 m s^{-1}, and the combination of high melting temperature with high kinetic energy leads to coatings of lower porosity than those produced by flame spraying. Oxidation of metallic coating materials is also less pronounced, since the carrier gas is inert. However, plasma spraying is usually

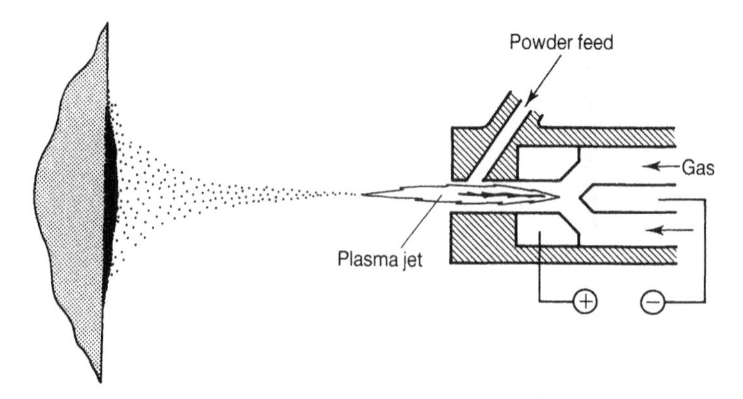

Fig. 8.11 Schematic diagram of a plasma spray gun

carried out in air, and entrapment of gas still causes some porosity; levels are typically 1 to 10%, with the higher levels being associated with ceramic materials of high melting point. Lower porosity can be achieved in the recently-developed *vacuum plasma spray* (or *low-pressure plasma spray*) process, in which spraying takes place in a partial vacuum. Very low levels of porosity and correspondingly stronger and stiffer coatings result. Figure 8.12 illustrates the microstructures of alumina coatings produced by plasma spraying in air and in vacuum; the difference in porosity is evident.

The thermal spraying processes can be used to apply coatings of a wide range of materials, including metals and cermets; refractory ceramic materials such as oxides can also be applied by the detonation-gun and plasma spraying processes which involve higher thermal and kinetic energies. In each case the properties of the coating depend strongly on the process conditions. The two most important considerations are the strength of the coating and the integrity of its bond with the substrate. Strong coatings are associated with complete interfusion between the molten droplets on striking the surface, and with low porosity. As we have seen, the porosity is determined to some extent by the nature of the coating process and some methods (notably the spray-fuse, detonation-gun and vacuum plasma spraying processes) are capable of producing coatings with lower porosity than others. The effect of coating microstructures can also be very significant. For example, pure alumina coatings deposited by vacuum plasma spraying have been found to show resistance to wear by solid particle erosion (with silica erodent) up to ten times that of a coating of the same composition plasma-sprayed in air. This difference can be largely ascribed to the difference in porosity between coatings produced by the two methods (see Fig. 8.12).

In order for a dense, strong coating to enhance the tribological properties of a substrate, it must remain firmly attached to the substrate and wear must occur within the coating rather than by its detachment from the substrate. For this reason, correct surface preparation before coating is very important. The bonding mechanism of weld-applied coatings is metallurgical, whereas for plasma-sprayed coatings it depends more on mechanical interlocking. Surface

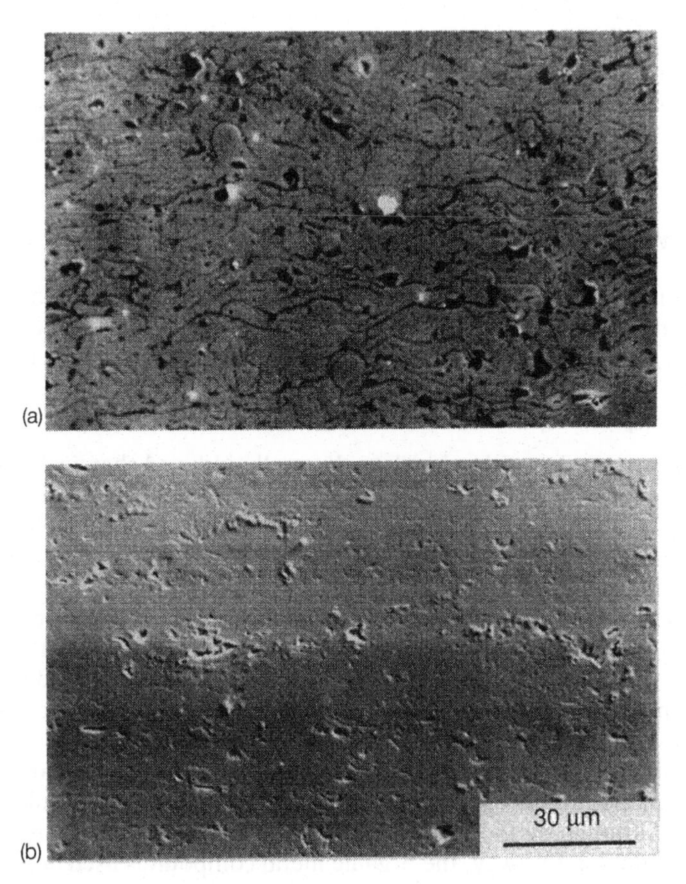

Fig. 8.12 Microstructures of alumina coatings produced by plasma spraying (a) in air, and (b) in a partial vacuum (courtesy of X X Zhang)

preparation by cleaning and roughening (for example, by grit-blasting) is particularly important before thermal spraying. Even a strong interfacial bond may still fail if sufficient stress is applied to it, and an important source of such stress is differential thermal contraction between the coating and substrate as they cool together from the deposition temperature. Residual stresses from this source limit the usable thickness of many ceramic coatings to 0.2 to 0.5 mm; for metallic coatings on metals the difference in expansion coefficients is usually less, and thicker coatings (up to several mm) are readily deposited by thermal spraying methods. In order to improve the interfacial adhesion of some sprayed coating materials (particularly ceramics) *bond-coats* are often used. These are intermediate strongly-adherent coatings which provide a strong mechanical key for the final sprayed coating. Nickel-aluminium powder is often used to form a bond-coat in thermal spraying; an exothermic reaction which occurs when the particles are heated in the spraying process raises their temperature still further and ensures good bonding. Molybdenum powder, which bonds readily to most substrates, is also used. In some thermal spraying processes the composition of the powder can be changed progressively during spraying by mixing, so that coatings of graded composition can be produced, with a steady decrease in the concentration of the bond-coat material into the coating away from the surface of the substrate.

8.3.3 Vapour phase processes

Methods in which a coating is formed from the vapour phase can be divided into two groups, described as *chemical vapour deposition* (CVD) and *physical vapour deposition* (PVD). We shall examine these two groups in turn.

Chemical vapour deposition

Chemical vapour deposition involves thermally-induced chemical reactions at the surface of a heated substrate, with reagents supplied in gaseous form. These reactions may involve the substrate material itself, but often do not. CVD coating processes employ temperatures typically in the range 600 to 1100 °C, although, as discussed below, efforts are being made to develop processes which will operate at lower temperatures. At these temperatures significant thermal effects may occur in the substrate material; steels, for example, will often be heated into the austenite phase region and the coating process must then be followed by suitable heat treatment to optimize the properties of the substrate.

The simplest CVD process involves the pyrolytic decomposition of a gaseous compound on the substrate to provide a coating of a solid reaction product. Although methods have been developed to deposit many metals as well as alumina in this way, the organic metal compounds used are hazardous and other methods tend to be preferred. Reaction of metal halides, for example, with hydrogen, nitrogen or methane is commonly used to provide coatings of pure metals or their nitrides or carbides. For example, the following reactions are used to produce solid coatings of tungsten metal, titanium carbide and titanium nitride respectively:

$$WF_6 + 3H_2 = W(\text{solid}) + 6HF(\text{gas})$$
$$TiCl_4 + CH_4 = TiC(\text{solid}) + 4HCl\ (\text{gas})$$
$$TiCl_4 + \tfrac{1}{2}N_2 + 2H_2 = TiN(\text{solid}) + 4HCl(\text{gas})$$

Figure 8.13 illustrates the arrangement of typical CVD equipment, as set up to deposit a coating of titanium carbide by the second of the reactions listed above. The gaseous hydrogen chloride product is removed in the carrier gas (in this case hydrogen) which is used to introduce the metal halide into the reaction vessel.

Alumina may be deposited by the reaction:

$$Al_2Cl_6 + 3CO_2 + 3H_2 = Al_2O_3(\text{solid}) + 3CO(\text{gas}) + 6HCl(\text{gas})$$

With mixtures of metal halides, hydrogen, oxygen, nitrogen, hydrocarbons and boron compounds, coatings of a wide range of metals and their oxides, nitrides, carbides and borides can be formed by CVD.

Although the reactions listed above involve only gaseous reagents, the substrate material in some cases also plays a significant role. For example, the rate-controlling step in the growth of a TiC layer on carbon steels or cemented carbides from $TiCl_4$ is the reaction

$$TiCl_4 + C + 2H_2 = TiC(solid) + 4HCl(gas)$$

which involves carbon from the substrate. This reaction can lead to decarburization of the substrate immediately beneath the coating, and must be taken into account in selecting a suitable substrate material.

Equipment for CVD coating usually consists of a heated reaction vessel, as shown in Fig.8.13, and associated gas-handling equipment; in most cases the substrate is heated by convection and radiation within the vessel. The process is controlled by varying the temperature of the workpiece and the composition and pressure of the gas mixture. Optimum deposition rate is achieved at a gas pressure below atmospheric; the rate depends strongly on the process conditions, but is typically 0.1 to 1 μm min^{-1}. Final thicknesses of up to 10 μm are commonly used, but thicker layers may be formed for some applications (e.g. corrosion resistance). Coatings formed by CVD often have a columnar microstructure, although the initial deposit is sometimes equiaxed; the grain size and microstructure depend strongly on the process conditions. In many cases, the most desirable microstructure may not be produced at a high deposition rate, and a compromise may have to be struck between a high growth rate, desirable for economic reasons, and a fine-grained microstructure with the best tribological properties.

Because of the high process temperature, appreciable interdiffusion can occur between coating and substrate, and CVD coatings therefore generally show strong adhesion. Diffusion can, however, lead to problems with the formation of brittle intermetallic compounds at the interface in some systems, which cause poor mechanical properties. Undercoats of other materials which

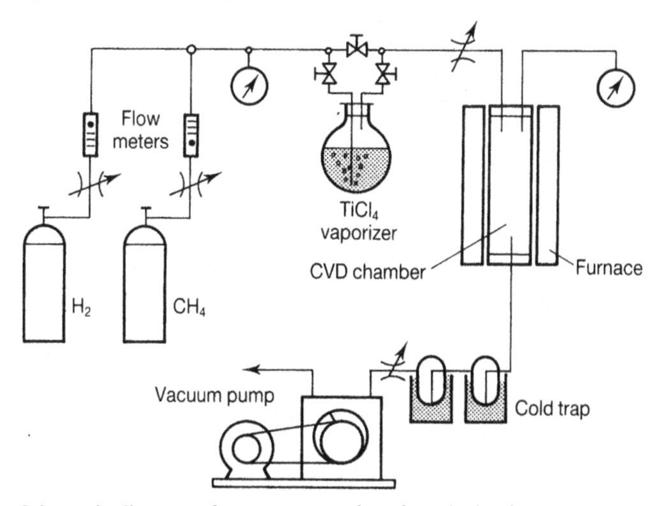

Fig. 8.13 Schematic diagram of apparatus used to deposit titanium carbide coatings by CVD (from Peterson M B, Ramalingham S and Rigney D A, in *Fundamentals of Friction and Wear of Materials*, Kigney D A (Ed.) ASM, pp. 331–372, 1981)

do not form such compounds with either the substrate or the final coating can be used as diffusion barriers to avoid this problem. For some applications a sequence of several CVD coatings can be applied, which not only provide chemical compatibility and reduce the problems of interdiffusion with the substrate, but also provide a gradation of mechanical, chemical and physical properties. For example, cemented carbide cutting tools coated with a three-layer sequence of titanium carbide, titanium carbonitride and titanium nitride show significantly longer lives than tools with a single layer coating; multilayer coatings with as many as ten components have also been used in such applications.

Future developments of CVD coating methods will probably aim to reduce the process temperature. One way in which this can be achieved is in *plasma-assisted CVD* (PACVD), in which an electrical discharge in a low-pressure gas is used to accelerate the kinetics of the CVD reaction. In this way process temperatures as low as 200 °C can be used, although higher temperatures may still be necessary to promote strong coating adhesion.

The plasma-assisted CVD method is one of many related techniques which can be used to deposit thin films of diamond, which through their very high hardness and potential for low friction (see Section 3.6) have attractive tribological properties. These methods involve the decomposition of a gaseous carbon-containing precursor in the presence of hydrogen and, often, oxygen: for example, a mixture of methane, hydrogen and carbon dioxide gases may be used. Considerable variation in film properties and growth rate

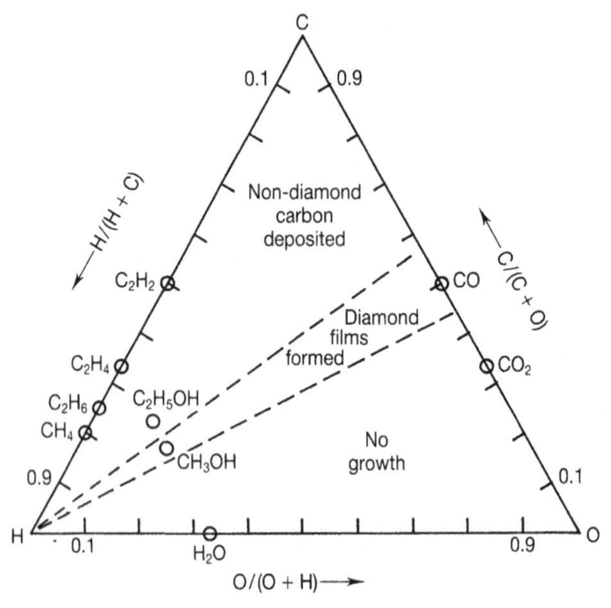

Fig. 8.14 Illustration of the domain of precursor gas composition, in terms of atomic C:H:O content, over which the growth of diamond films by energy-enhanced CVD methods has been achieved (from Bachmann P K, Leers D and Lydtin H, *Diamond and Related Materials* **1**, 1–12, 1991)

is seen with gas composition and process conditions, and the optimum process is still being sought. However, it seems that the precursor composition must lie within a well-defined domain of carbon, hydrogen and oxygen content for diamond films to be produced, as shown in Fig. 8.14

Physical vapour deposition

In the processes collectively known as physical vapour deposition (PVD), the coating material is transported to the surface in atomic, molecular or ionic form, derived by physical rather than chemical means from a solid, liquid or gaseous source. Chemical reactions may, but need not, occur at the surface of the substrate, which is generally much cooler than in CVD coating (typically from 50 to 500 °C). The fact that PVD processes are carried out at relatively low temperatures makes them very attractive, since the coating process may then not influence the microstructure and properties of the underlying substrate material.

The simplest PVD process is *evaporation*, which has been used for many years to produce coatings on glass lenses and other optical components. The principle of the method is illustrated in Fig. 8.15(a). The coating is formed by the evaporation of material from a molten source, often heated by an electron beam; the process is carried out in vacuum ($\sim 10^{-3}$ Pa), and is restricted to coating materials of relatively low boiling point which do not dissociate on heating, such as metals. If a suitable reactive gas is present (e.g. methane, oxygen or nitrogen), chemical reactions can be induced with the evaporated species on the substrate surface, and coatings of compounds can be laid down. The process is then known as *reactive evaporation*.

In evaporative processes the kinetic energy of the atoms striking the substrate results only from their thermal energy. On a cool substrate they lose this energy quickly, and little mixing occurs between the surface atoms of the substrate and those of the coating. Adhesion is therefore weak, and is further

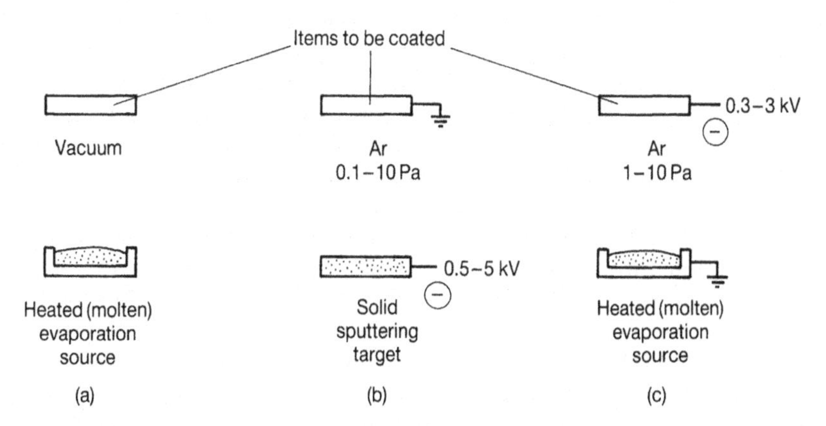

Fig. 8.15 Schematic illustration of the principal methods of physical vapour deposition (PVD): (a) evaporation; (b) sputtering; (c) ion plating. All these methods may be modified by the incorporation of a reactive gas into the system

reduced by any contaminants (adsorbed gases etc) on the surface. Some improvement in adhesion can be achieved by heating the substrate during or after coating to allow some interdiffusion, but the bond strengths of evaporated coatings are often too low for tribological purposes. Another limitation of evaporative methods is that the coating is transported along a line of sight from source to substrate. An even coating is difficult to achieve and the coating of complex shapes is sometimes impossible.

In *sputtering* the energy to transport material from the source to the substrate is supplied by energetic heavy gas ions. The process is illustrated in Fig. 8.15(b). Positive ions are formed in a glow discharge in a low pressure gas (usually argon at 0.1 to 10 Pa) and strike the solid source material, which is negatively biased, with energies up to a few keV. This ion bombardment causes atoms to be sputtered from the target which then strike the substrate a short distance away. With a direct current source the process is limited to coatings of conducting materials, but the use of an alternating (radio-frequency) source removes this limitation. The diode arrangement shown in Fig. 8.15(b) is the simplest; higher sputtering rates can be gained by increasing the ionization efficiency, and magnetron sources are often used for this purpose. Diode systems give coating rates typically of 1 to 100 nm min^{-1}, while with a magnetron source the rate may be as high as 2 μm min^{-1}.

Sputtering is a versatile method, which can be used to form coatings of a wide range of materials. The process may also be modified by the incorporation of a reactive gas (*reactive sputtering*), and this process is widely used to deposit oxides, nitrides and carbides. Because atoms sputtered from the source have much higher energies (typically tens of eV) than the thermal energies of evaporated atoms (fractions of 1 eV), they undergo more interdiffusion and mixing with the surface atoms of the substrate, giving stronger adhesion than in evaporated coatings.

The third important PVD process is *ion plating*, in which atoms or molecules of the coating material are evaporated from a hot source into a glow discharge, usually in argon at a pressure of 0.1 to 10 Pa. Figure 8.15(c) illustrates the process. The vapour source may be heated resistively or by an electron beam, or may be provided by an arc struck against a solid source material. Some atoms of the vapour become positively ionized in the plasma and are accelerated towards the substrate, which is held at a negative potential of 2 to 5 kV. Others move under their thermal velocities, enhanced by collisions with the energetic argon ions, and also strike the substrate, which is also being bombarded continuously by argon ions. The high energies of the atoms at the substrate surface, together with the scattering provided by collisions with the argon ions, provide a uniformly distributed coating with good adhesion through mixing at the interface. Cleaning of the substrate before coating is readily carried out by sputtering in the glow discharge. Deposition rates for metals of several μm min^{-1} can be achieved by ion plating. As with the other PVD processes described above, incorporation of a reactive gas into the system allows compounds to be be formed at the surface. The process is then termed *reactive ion plating*. This process is used extensively as a commercial process to form coatings of titanium nitride (see Section 7.6.2), with the titanium metal being evaporated into a glow discharge in a mixture of nitrogen and argon gases surrounding the workpiece, which reaches a temperature of about 400 °C. The process can therefore be applied, for example, to tool steels and other hard substrate materials which are microstructurally stable at this temperature, without causing the substrate to become softer.

We have seen that as the energy of the atoms or ions striking the surface is raised, so is the extent of interdiffusion and mixing between the atoms of the coating and the surface layers of the substrate. If ions of high energy strike a suface, they can penetrate to a macroscopic distance, and change the properties of the material to this depth. This provides the principle of the *ion implantation* process. Although it is better described as a method of surface modification rather than coating, it is convenient to discuss it here since the method is physical rather than chemical in action. Ions commonly used for implantation in surface engineering include N^+, N^{2+}, C^+ and B^+, and also metal ions: e.g. Ti^+, Al^+ and Y^+. Their energies are usually in the range from 50 to 200 keV, giving penetration depths of less than 1 μm. Figure 8.16, for example, shows how the concentration of 100 keV nitrogen ions implanted into iron varies with depth beneath the surface. The peak concentration, at this overall dose of 10^{17} ions cm^{-2}, is >10 at.%. Doses of this level are about the minimum which give practically useful changes in tribological properties, and are several orders of magnitude greater than those used to modify the electronic properties of semiconductors. Because of the shallow depth of material affected by ion implantation, the process causes negligible change in the dimensions or surface finish of the substrate. It can be applied to ceramics and cermets as well as to metals.

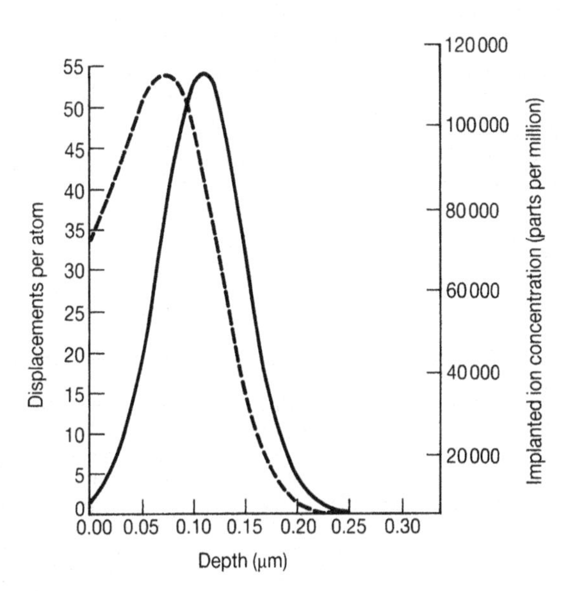

Fig. 8.16 Concentration of implanted ions (solid curve) and number of displacements per atom (broken curve) plotted against depth beneath the surface for pure iron implanted with 100 keV nitrogen ions to a dose of 10^{17} ions cm^{-2} (from Dearnaley G, *J. Metals* **34** (9), 18–28, 1982)

Implantation of energetic ions alters the structure of the surface material in two ways: through the introduction of the implanted species, which may form a solid solution or compound, and also through radiation damage—the introduction of lattice defects through the displacement of matrix atoms by ion-induced collisions. The extent of radiation damage, in terms of the average number of displacements suffered by each atom of the substrate, is also illustrated in Fig. 8.16. Ion implantation leads in many materials to a significant decrease in sliding wear rate, for which several reasons have been suggested. Implantation of some metals by certain ions also leads to a reduction in friction, which may be directly responsible for the reduced wear rate by lowering the tangential forces transmitted on sliding. For example, implantation by Ti^+ ions reduces the coefficient of friction for dry sliding of a hard bearing steel on itself from 0.6 to 0.3, probably by modifying the surface composition, its oxidation characteristics and the nature of the oxide formed. It also leads to a lower wear rate. However, implantation of N^+ ions also reduces the rate of wear of relatively ductile steels (though not hard martensitic steels), despite causing no reduction in friction. In this case the main source of the increased wear resistance is probably the changes in near-surface mechanical properties: in particular, hardness and strain-hardening rate. Microhardness measurements at very small loads have revealed increases in the surface hardness of N^+-implanted steels of from 50 to 100%. Steels containing nitride formers (e.g. aluminium, chromium or vanadium) show a greater effect than others. A further effect of ion implantation, which may contribute to tribological benefits, is the introduction of residual compressive stress in the surface regions.

The processes discussed above and grouped together as PVD methods have a good deal in common. Intermediate techniques can be devised which incorporate features of more than one of the simpler processes described here, and it is possible to combine different processes, either simultaneously or sequentially, to treat a single substrate in one vacuum chamber. In this way, the properties of coatings and interfaces can be tailored for particular applications. Considerable further development of such processes will undoubtedly occur.

Further reading

ASM Handbook, 10th edn., vol. 3, Friction, lubrication and wear technology, ASM International, 1991

Bhushan B and Gupta B K, *Handbook of Tribology: Materials, Coatings and Surface Treatments*, McGraw Hill, 1991

Budinski K J, *Surface Engineering for Wear Resistance*, Prentice Hall, 1988

Child H C, *Surface Hardening of Steel*, Engineering Design Guide No. 37, Oxford University Press, 1980

Gabel M K and Donovan D M, Wear resistant coatings and treatments, in Peterson M B and Winer W O (Eds), *Wear Control Handbook*, ASME, New York, pp. 343–371, 1980

Peterson M B and Ramalingam S, Coatings for tribological applications, in Rigney D A (Ed), *Fundamentals of Friction and Wear of Materials*, ASM, Metals Park, pp. 331–372, 1981

Ramalingam S, New coating technologies for tribological applications, in Peterson M B and Winer W O (Eds), *Wear Control Handbook*, ASME, New York, pp. 385–411, 1980

Towler B, *Flame Deposition*, Engineering Design Guide No. 25, Oxford University Press, 1978

UK Department of Trade and Industry, *Wear Resistant Surfaces in Engineering*, HMSO, 1986

9

Materials for bearings

9.1 INTRODUCTION

A *bearing* is a device which permits two parts of a mechanism to move freely relative to each other in either one or two dimensions, while constraining them in the remaining dimensions. Bearings form vital components in almost all mechanical systems, and are of central concern to the tribologist.

The simplest arrangement is the *linear bearing*, which allows linear motion but can nevertheless support substantial normal loads. Figure 9.1(a) represents a linear bearing in idealized form. More common in engineering use is the *journal bearing*, which permits a cylindrical shaft (the *journal*) to rotate freely, while transmitting radial loads and maintaining the axis of the shaft in a fixed position (Fig. 9.1(b)). A *thrust bearing* (Fig. 9.1(c)) also allows rotation of a shaft, but transmits a load parallel to the axis of rotation. Some designs of shaft bearing can carry both radial and axial loads.

The most important requirement of a bearing is that it operates with low friction. This is usually achieved in one of two ways: with smooth lubricated sliding surfaces in a *plain bearing*, or by interposing balls or rollers between the moving surfaces in a *rolling element bearing*.

Both linear and journal bearings (for axial or radial loads) can be constructed as either plain bearings or rolling bearings. Most types of bearing will function satisfactorily only over limited ranges of operating conditions, and the load carried by the bearing and the sliding speed at which it operates are the two most important factors determining its life. Figure 9.2 illustrates how the limiting operating conditions depend on rotational speed and radial load for journal bearings of various types, for one particular shaft diameter. At low speeds, marginally lubricated and 'dry' plain bearings perform as well as rolling element bearings, but as speed is increased, rolling element bearings offer progressively greater load-carrying capacity. Once the sliding speed is sufficiently high to maintain a hydrodynamic film, hydrodynamically-lubricated plain bearings will carry greater loads than rolling bearings, and can be used at considerably higher speeds, ultimately limited by the strength of the rotating shaft itself.

The choice of bearing type for a given application will usually be based on many factors apart from speed and load, and is the subject of detailed

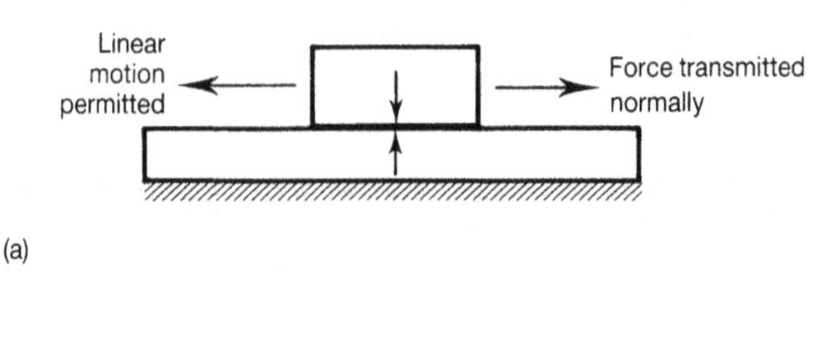

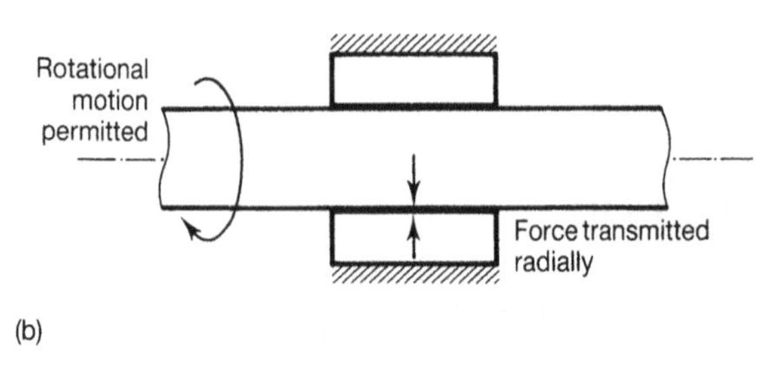

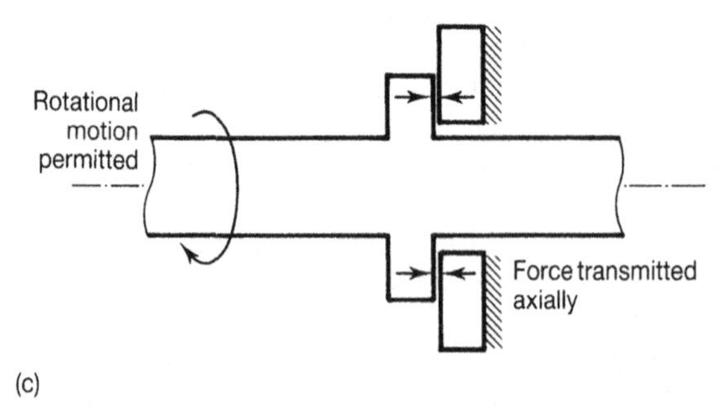

Fig. 9.1 Schematic illustrations of three different types of bearing: (a) linear bearing; (b) journal (shaft) bearing; (c) thrust bearing

discussion in some of the references listed at the end of this chapter. It is sufficient here to note that among other factors which may also be important in the selection and design of bearings are size, weight, cost, precision of shaft location, lubricant requirements and frictional drag. The differing demands of the wide range of applications for which bearings are used has resulted in a correspondingly wide range of types of bearing. In this chapter we shall review these types, paying particular attention to the materials employed in their construction.

9.2 ROLLING BEARINGS

9.2.1 Introduction

In a rolling element bearing, the load is carried by a set of balls or rollers, located between a pair of inner and outer tracks or *races*. The rolling elements

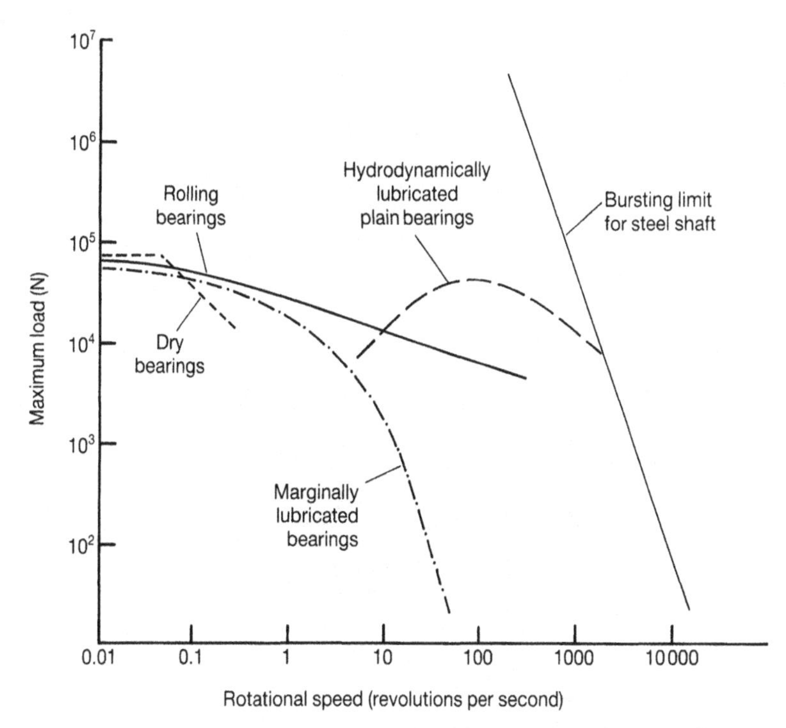

Fig. 9.2 Diagram showing the limiting values of load and rotational speed for journal bearings of various types, for a shaft diameter of 50 mm and a bearing life of 10^4 hours. The curve relating to hydrodynamically lubricated bearings applies to a bearing of length 50 mm (data from Neale M J (Ed.), *Tribology Handbook*, Butterworths, 1973)

are often held in place, regularly spaced around the tracks, by means of a *cage*, which is free to move with the balls or rollers between the tracks. Figure 9.3 shows the design of typical single-row radial ball and roller bearings. These are the simplest designs. Parallel roller bearings can carry high radial loads but essentially no load in the axial direction, whereas radial ball bearings can to some extent withstand loads in both directions. Many other designs of bearing are used, which provide for a wide range of loading conditions : ball thrust races, for example, can carry large axial loads, while tapered roller bearings, used as opposed pairs, combine good radial and axial load capacity. Rolling bearings are manufactured to standardized dimensions, to very high precision and at moderate cost as a result of volume production.

When a purely radial load is applied to one of the types of bearing illustrated in Fig. 9.3, the load is borne to some extent by all the balls or rollers in the load-carrying half of the bearing. The most heavily loaded rolling element carries about $5/Z$ times the applied load, where Z is the total number of balls or rollers in the bearing ($Z>5$). Figure 9.4 shows the geometry of the contact between a ball and its track. For both balls and rollers the contact with the track is counterformal, and the contact pressures resulting from the applied loads are therefore high. These pressures, together

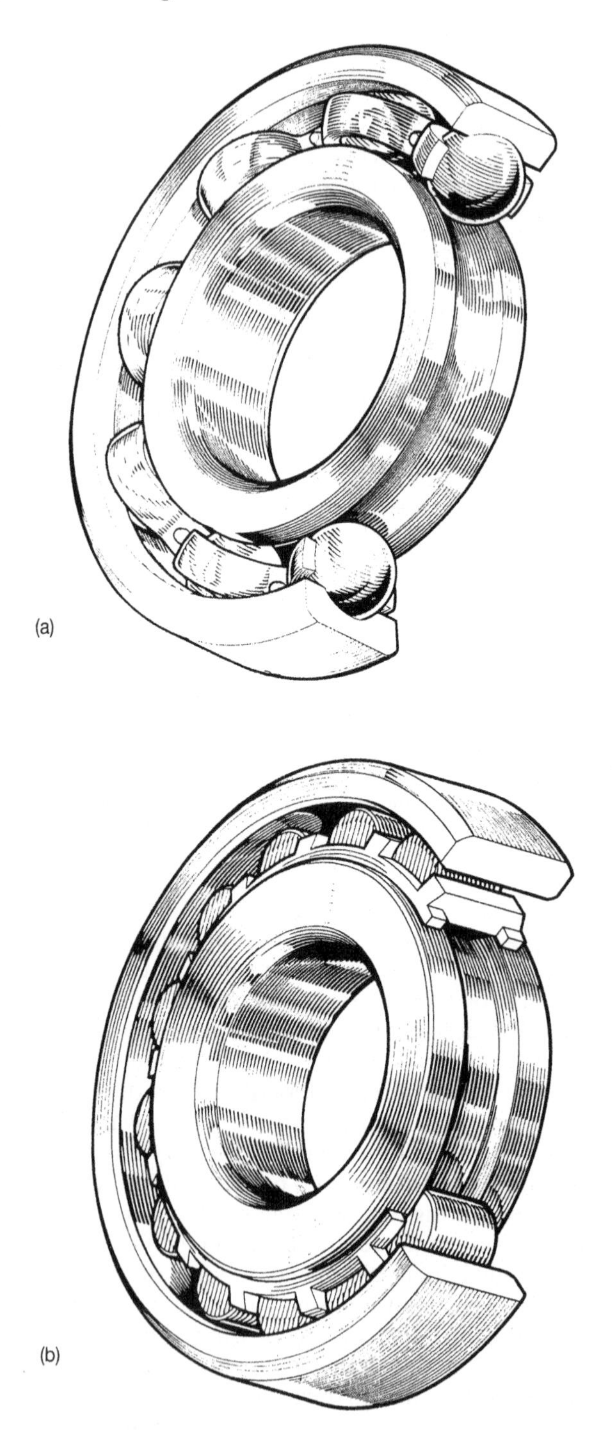

Fig. 9.3 Typical constructions of simple rolling element bearings: (a) single-row radial ball bearing; (b) single row cylindrical roller bearing (courtesy of RHP Industrial Bearings, Newark, UK)

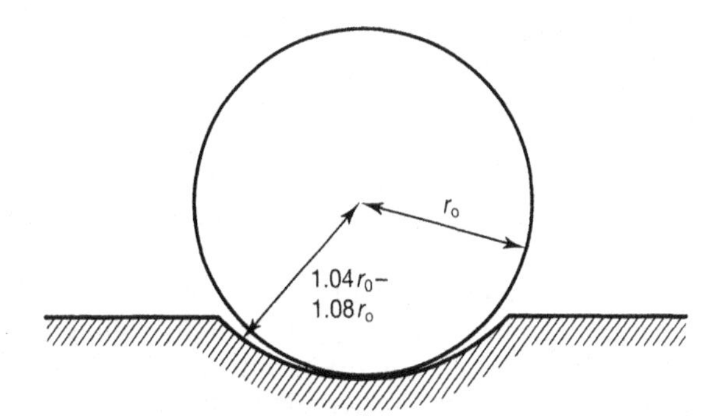

Fig. 9.4 Illustration of contact between a bearing ball and its track. Typically, the radius of curvature of the track is 4 to 8% greater than that of the ball

with the associated subsurface stresses, can be calculated from Hertzian elasticity theory, provided that plastic flow does not occur.

9.2.2 Materials for rolling elements and races

A primary requirement for the operation of a rolling element bearing is that deformation of its component parts should at all times be elastic; in view of the high stresses which result from contact between the balls or rollers and the tracks, this means that the material chosen for them must have a high yield stress. A high resistance to fatigue failure is also needed.

In most applications, steels are used for both the balls or rollers and the tracks, with typical hardnesses of 700 to 800 HV (60 to 64 Rockwell C). The rolling elements may be up to 10% harder than the tracks. The standard alloy, used for general applications, is a through-hardened 1% carbon, 1.5% chromium steel, designated 534A99 or 535A99 (to British Standard BS970: previous designation En 31). Its US designation is 52100 (AISI). The high yield stress of this steel after heat treatment results from a microstructure of high carbon martensite, containing primary carbides. Bearings fabricated from this alloy are suitable for service at up to 125 °C; for non-critical applications they can be used at up to 200 °C. Case-hardened steels (typically chromium-nickel or manganese-chromium low alloy steels with 0.15% carbon) are sometimes used where heavy sections render through-hardening impossible, or where the higher toughness of the core of a case-hardened component is beneficial in resisting shock loading.

For high temperature applications, as in gas turbine engines, high speed tool steels are adopted: commonly in the UK a high tungsten alloy, 18-4-1 (0.7% C, 0.3% Mn, 0.25% Si, 4% Cr, 1% V, 18% W, designated BT1 to British Standard BS4659), and in the USA a molybdenum steel, AISI M50 (0.8% C, 0.3% Mn, 0.25% Si, 4% Cr, 1% V, 4.25% Mo). With these alloys, bearings can operate suitably lubricated at up to 500 °C; both these tool steels give similar performance as rolling bearing materials.

Bearings required to operate in corrosive environments may be made from stainless steels: AISI 440C (1% C, 1% Mn, 1% Si, 18% Cr, 0.75% Mo), a martensitic steel, is a common choice, and can be hardened almost to the strengths attained by standard bearing steels.

For very light loads, bearings with thermoplastic polymer races, typically injection moulded from acetal or polypropylene, can be used with stainless steel balls. These can be significantly cheaper than bearings with steel races as well as being lighter, operating more quietly, needing no lubrication and resisting corrosion.

Excellent corrosion resistance in many environments, together with high load-carrying capacity, can be achieved with ceramic rolling bearings, which are also of considerable interest for operation at very high temperatures. Development of the materials and fabrication methods for such bearings, and of the necessary high temperature lubricants, is still in progress. Ceramic bearings should be usable at temperatures up to 1200 °C. Hot pressed or hot isostatically pressed (HIPed) silicon nitride (Si_3N_4) is currently the most widely favoured ceramic, and with a good surface finish and careful fabrication this material can show a very good fatigue life, significantly better than M50 tool steel. The lower density of ceramic materials (typically 40% that of tool steel) results in another advantage for high speed use: a substantial fraction of the loading on the outer race of a high speed bearing is due to centrifugal forces, which are significantly lower for a material of lower density.

The fatigue behaviour of bearing steels under rolling contact is well characterized, and rolling bearings tend to be designed on the basis of fatigue lifetime predictions, on the assumption that the bearing life will be limited by fatigue cracks which initiate at subsurface flaws and propagate under the cyclic stress field associated with rolling contact. Substantial improvements in the fatigue lives of rolling bearings have resulted from improved steelmaking practice, in particular by the reduction of oxide and other inclusion contents, since these inclusions act as nuclei for subsurface fatigue cracking. Striking benefits have resulted from the introduction of vacuum induction remelting (VIR), vacuum arc refining (VAR) and electroflux remelting (EFR). These improvements are illustrated in Fig. 9.5. Vacuum degassing is now routinely applied to bearing steels, and results in twice the fatigue life of an air-melted steel at little extra cost; the use of the more expensive techniques of VIM/VAR and EFR is restricted to steels for critical applications such as gas turbine shaft bearings, and bearings used in navigational gyroscopes. Over the past 20 years or so, there has been such an improvement in the quality of bearing steels that in most applications bearings now only seldom fail by rolling contact fatigue; more commonly, their life is determined by failure mechanisms involving surface phenomena, which can be collectively termed 'surface distress'. These include pitting, surface-initiated fatigue and various modes of wear, and can be ascribed to defects in the original surface finish of the components, or to poor lubrication, or to contamination by abrasive particles or wear debris. Such failure modes are less well understood than rolling contact fatigue (which originates from subsurface flaws).

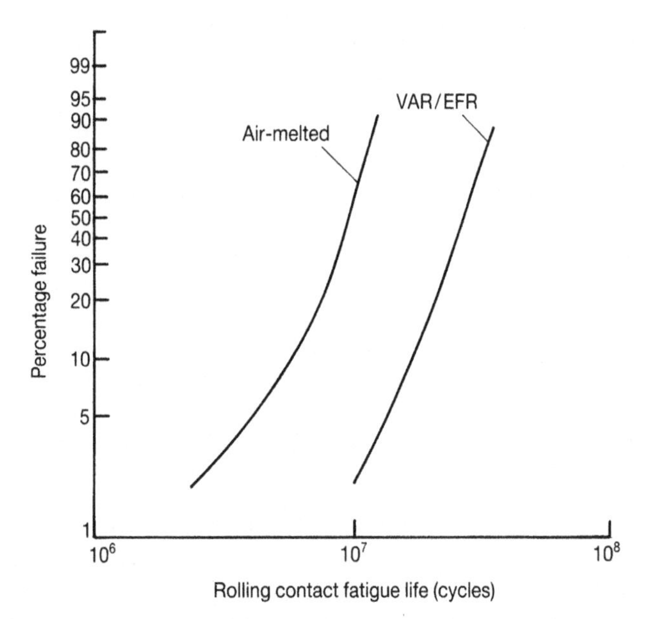

Fig. 9.5 Influence of steel processing methods on the rolling contact fatigue lives of rolling bearings made from AISI 52100 steel. The reduction of oxide and other inclusion contents by the use of vacuum arc refining (VAR), electroflux remelting (EFR) and vacuum induction remelting (VIR) has led to significant improvements in fatigue life (from Oakes G and Barraclough K C, in Meetham G W (Ed.), *The Development of Gas Turbine Materials*, Applied Science Publishers, 1981)

9.2.3 Materials for bearing cages

The cage or separator which retains the balls or rollers between the races is an important component in most designs of rolling bearing (see Fig. 9.3). If a cage is not present, neighbouring balls or rollers will rub against each other with a very high local sliding speed, and damage may result. By separating the rolling elements, the cage can lead to a substantial increase in the safe operating speed of a bearing.

Since the cage rubs against both the inner and outer races and the rolling elements, it must be made from a material which can slide against these components under poorly lubricated conditions without a high wear rate. Mild steel pressings, rivetted or spot-welded together, are commonly used in standard bearings, sometimes with a tin or silver plating or a phosphate coating to lower the friction. Cages machined from 60/40 brass, bronze, or aluminium alloy are also sometimes used. Polymer materials can provide the advantages of low running noise, lightness and low wear rates. Fabric-reinforced phenolic resin is used in some applications. Cages in thermoplastic polymers can be fabricated very economically by injection moulding. Nylon 66, often reinforced with glass fibres, has been extensively used, although limited to operating temperatures below 120 °C and above −40 °C. Polyethersulphone, also reinforced by glass fibres, offers good performance at temperatures up to 180 °C.

9.2.4 Friction and lubrication in rolling bearings

The frictional torque imposed by a rolling bearing supporting a shaft is very low: expressed as a coefficient of friction, it has a typical value of 0.0015 for a radial ball bearing with a single row of balls, and can be as low as 0.0010 for a cylindrical roller bearing with short rollers. A rolling bearing is therefore a very efficient machine component. Only a well-designed hydrodynamically lubricated plain bearing operating near the minimum of the Stribeck curve (see Fig. 4.6) can offer a comparably low coefficient of friction; the rolling bearing, however, maintains its low friction even from a standstill, whereas a plain bearing will exhibit a much higher starting torque.

The frictional drag in a rolling bearing originates from several sources. The most important, in a lubricated bearing, is the energy dissipated in the thin lubricant films between the rolling elements and the tracks. At moderate speeds in a cylindrical roller bearing lubricated with oil, about 60% of the total energy lost is dissipated in these EHL films. Nearly all the remaining energy is accounted for by the lubricated sliding of the cage against the other components of the bearing. Only a very small fraction (of the order of 1%) of the energy is lost in the hysteresis associated with the cyclic elastic deformation of the balls or rollers and the track (as illustrated in Fig. 3.29).

Lubrication of rolling bearings is essential, mainly in order to form protective films between the various components and thus prevent excessive wear. Sliding contact occurs between the cage and the rolling elements and races; it also occurs to some extent between the rollers or balls and the races. Although under zero load a cylinder or sphere will roll over an elastic plane surface without slipping, when it supports a load some local slippage occurs in the following ways.

Cylindrical rollers and spheres will both experience *Reynolds slip* (named after Osborne Reynolds, 1842–1912), due to the progressive stretching of the surface within the contact region. Figure 9.6 illustrates a cross-section through a cylinder, rolling over an originally plane surface under load. The

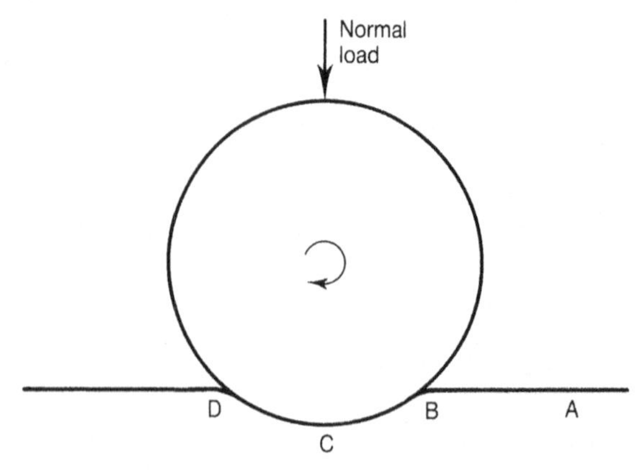

Fig. 9.6 Cross-section through a sphere or cylinder rolling over an elastic surface, illustrating the origin of Reynolds slip

surface between points B and D has been stretched to accommodate the cylinder. Reynolds envisaged that the strain in the surface would be greater at point C than at B or D, and that there would therefore have to be some differential movement between the surface of the plane and the surface of the cylinder as the cylinder rolled. He suggested that frictional energy dissipation associated with this slip was the cause of rolling resistance. It is now clear, however, that Reynolds slip is of minimal importance even in the rolling of steel spheres on rubber where the surface strains can be quite high; in practical rolling bearings it plays an even smaller role.

The second type of slip occurs only with spherical rolling elements and not with cylindrical rollers. Figure 9.7 shows in plan view and in section a sphere rolling under load in a cylindrical groove. The contact zone is elliptical. If the sphere rotates through one complete revolution, then point A on its surface will have travelled a distance $2\pi r$, while point B has travelled the smaller distance $2\pi r'$. Since the sphere moves as a rigid body, slip must therefore have occurred within the contact region in order to accommodate this difference; this differential slippage is called *Heathcote slip*. While this simple picture illustrates the origin of Heathcote slip, more detailed analysis is

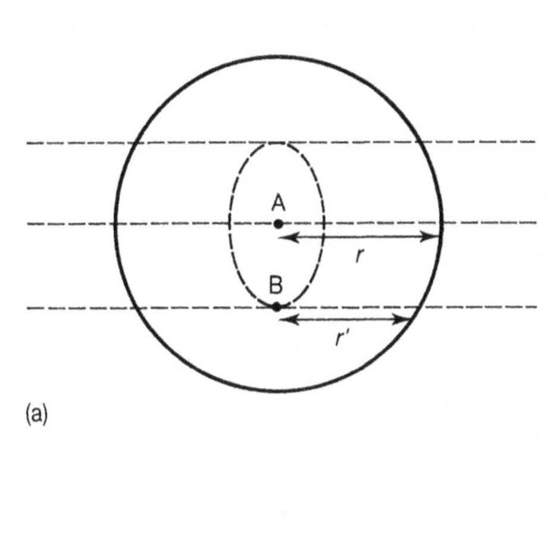

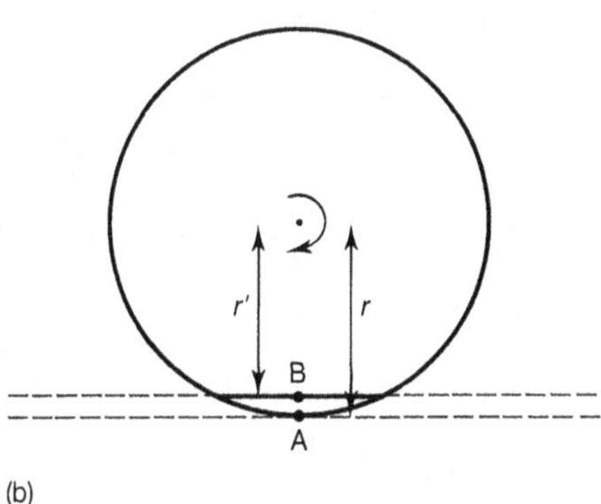

Fig. 9.7 (a) Plan view and (b) section through a sphere rolling over a bearing track, to illustrate the origin of Heathcote slip

needed to establish exactly which areas within the contact patch do indeed slip, and in which areas the displacements are accounted for by elastic strains.

Friction in the areas undergoing Heathcote slip makes a considerably greater contribution to rolling resistance than does Reynolds slip, but neither is important in the overall energy balance in a rolling bearing. Slip in the contact region will, however, cause surface damage which can lead to severe wear or early failure through surface-initiated fatigue if the bearing is unlubricated; an important function of the lubricant in a rolling bearing is therefore to provide a film thick enough to separate the surfaces, usually in the elastohydrodynamic regime, and thereby reduce wear and surface damage. Grease and oil (see Section 4.3) are both used to lubricate rolling bearings. Oil offers lower frictional drag and also transfers heat away from the bearing, a valuable function at high speeds, while grease needs less complex means of sealing and replenishment, and is widely used. Both oils and greases give some protection against corrosion. Suitable synthetic oils can be used with tool steel bearings at high temperatures, while for even higher temperatures, with ceramic bearings, solid lubricants are being investigated (see Section 4.7).

9.3 FLUID FILM LUBRICATED BEARINGS

9.3.1 Introduction

Plain bearings may be broadly classified into two types: those in which a lubricant (usually oil) is supplied continuously in order to produce a hydrodynamic (or hydrostatic) lubricating film during operation, and those operating under more marginal conditions of lubrication, where the lubricant is either oil or grease supplied intermittently and retained in the bearing structure, or a solid lubricant incorporated into or even constituting the bearing material. We shall consider the first type, the fluid film lubricated bearing, in this section, while marginally lubricated and dry bearings are discussed in Section 9.4.

As we have seen in considering hydrodynamic lubrication (Section 4.4), in a journal bearing operating under fully hydrodynamic (or hydrostatic) conditions the journal and shaft should remain separated by the lubricant film, and there should ideally be no contact between them. It might therefore be thought that the materials of the bearing housing and shaft would be unimportant. When the shaft is not rotating, however, the lubricant film is squeezed out from the bearing contact, and the bearing system must be able to operate under very poorly lubricated conditions on starting, stopping, and possible extreme overloading. The bearing material must therefore exhibit *compatibility* with the journal material, to tolerate limited direct contact without gross seizure. Furthermore, the lubricant will inevitably become contaminated with solid particles, perhaps from atmospheric dust, or from wear debris. If these particles are larger than the thinnest lubricant film encountered in the bearing, then abrasive wear may occur. Plain journal bearing materials can tolerate some abrasive contamination if they have good *embeddability*: that is, if they will allow abrasive particles to embed into their surface, so that they do not circulate repeatedly with the lubricant causing abrasive wear of both journal and bearing. In many applications, journal bearings are subject to repetitive loading which results in high cyclic pressures in the lubricant film; this can lead to fatigue failure of the bearing material, which should therefore possess good *fatigue strength* as well as high *compressive strength*. Since bearings, shafts and machine structures are all built to finite tolerances, and all will distort to some extent under load, there may be some misalignment between a journal and its housing. A good plain bearing material will tolerate some misalignment if it shows *conformability*. Finally, bearing materials should resist *corrosion* by the lubricant, even though it may be contaminated with water, oxygen, acidic oxidation products, or, for example, with combustion products in an internal combustion engine.

These properties of compatibility, embeddability, fatigue resistance, compressive strength, conformability and corrosion resistance are needed by all plain bearings, although their relative importance depends considerably on the conditions of load and speed under which the bearing is required to operate. No single material offers ideal properties in all these areas, since in some cases they are mutually incompatible: a high compressive strength, for example, implies a high hardness, which in turn implies poor embeddability. The choice of a bearing material involves a compromise, and different materials are therefore suitable for different applications.

We can separate applications of lubricated bearings into two types: those of low severity, in which the average bearing pressure is low or moderate and in which the bearing often carries a steady load, and those characterized by high pressures and large cyclic loads. Conditions of the first type are found in the shaft bearings of large rolling mills, steam turbines and low-speed diesel engines, often for marine use, while the second, more severe conditions, are associated with the crankshaft bearings of high speed petrol (gasoline) and diesel engines.

In nearly all applications the journal running in the bearing is steel; it is a more expensive component to replace than the bearing sleeve, and wear must be avoided. It is therefore often hardened, particularly if one of the harder bearing materials is used. Small shafts, for example, can be nitrided, which increases both the surface hardness of the journal and its fatigue resistance (see Section 8.2.3).

9.3.2 Low stress applications

In applications involving low bearing pressures and large bearings, some degree of misalignment is inevitable, and it is particularly important that seizure of the bearing on starting and stopping is avoided. Stresses, however, are low. Compatibility and conformability are therefore more important than strength in the bearing material, and the *whitemetals* are very widely used. Whitemetals, also known as *Babbitts* after Isaac Babbitt who patented the use of pewter as a bearing metal in 1839, are *tin-based* and *lead-based* alloys with unrivalled compatibility, conformability and embeddability.

Tin-based whitemetals, historically the earliest compositions, are alloys of tin with antimony and copper. A typical composition is 89% tin, 7.5% antimony and 3.5% copper (British Standard BS3332/1, ASTM B23 Alloy 2); several other compositions are also used and are covered by national standards. Figure 9.8 shows the microstructure of a typical cast tin-based whitemetal. Alloys containing up to 8% copper and less than about 8% antimony form needles of Cu_6Sn_5, often in a characteristic star-like formation, sometimes together with some particles of SbSn, in a solid-solution (tin–antimony) matrix. The proportion of the copper–tin constituent increases with copper content. Primary precipitation of the copper–tin needles tends to form a network which prevents the antimony–tin particles from segregating. Higher antimony contents (over about 8%) lead to the formation of primary cuboids of SbSn, although the precise microstructure depends on cooling rate and will also, due to segregation effects, vary with location within the casting. Whitemetals with a higher antimony content (e.g. 87% Sn, 9% Sb, 4% Cu: BS3332/2) are somewhat stronger but less ductile. The lead content in all tin-based whitemetals is kept low (typically below 0.5%) to avoid the formation of the low melting point lead–tin eutectic (M.P. 183 °C).

Lead-based whitemetals usually contain tin and antimony, with arsenic in some compositions. A typical arsenic-free alloy contains 10% antimony and 6% tin (ASTM B23 Alloy 13, SAE 13); its microstructure comprises primary cuboids of SbSn, embedded in a matrix of lead–antimony–tin. Segregation of the antimony–tin cuboids can occur in casting. The addition of arsenic strengthens the alloy, particularly at high temperatures. A typical arsenic-containing lead-based alloy contains 16% antimony, 1% arsenic and a nominal 1% tin (ASTM B23 Alloy 15). Its microstructure contains fine particles of antimony in a solid-solution matrix. Some lead-based whitemetals are also used, to a limited extent, which contain calcium and traces of alkali metals (e.g. 1–1.5% Sn, 0.5–0.75% Ca, traces of sodium).

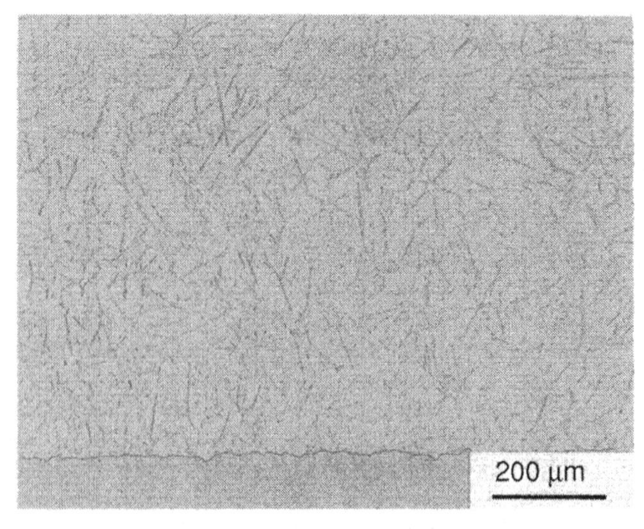

Fig. 9.8 Microstructure of typical tin-based whitemetal bearing alloy cast on to a steel backing (courtesy of Glacier Vandervell Ltd, Kilmarnock, Scotland)

Both tin-based and lead-based whitemetals commonly contain hard particles (e.g. antimony–tin cuboids or copper–tin needles) in much softer solid-solution matrices. The coefficients of friction for the unlubricated sliding of these materials against steel do not, however, appear to be influenced by the presence of the particles. Measured friction values against steels are quite high without lubrication: typically $\mu = 0.5$ to 0.8.

Whitemetals show very good compatibility and conformability, with the tin-based alloys being somewhat more resistant to seizure. Tin-based whitemetals can be susceptible to thermal fatigue if subjected to wide-range thermal cycling; this results from the anisotropy of the thermal expansion of tin (which has a body-centred tetragonal structure) and does not occur in lead-based alloys. Lead-based whitemetals, on the other hand, can corrode severely in the presence of acids (although to some extent protected by alloying with tin and antimony), whereas tin-based whitemetals are corrosion-resistant.

The use of whitemetal bearings is limited largely by their strength and, particularly in applications involving cyclic loading, by their fatigue strength. Although the strengths of whitemetals depend to some degree on their composition, they vary more strongly with the thickness of the bearing metal employed. Whitemetal bearings are cast, usually by gravity or centrifugally; replaceable steel bearing shells lined with whitemetal are commonly used, although in some applications in heavy machinery the alloy may be directly cast *in situ*. As a result of the plastic constraint introduced by the backing, a thin layer of whitemetal bonded as a lining on to a steel backing has a considerably higher compressive strength than a thicker lining. Similar effects are seen on the fatigue strength: Fig. 9.9 shows the increased fatigue strength obtained as the thickness of the whitemetal layer is reduced.

In very demanding applications, very thin linings of whitemetal on steel or bronze backings are used. However, as the lining becomes thinner, so its tolerance to abrasive particles and to shaft misalignment diminishes. In

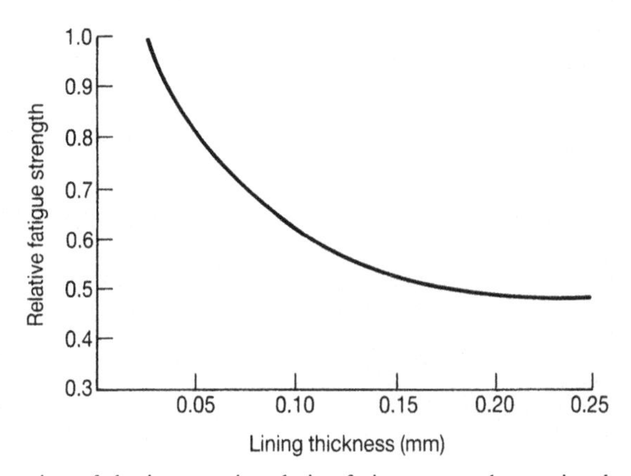

Fig. 9.9 Illustration of the increase in relative fatigue strength associated with a reduction in thickness of a whitemetal bearing alloy supported on a steel backing (data from Neale M J (Ed.), *Tribology Handbook*, Butterworths, 1973)

situations where the bearing pressures are high, and where there is a large dynamic component to the loading, whitemetal bearings are inadequate and the materials described in the next section must be used.

9.3.3 High stress applications

Perhaps the most demanding application in which lubricated plain bearings operate is as crankshaft main and big-end (US crankpin) bearings in the modern high speed internal combustion engine. The environment is severe: they carry large cyclic loads at high rotational speeds. To achieve good thermodynamic efficiency the running temperature of the engine is high; the bulk lubricating oil temperature may be up to 130 °C, and still higher temperatures will be attained within the hydrodynamic film. The maximum oil film pressures generated on the power stroke can be over 200 MPa, and good compressive strength and fatigue resistance are therefore needed in the bearing material. Embeddability, conformability and compatibility are also important, since the oil film can typically reach a minimum thickness during running of only 5 μm, a dimension which may be less than the size of the abrasive contamination, or the elastic deflections of the crankshaft. To meet these stringent requirements, composite bearings are used, as shown in Fig. 9.10. The bearing, known as a *thin wall bearing*, consists of a thin steel shell, typically 2 mm thick, lined with one or more layers of bearing material with a total thickness of about 300 μm. The shell is split into two to allow assembly, and is formed from low-carbon steel strip to standard dimensions for interchangeability. The increased strength resulting from the use of a thin lining on a more rigid backing has been mentioned in the previous section. Bearing alloys commonly used in these applications are *copper-based* (copper–lead or leaded bronze) and *aluminium-based* (aluminium–tin, aluminium–lead and aluminium–silicon).

Copper–lead bearing alloys typically contain 25 to 40% lead, and are fabricated by casting or sintering on to the steel bearing shell. Copper and lead are miscible in the liquid state in these proportions, but their solid solubility is effectively zero. Continuously cast material contains copper

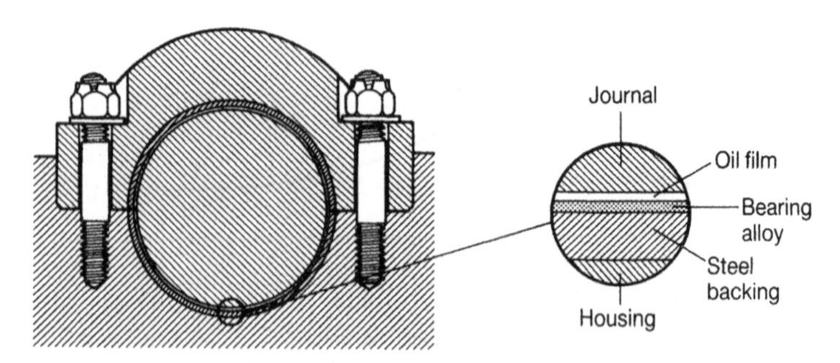

Fig. 9.10 Typical construction of a thin wall crankshaft bearing in a reciprocating internal-combustion engine. The bearing consists of a thin layer of bearing alloy supported on a replaceable steel shell

dendrites which have grown outwards from the steel backing, separated by an interdendritic lead phase (see Fig. 9.11(a)). In production by the powder route, copper–lead particles are made by gas-atomization of the melt (Fig. 9.11(b)). These are then spread on to the copper-plated steel backing strip, sintered in a reducing atmosphere, rolled and resintered to improve the interparticle bonding and eliminate porosity. The resulting structure is more equiaxed than that produced by casting (Fig. 9.11(c)). Whether the alloy is cast or sintered, the structure consists of a relatively strong continuous copper phase interspersed with regions of much softer lead. The alloy in this form exhibits low sliding friction against steel in the dry state ($\mu \approx 0.2$), due to the lubricating action of the thin film of lead which becomes smeared out over the surface (see Section 4.7).

Thin wall bearings lined only with copper–lead are used in some undemanding applications. Substantial improvements in bearing performance, however, can be gained from the application of a further thin *overlay coating*, typically 25 μm thick, on top of the copper–lead lining. When copper–lead bearings are used in internal combustion engines, they are usually overlay coated. Figures 9.11(a) and (c) show bearing materials with overlay coatings.

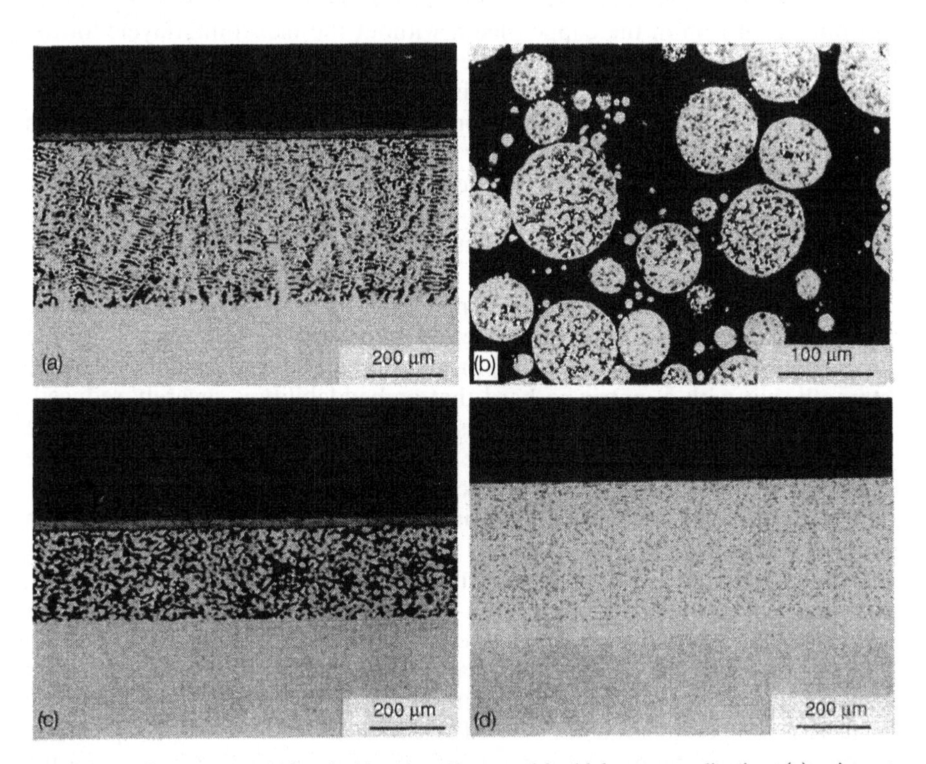

Fig. 9.11 Microstructures of typical bearing alloys used for high stress applications (a) strip-cast copper–lead on a steel backing; (b) particles of copper–lead alloy produced by gas atomization from the melt; (c) copper–lead alloy produced by sintering from gas-atomized particles; (d) reticular aluminium–20% tin alloy roll-bonded with a pure aluminium interlayer on to a steel backing. The materials in (a) and (c) both have thin overlay coatings (courtesy of Glacier Vandervell Ltd, Kilmarnock, Scotland)

The overlay is chosen to be much softer than the copper–lead; common compositions are lead–10% tin, with a possible copper content of up to 3%, or lead–8% indium. These coatings are applied by electroplating. They improve the compatibility and embeddability of the bearing, and also appreciably increase its fatigue strength. A further advantage conferred by overlay plating is corrosion protection; the almost pure lead phase which is exposed in uncoated copper–lead bearings can suffer severe corrosion from acidic oil, and this is prevented by a lead–tin overlay. Overlays on commercial bearings are frequently separated from the copper–lead lining by a very thin electroplated nickel layer, which is found to increase the bearing life. Explanations for this effect are equivocal: it has been suggested that the nickel forms a barrier to diffusion of tin from the overlay, thereby preventing the formation of brittle copper–tin compounds (Cu_6Sn_5 or Cu_3Sn) at the interface. The tin content of the overlay (in the case of lead–tin) is responsible for its corrosion resistance, and it is therefore possible that by preventing diffusion of tin out of the overlay the nickel barrier enables the original corrosion resistance to be maintained. It has also been proposed that the nickel layer is particularly beneficial when the copper–lead lining has been cast on to the steel shell, since this fabrication route leads to significant dissolution of iron in the copper–lead; without the nickel interlayer, formation of a brittle copper–tin–iron intermetallic might occur, weakening the bond between the overlay and the lining.

A stronger bearing material than copper–lead, with greater fatigue strength at the expense of lower compatibility and embeddability, is obtained by the addition of some tin to form the *lead bronzes*. Lead bronze linings are fabricated by the same methods as the copper–lead alloys, and have similar structures; the tin is present preferentially in the copper-rich phase. At 1 to 2% tin (typical composition 72% Cu, 26% Pb, 2% Sn), useful solid solution strengthening is obtained, but the lining is still soft enough to be run (with an overlay plating) against an unhardened steel shaft. Higher levels of tin (typically 74% Cu, 22%Pb, 4% Sn) produce greater improvements in fatigue strength, but with a matrix so hard that embeddability is severely reduced. Tolerance to abrasive contaminants is now determined almost solely by the overlay, and a hardened steel shaft (e.g. induction-hardened or nitrided, see Section 8.2) must be used to avoid excessive wear.

The second class of bearing materials widely used in internal combustion engines is *aluminium-based*. A highly successful alloy is aluminium–20% tin. Tin and aluminium are virtually insoluble in the solid state, and the structure, like that of copper–lead, consists of a very soft phase dispersed in a stronger matrix. Aluminium–tin bearings are fabricated by casting, followed by rolling. The strip of bearing material is then roll-bonded on to the steel backing strip, usually with a thin intervening layer of pure aluminium to achieve a good bond strength. As cast, the alloy is weakened by the presence of tin at the aluminium grain boundaries; the optimum structure for bearing performance is developed by subsequent heat treatment which allows the aluminium to recrystallize and the tin to form a three-dimensional network through the aluminium matrix. This type of structure, illustrated in Fig. 9.11(d), is called *reticular* (net-like). Reticular aluminium–tin bearings are, in several respects, superior to copper–30% lead. They have better embeddability and seizure resistance, and will function satisfactorily without an overlay coating; furthermore, they are very resistant to corrosion. The alloy is sometimes strengthened by the addition of copper (~1%), which hardens the aluminium phase.

A lower tin content is used in the aluminium–6% tin alloys, which have good fatigue resistance but poor compatibility. They must therefore be used with overlays. A typical composition contains 6% tin, 1% copper and 1% nickel; the copper provides some solid-solution strengthening of the matrix, while the nickel forms $NiAl_3$ particles. In order to electroplate an overlay on this alloy, a multistage process must be used, involving a thin electroless deposit of zinc, followed by electroplated layers of nickel and of lead–tin.

The use of lead instead of tin in aluminium alloys, at about 8% by weight, results in a bearing material with good compatibility which will operate without an overlay; a small percentage of tin, which dissolves in the lead phase, eliminates corrosion problems, and hardening of the aluminium can be produced by small additions of silicon or copper. A typical composition contains 8% lead, 5% silicon and 2% tin.

Several other aluminium alloys are also used in thin wall bearings: aluminium–silicon (typically 11% Si, 1% Cu) has good fatigue strength but needs an overlay, as do aluminium–cadmium alloys (typically 3% Cd) and aluminium–cadmium–silicon (1% Cd, 4% Si).

9.4 MARGINALLY LUBRICATED AND DRY BEARINGS

9.4.1 Introduction

In many general engineering applications, adequate bearing performance can be obtained from plain bearings which do not operate under fully hydrodynamic conditions. These bearings, which may be only intermittently lubricated, are termed *marginally lubricated*, the mechanisms of lubrication being partly hydrodynamic, but also involving contributions from elastohydrodynamic, boundary and sometimes solid lubrication. For some uses, plain bearings containing solid lubricants are used without any external supply of lubricant; these are called *dry bearings*. Some marginally lubricated and dry bearings can carry loads as high as rolling element bearings and fluid film plain bearings, but they are more limited in speed, as shown in Fig. 9.2; their major advantages lie in their small dimensions and low cost, both for the bearing itself and for the means of supplying the lubricant.

Marginally lubricated bearings are constructed either as solid components or as porous metal parts. Dry bearings are nearly always solid. A frequently used guide to the capacity of a marginally lubricated or dry bearing is the value of the product PV, where P is the mean bearing pressure in MPa (i.e. load divided by projected bearing area) and V is the sliding velocity in m s^{-1}; maximum values of PV are often quoted by manufacturers of proprietary bearing materials. In the absence of more detailed design information, the PV value is useful, but can be misleading; a better guide is provided by a plot of P

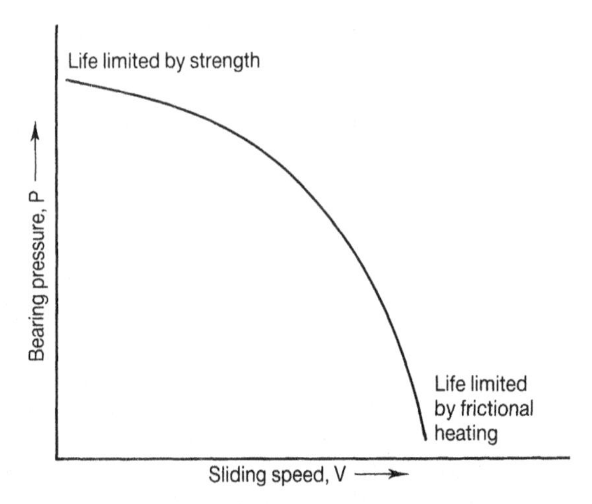

Fig. 9.12 Typical relationship between bearing pressure P and operating speed V for a fixed lifetime, for a marginally lubricated or dry bearing

against V for some specified lifetime or total wear, as shown schematically in Fig. 9.12. At high loads, the life of the bearing will be limited by its strength, whereas at high speeds it is limited by frictional heating. Only in the intermediate regime does the specification of a maximum PV value provide a reliable criterion for design.

9.4.2 Marginally lubricated bearings

A wide range of solid bearing materials is used under marginally lubricated conditions; selection must depend on many factors, including the load, speed and nature of the movement (whether continuous or oscillatory), the means of lubrication, cost and methods of fabrication and installation.

Solid bronze bushes, often leaded, are commonly used; the material is readily available in bar form and is therefore suited to low-volume production of bearing bushes by simple machining operations. The load capacity of leaded bronze bearings decreases with increasing lead content, while their conformability and embeddability increase. The higher strength alloys with lower lead content tend to be favoured for low speed, grease lubricated operation, while higher lead contents are used for higher speeds with oil lubrication. Typical compositions are 89% copper–10% tin–0.5% phosphorus (unleaded phosphor bronze), 80% copper–10% tin–10% lead (medium-lead tin bronze) and 70% copper–5% tin–25% lead (high-lead tin bronze); many other compositions are also used, including the copper–lead alloys and lead bronzes described in Section 9.3.3 above. The lead phase present in the leaded bronzes results in a low coefficient of dry friction (see Sections 3.5.4 and 4.7), and is therefore beneficial under marginally lubricated conditions. Further improvements in bearing capacity under conditions of poor lubrication can be achieved by incorporating graphite as a solid lubricant into the bearing material. Proprietary materials are available containing from 4 to

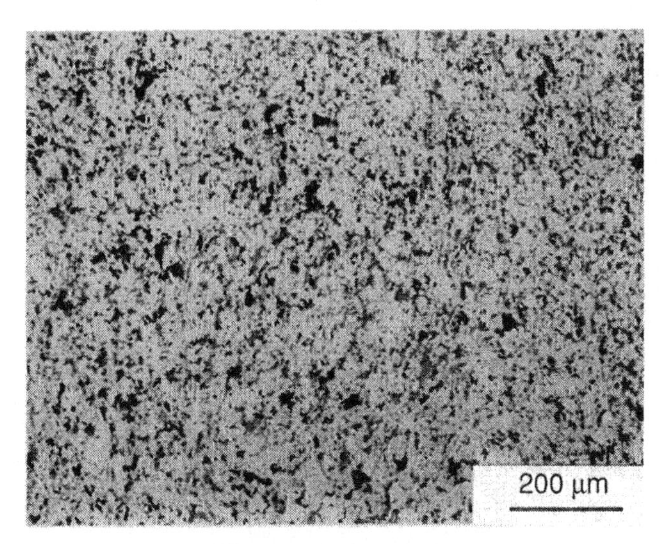

Fig. 9.13 Microstructure of a graphited tin bronze bearing material, produced by powder metallurgical methods (courtesy of Glacier Vandervell Ltd, Kilmarnock, Scotland)

14% by weight (10 to 40% by volume) of graphite, finely dispersed in a leaded or unleaded bronze matrix (see Fig. 9.13). These materials are fabricated by powder metallurgical techniques, and can be used at up to 200 °C. For higher temperature use, where conventional lubricants are unusable, similar composite materials containing graphite and iron or graphite and nickel are available, which can be used without external lubrication at up to 600 °C.

Other bearing materials described in Section 9.3, such as the aluminium–tin alloys and the tin-based and lead-based whitemetals, are also sometimes used for marginally lubricated applications. All these materials can be used in the form of solid bushes, or at higher loads as relatively thin linings on steel or bronze backings.

Porous bronze or iron bushes, impregnated with oil, are widely used; these are pressed to shape from powder and then sintered, to give an interconnected porosity of 10 to 25%. They are then impregnated with oil under vacuum before use. Bronze is a common material, often containing some graphite as a solid lubricant (e.g. 89% Cu, 10% Sn, 1% graphite); porous bushes in this material can be run against unhardened steel shafts and, on account of their good thermal conductivity, function well at high rotational speeds. Porous iron bushes (e.g. pure iron, with 2 to 25% copper) are stronger, but have lower thermal conductivity; they are therefore more suitable for higher load, lower speed applications. Because the material is harder, a hardened steel shaft must be used to avoid heavy wear. The mechanism of lubrication in an oil-impregnated porous bush is, ideally, hydrodynamic; the hydrodynamic pressure distribution in the bearing when it is running leads to continuous circulation of oil through the pores. They are therefore better suited to running at high speeds (and correspondingly low loads) than to lower speeds where the lubrication will be poorer. As the

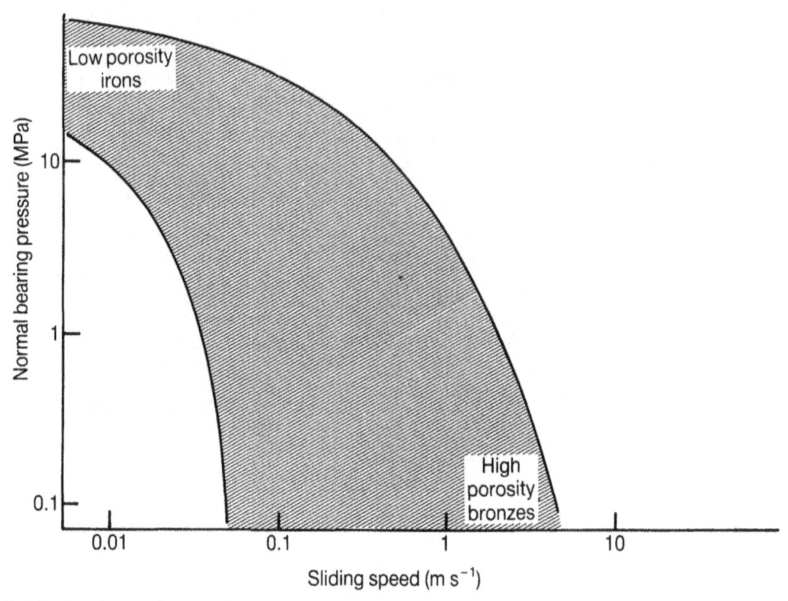

Fig. 9.14 Regime of speed and pressure over which porous metal bushes are useful as marginally lubricated bearings (data from Waterman N A and Ashby M F (Eds.), *Elsevier Materials Selector*, Vol. 1, Elsevier, 1991)

porosity of the bush is decreased, so its load capacity increases; on the other hand, high porosity produces a large inbuilt reservoir of oil and allows it to circulate more readily, resulting in better lubrication. Figure 9.14 illustrates the regimes of speed and load over which porous metal bushes are useful. A maximum PV value of 1.75 MPa m s^{-1} is sometimes specified for the use of this type of bearing.

Solid polymer bushes are used in some marginally lubricated, lightly loaded applications. Unfilled thermoplastics, particularly polyamides (nylons) and polyacetals, are suitable only for low loads; PV limits of 0.05 MPa m $^{-1}$ are typical. Better performance under marginal lubrication results from the incorporation into these polymers of a solid lubricant. Polytetrafluoroethylene (PTFE) or molybdenum disulphide (MoS_2) are common additives, while the bulk strengths of these materials can be increased by the use of inert fillers such as glass fibres. Reinforced thermosetting polymers (e.g. polyester or phenolic resins) are also used in some applications. Solid thermoplastics bearings can be injection moulded and are therefore cheap to produce; the performance of these materials can be considerably improved, at the expense of more complex fabrication, by bonding them as a thin layer to a metallic backing. Commercial composite bearings of this type are made by sintering a porous layer of bronze powder (typically 0.2 to 0.3 mm thick) on to a steel backing strip, previously copper plated to provide a good bond, and then infiltrating the bronze layer with a thermoplastic material. Polyacetal is a common choice; a mixture of polyvinylidene fluoride (PVDF), PTFE and lead powder is also used, and more recently, a composite of polyetheretherketone (PEEK), PTFE and graphite. The microstructure of a

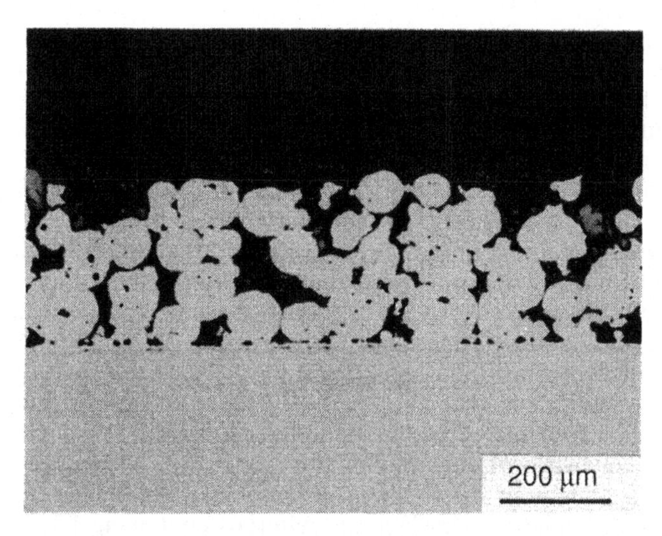

Fig. 9.15 Microstructure of a typical marginally-lubricated metal–polymer composite bearing. A porous bronze network sintered on to a steel backing strip is infiltrated with a composite material containing a thermoplastic polymer and solid lubricants (courtesy of Glacier Vandervell Ltd, Kilmarnock, Scotland)

typical material of this type is shown in Fig. 9.15. Such bearings can operate at PV values up to 2 MPa m s^{-1}, and show good embeddability. Lubrication is usually by grease, often placed during initial assembly into indentations formed in the bearing surface.

9.4.3 Dry bearings

All the bearing materials discussed in the previous section are lubricated to some extent by externally supplied grease or oil. They may be lubricated only before use, with the lubricant retained within the bearing sufficing for its whole life, or they may be relubricated at intervals. In some applications, for example in the food processing industry, or for a bearing operating in vacuum or at high temperature, the use of oil or grease may not be possible, or the avoidance of periodic lubrication may be attractive. For these and other reasons *dry bearings* are used. These incorporate a lubricant in their composition, and no external supply of lubricant is required. PTFE, molybdenum disulphide and graphite are commonly used solid lubricants (see Section 4.7), and dry bearing materials are usually composites containing one or more of these lubricants.

PTFE is a good solid lubricant, but extremely soft, and must be reinforced for use as a bulk bearing material. Common fillers are glass fibres, bronze particles, MoS_2 and graphite; the composite material must be prepared by powder processing methods since PTFE cannot be processed as a melt. Solid bearings made from these materials (e.g. a composite of PTFE, graphite and bronze powder) exhibit low friction and good wear behaviour when not externally lubricated; however, the maximum loading is low, with limiting *PV* values of about 0.3 MPa m s^{-1} for reasonable life. Substantially better performance is given by composite thin films on steel backing shells, with structures very similar to that shown in Fig. 9.15. A typical material, widely used and hard to improve upon for moderate loads (up to ~35 MPa) and temperatures (up to 250 °C), consists of an intimate mixture of PTFE and lead powder, impregnated into a porous bronze layer sintered to a steel strip. Here the normal load is largely supported by the bronze particles, and sliding occurs in a thin film of PTFE at the surface. The lead is thought to play a role, together with copper from the bronze, in the formation of a stable transfer film of PTFE on the steel counterface (see Section 5.11.3). Low values of μ are attained, typically 0.03 to 0.1, although these composite materials, like bulk PTFE, do not obey Amontons' Laws. There is a substantial dependence of μ on both load and sliding speed and the lowest coefficient of friction is found at low sliding speed and at high load. Sliding speeds are usually limited to 1 to 2 m s^{-1} to avoid excessive frictional heating. However, as a result of the compressive strength imparted by the bronze matrix, composite bearings of this type can be used at *PV* ratings of up to about 2 MPa m s^{-1}.

Many other composite dry bearing materials containing PTFE are used. Some contain other solid lubricants as well (e.g. graphite or MoS_2) and a wide variety of other constituents have been employed (e.g. epoxy resins, polyimide in bulk or fibre form, glass or carbon fibres, phenolic resins). This is an area of considerable current research interest, where new composite bearing materials are being sought to exploit the properties of other lubricants (e.g. graphite fluoride—see Section 4.7) and polymers (e.g. polyetheretherketone—PEEK).

For operation at high temperatures, where organic materials would decompose, graphite and molybdenum disulphide are useful lubricants. The graphite–bronze, graphite–iron and graphite–nickel composites discussed above for marginally lubricated conditions (Section 9.4.2) will also operate dry at high temperatures, and bearings of solid graphite are also sometimes used. Many other proprietary composite materials have been developed for high temperature use; some for example, incorporate gold as a solid lubricant, while others use molybdenum disulphide in a refractory metal matrix (e.g. molybdenum, tungsten, tantalum or niobium). Materials of this latter type can be used in air at up to 400 °C, their operation being limited by oxidation, while in vacuum or inert gas they will function at up to 1200 °C. Other metal-matrix composites, and carbon–carbon composites, also offer exciting potential for the future.

Further reading

ASM Handbook, 10th edn, vol. 3, *Friction, Lubrication and Wear Technology*, ASM, 1992
Barwell F T, *Bearing Systems: Principles and Practice*, Oxford University Press, 1979
Cullum R D, *Handbook of Engineering Design*, Butterworths, 1988
Jones M H and Scott D (Eds), *Industrial Tribology*, Tribology Series No. 8, Elsevier, 1983
Kirkham R O, Park G, Groves W S and Stobo J J, Plain bearing alloys, *Metals Forum* **6**, 149–160, 1983
Neale M J (Ed.), *Tribology Handbook*, Butterworths, 1973
Nisbet T S, *Rolling Bearings*, Engineering Design Guide No. 4, Oxford University Press, 1974
Peterson M B and Winer W O (Eds), *Wear Control Handbook*, ASME, 1980
Pratt G. C., Materials for plain bearings, *International Metallurgical Reviews* **18**, 1973
Waterman N A and Ashbey M F (Eds), *Elsevier Materials Selector*, vol. 1, Elsevier, 1991

Author index

Subject index

Lightning Source UK Ltd.
Milton Keynes UK
UKOW05f0058041216
289121UK00002B/35/P

9 780340 561843